Quantitative Methods in Transportation

T0138850

Quantitative Methods in Transportation

Dušan Teodorović and Miloš Nikolić

CRC Press
Taylor & Francis Group
Boca Raton London New York

CRC Press is an imprint of the
Taylor & Francis Group, an **informa** business

First edition published 2020
by CRC Press
6000 Broken Sound Parkway NW, Suite 300, Boca Raton, FL 33487-2742

and by CRC Press
2 Park Square, Milton Park, Abingdon, Oxon, OX14 4RN

© 2021 Taylor & Francis Group, LLC

CRC Press is an imprint of Taylor & Francis Group, LLC

Library of Congress Cataloging-in-Publication Data

Names: Teodorović, D. (Dušan), 1951- author. I Nikolić, Miloš, 1984- author.
Title: Quantitative methods in transportation / Dušan Teodorović and Miloš Nikolić.
Description: First edition. I Boca Raton, FL : CRC Press, 2021. I Includes bibliographical references and index.
Identifiers: LCCN 2020011458 (print) I LCCN 2020011459 (ebook) I ISBN 9780367250546 (hardback) I ISBN 9780367250539 (paperback) I ISBN 9780429286919 (ebook)
Subjects: LCSH: Transportation engineering--Mathematical models. I Transportation engineering--Statistical methods. I Transportation--Mathematical models. I Transportation engineering--Planning.
Classification: LCC TA1145 .T46 2021 (print) I LCC TA1145 (ebook) I DDC 388.3/10721--dc23
LC record available at https://lccn.loc.gov/2020011458
LC ebook record available at https://lccn.loc.gov/2020011459

ISBN: 978-0-367-25054-6 (hbk)
ISBN: 978-0-367-25053-9 (pbk)
ISBN: 978-0-429-28691-9 (ebk)

Typeset in Sabon
by Deanta Global Publishing Services, Chennai, India

Visit the eResources: https://www.routledge.com/9780367250539

To my wife Ljiljana, my children, and grandchildren

Dušan Teodorović

To my parents and my wife Jelena

Miloš Nikolić

Contents

Preface

We met, for the first time, in 2007, at the University of Belgrade, Serbia, where the first author has been professor, and the second author, an undergraduate student. For many years, we have been investigating traffic phenomena. At the same time, as professors, we have been teaching quantitative methods in transportation within different courses at different universities. Eventually, we thought that it was time to write a book about quantitative methods in transportation. We believe that every new book in a specific area opens new views to the persons who read.

The book is aimed for use by students at the senior undergraduate level, and at the graduate level, for those interested in transportation engineering, civil engineering, city and regional planning, urban geography, and economics. We hope that the book would also be helpful for planners and engineers working in the area of transportation. Our teaching experience had shown that the majority of students learn best by means of examples, so we have offered numerous examples in this book.

Why Quantitative Methods in Transportation? Currently, large-scale transportation networks, their density, high level of congestion, and high transportation costs require a comprehensive approach to transportation planning, logistics, and traffic control problems. As a rule, we have limited financial resources with which to expand our transportation networks, and/or to improve the level of service in transportation. Which street to widen? Should a new bridge be built? Is it necessary to purchase additional buses for the public transport operator? How to further develop the metro line network? How to synchronize timetables in public transport? What is the best topology of the air carrier's network? How can daily airline schedule disruptions be mitigated? Planners, traffic engineers, managers, and dispatchers try to find answers to these and similar questions.

In essence, transportation engineering methods and techniques are quantitative. By these methods, traffic engineers try to optimize traffic and transportation operations. They are quantitative, as wrong decisions in the transportation arena are environmentally harmful and highly costly. During the past nearly one hundred years, quantitative methods have been utilized to support decision making in transportation planning, logistics, and traffic control. Traffic network equilibrium problems have been studied since the 19th and 20th centuries. With increasing levels of traffic congestion, by the middle of the 20th century, traffic problems required more sophisticated analysis.

Some of the methods widely used in transportation planning, logistics, and traffic control are mathematical programming, heuristic and meta-heuristic algorithms, simulation, queueing theory, and statistical analysis.

The following is a brief description of the book chapters.

Chapter 1 introduces the reader to the field of applied mathematical programming in traffic and transportation. The chapter offers basic definitions and describes the most important mathematical programming techniques, namely linear programming, integer programming, mixed-integer programming, and dynamic programming.

Chapter 2 covers the transportation network basics, as well as shortest paths problems. We provide numerous examples in this chapter.

Chapter 3 deals with the multi-attribute decision-making (MADM) methods and data envelopment analysis (DEA). On a regular basis, government, industry, and/or traffic authorities have to evaluate sets of transportation projects (alternatives). The ranking of the alternatives is usually done according to a number of criteria that, as a rule, are mutually conflicting. This chapter covers the most important MADM methods, as well as DEA, that are increasingly being used for measuring efficiency in transportation.

Chapter 4 covers basic elements of probability theory. The many independent random factors have an effect on various traffic phenomena (travel time, the total number of cars on a specific link, the total number of passengers on a specific flight, demand at nodes of a distribution system, etc.). Consequently, there is a call for a probabilistic analysis of traffic phenomena.

Chapter 5 describes basic statistical methods. In traffic engineering, we perform measurements or counts. We use statistics to explain the collected data with numbers, pie charts, bar graphs, and histograms. Statistics help us to analyze data, to reach conclusions, to find answers to engineering questions, and to make appropriate decisions.

Chapter 6 describes the basic simulation techniques. Simulation models imitate the behavior of real-world systems, most frequently *via* computers. These models study interactions among elements of the transportation system, as well as the system's operating characteristics. Based on the results obtained from simulations, engineers come to various conclusions and take appropriate actions.

Chapter 7 deals with queueing theory. Queueing in various transportation systems happens on a daily basis. Clients' total waiting time depends on the number of other requests for service, as well as on the number of servers. Queueing theory is the mathematical analysis of queues. Queueing models help us to predict queue lengths, as well as waiting times, in queues.

Chapter 8 covers the basics of the metaheuristic algorithms (simulated annealing, tabu search, genetic algorithms, ant colony optimization, bee colony optimization). In many cases, the optimal solution to the transportation problems under consideration (vehicle routing and scheduling, crew scheduling, transportation network design, location problems, etc.) cannot be found in acceptable central processing unit (CPU) time. In order to conquer these difficulties, various heuristic and metaheuristic algorithms have been proposed over the past five decades. Many of the heuristic algorithms developed have been able to generate sufficiently good solution(s) in an acceptable amount of CPU time.

This book offers various quantitative techniques related to transportation planning, logistics, and traffic control. We hope that these techniques will be used by new generations of planners and engineers to make future transportation systems safer, more cost effective, and more environmentally friendly.

Dušan Teodorović

Miloš Nikolić

Authors

Dušan Teodorović is Professor at the University of Belgrade, Serbia, and Professor Emeritus at Virginia Tech. He has been elected a member of the Serbian Academy of Sciences and Arts and the European Academy of Sciences and Arts. He is Editor of the *Routledge Handbook of Transportation*.

Miloš Nikolić is Assistant Professor at the University of Belgrade, Serbia, and he has been a visiting scholar at the University of California, Berkeley.

Chapter 1

Mathematical programming

A lot of real-life traffic and transportation problems can be quite simply formulated in words. In the next step, engineers typically transform such a verbal problem description into a mathematical description. Key components of the mathematical description of the problem are *variables, constraints*, and the *objective*. Variables are also occasionally called *unknowns*. Certain variables are under the control of the analyst. There are also variables that are not under the control of the analyst. Constraints could be physical, caused by some engineering rules, laws, or guidelines, or by a variety of financial causes. Nobody could allow more than 100 passengers for a planned flight, for example, if the capacity of the aircraft is 100 seats. This is an example of a physical constraint. Financial constraints are typically related to a range of investment decisions. For example, no one could put in road development costs of more than $15,000,000 if the existing budget equals $15,000,000. Variable values could be feasible or infeasible. Variable values are feasible when they satisfy all the defined constraints. An objective corresponds to the end-result which the decision maker wants to achieve by selecting a certain program of action. The typical objectives in profit-oriented organizations are revenue maximization, cost minimization, or profit maximization. Providing the best level of service (travel time, waiting time, ...) to the clients represents a common objective in a nonprofit organization.

A mathematical description of a real-world problem is called a *mathematical model* of the real-world problem. We denote the n decision variables of the studied problem by $x_j, j = 1, 2, ..., n$. Let us assume that the observed system is subject to m constraints. The general mathematical model can be written in the following way:

Optimize

$$y = f(x_1, x_2, ..., x_n)$$

subject to:

$$g_i(x_1, x_2, ..., x_n) \leq a_i \qquad i = 1, 2, ..., m$$

$$x_j \geq 0 \qquad j = 1, 2, \ldots, n$$

where

$y = f(x_1, x_2, \ldots, x_n)$ - objective function

$g_i(x_1, x_2, \ldots x_n) \leq a_i \qquad i = 1, 2, \ldots, m$ - constraints

$x_j \geq 0 \quad j = 1, 2, \ldots, n$ - constraint (non-negativity restrictions)

Optimization looks for the best value (*optimal value*) of the objective function. Optimization frequently proposes the maximization or minimization of the objective function. *Optimal solution* to the model is the finding of a set of variable values (feasible) that produce the optimal value of the objective function.

An *algorithm* represents some quantitative method used by an analyst to solve the defined mathematical model. Algorithms are composed of a set of instructions, which are usually followed in a defined, step-by-step procedure. An algorithm produces an optimal (the best) solution to a defined model. The *optimal solution* to the model is the discovery of a set of variable values (feasible) that generate the optimal value of the objective function. Depending on a defined objective function, the optimal solution corresponds to maximum revenue, minimum cost, maximum profit, etc.

1.1 LINEAR PROGRAMMING IN TRAFFIC AND TRANSPORTATION

In a lot of cases, all the variables in a problem are continuous variables. In addition, there is typically only one objective function. Often, the objective function and all constraints are *linear*, meaning that any term is either a constant or a constant multiplied by a variable. Any mathematical model that has one objective function, all continuous variables, a linear objective function, and all linear constraints is called a *linear program* (LP). It has been shown, over many years, that many real-life problems can be formulated as linear programs. Linear programs are frequently solved using broadly spread simplex algorithms.

Example 1.1

There are two types of goods that could be transported from airport A to airport B. The unit transportation cost of the first type of goods is \$200/t, whereas the unit transportation cost of the second type of goods equals \$150/t. The total aircraft cargo capacity equals 40 t and the aircraft cargo volume equals 240 m³. The volumes per ton of goods are 8 m³/t and 4 m³/t for the first and second types of goods, respectively. Determine the quantity of the first type of goods and the quantity of the second type of goods that would maximize the air carrier's revenue.

SOLUTION:

Let us introduce the following notation:

x_1 – the quantity of first type of goods that should be transported
x_2 – the quantity of second type of goods that should be transported

The linear program reads:
  Maximize

$$F = 200 \cdot x_1 + 160 \cdot x_2$$

subject to:

$$x_1 + x_2 \leq 40$$

$$8 \cdot x_1 + 4 \cdot x_2 \leq 240$$

$$x_1 \geq 0, x_2 \geq 0$$

This problem can be solved graphically, as we have two decision variables. Our axes are x_1 and x_2. We take into consideration only positive values of the decision variables since $x_1 \geq 0, x_2 \geq 0$. Each constraint is represented by one straight line. We replace signs "≤" or "≥" by "=" for each constraint. In our example, the constraints $x_1 + x_2 \leq 40$ and $8 \cdot x_1 + 4 \cdot x_2 \leq 40$ are represented by straight-line equations $x_1 + x_2 = 40$ and $8 \cdot x_1 + 5 \cdot x_2 = 240$ (Figure 1.1). These straight lines are plotted in Figure 1.1. In the next step, we determine the feasible solution space, taking into account each constraint. This task can readily be completed by taking one point at a time and checking whether that point belongs to the feasible area. For example, let us consider the point with coordinates $x_1 = 0$ and $x_2 = 0$. By substituting these values into the first constraint, we obtain: $x_1 + x_2 \leq 40 \Rightarrow 0 + 0 \leq 40$. We conclude that this point belongs to the feasible solution space of the first constraint. The directions of arrows in Figure 1.1 designate the feasible region in which each constraint is satisfied, when the inequality is made active. Each point inside or on the border line of the feasible region complies with all the constraints. Every point that satisfies all the constraints is called a feasible point. We also represent the objective function by the straight line. We set that the objective function equals the arbitrary values. For example, we can set that $F=0$, e.g. $200 \cdot x_1 + 160 \cdot x_2 = 0$. For this equation, we plot the straight line. We do the same for some other arbitrary values, and we detect the direction in which our objective function changes. The optimal solution is the point that belongs to the feasible solution space and has the highest objective function value. We note that the highest objective function value has the point with the coordinates $x_1 = 20$ and $x_2 = 20$ (one of the corner points). For these values of the decision variables, the objective function equals $F=\$7200$ (Figure 1.2).

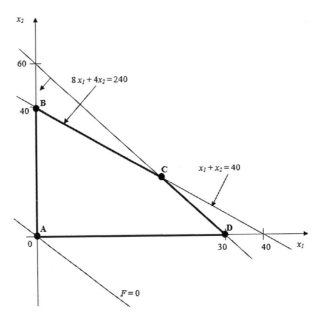

Figure 1.1 Solution space

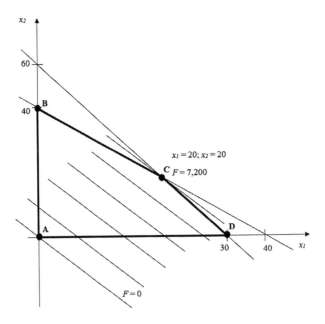

Figure 1.2 Optimal solution

We have used a graphical method for solving the given linear program. This method is suitable when there are two variables but is inconvenient for linear programs that have more decision variables. The other widely used solution approach is the simplex method.

1.2 THE SIMPLEX METHOD

Graphical solutions to many problems show that the optimal solution is associated with a corner point of the solution space every time. Unfortunately, the graphical method is appropriate only for problems that include two variables. The simplex method can be used for any number of decision variables. The simplex method is based on the idea that the optimum solution is, at all times, associated with a corner point of the solution space, as illustrated by the previous example.

The simplex method starts from one feasible corner point and improves the solution in iterations. The method moves from one corner point to an adjacent corner point. The algorithm stops when the optimal solution is reached To demonstrate how this method works, let us consider again the previous example.

The simplex method assumes that the solution space is represented by the standard form. The standard form assumes the following: (a) the objective function is minimization or maximization; (b) all the constraints are equations; (c) all the variables are nonnegative. This means that all constraints of the type $\leq$ or $\geq$ are converted to equations. This is done by adding a slack variable or by subtracting a surplus variable to the left side of the constraint.

Let us consider our first constraint:

$$x_1 + x_2 \leq 40$$

By adding the slack variable, $x_3 \geq 0$, we convert this constraint into the following equation:

$$x_1 + x_2 + x_3 = 40$$

Similarly, by adding the slack variable $x_4 \geq 0$, we convert the constraint:

$$8 \cdot x_1 + 4 \cdot x_2 \leq 240$$

into the following equation:

$$8 \cdot x_1 + 4 \cdot x_2 + x_4 = 240$$

Let us solve the considered linear program by the simplex method. The standard form reads:

$$F - 200 \cdot x_1 - 160 \cdot x_2 = 0$$

$$x_1 + x_2 + x_3 = 40$$

$$8 \cdot x_1 + 4 \cdot x_2 + x_4 = 240$$

$$x_1 \geq 0, \quad x_2 \geq 0, \quad x_3 \geq 0, \quad x_4 \geq 0$$

where x_3 and x_4 are nonnegative slack variables.

Let us, for example, consider the case where $x_3 = 0$. This is equivalent to $x_1 + x_2 = 40$. This equation represents the edge BC in Figure 1.2. By exploring the extreme points in Figure 1.2, we conclude that every extreme point has two variables equal to zero (zero variables). It is important to note that the difference between the number of unknowns (four in our case) and the number of equations (two in our case) is also equal to two. The number of zero variables at extreme points is always equal to the difference between the number of unknowns and the number of equations.

The extreme points and the corresponding zero and nonzero variables are given in Table 1.1.

In the simplex method, the variables equal to zero are called non-basic variables. The other variables are called basic variables.

Let us insert the equations obtained into the simplex table (Table 1.2). In the leftmost column are inserted the basic variables. In this example, the basic variables are x_3 and x_4. In this solution, their values are given in the column "Solution": $x_3 = 40$ and $x_4 = 240$. All other variables are non-basic.

Table 1.1 The extreme points, and the corresponding zero and nonzero variables

Extreme point	Zero variables	Nonzero variables
A	x_1, x_2	x_3, x_4
B	x_1, x_3	x_2, x_4
C	x_3, x_4	x_1, x_2
D	x_2, x_4	x_1, x_3

Table 1.2 Simplex table with the initial feasible solution

	F	x_1	x_2	x_3	x_4	Solution
F	1	−200	−160	0	0	0
x_3	0	1	1	1	0	40
x_4	0	8	4	0	1	240

The non-basic variables, in our example, are x_1 and x_2 and they equal zero $(x_1 = 0, x_2 = 0)$. For this solution, the objective function equals 0. This is also given in the column "Solution" (in the first row).

The solution obtained is feasible if the following condition is satisfied. Let us note the basic variables. In our examples, they are x_3 and x_4. All the elements in the columns of variables x_3 and x_4 must be zero, except for elements that are at the intersections with rows of variables x_3 and x_4. In our example, we note that this condition is satisfied. We will demonstrate in the next example what we have to do if this condition is not satisfied.

Once again, our first feasible solution is $x_1 = 0, x_2 = 0, x_3 = 40$, and $x_4 = 240$. For this solution, the objective function value equals 0. The question in this moment is: is this solution the optimal one? The answer is no. How do we know that the solution is not optimal? We note the elements in the row of the objective function (row denoted by F). If there were no negative elements, then the solution would be optimal. In our example, two elements are negative (-200 and -160) and, because of that, this solution is not optimal (In our example, we are maximizing the objective function. Obviously, the objective function value could be improved by increasing the x_1 or x_2 to greater than zero). Generally speaking, when the solution is not optimal, it can be improved. We try to make improvements to the objective function value in the following way. One of the non-basic variables becomes the basic variable. We term the column of the non-basic variable that will become the basic variable the "the pivot column". We select the pivot column in the following way. We select the decision variable that has the highest negative value in row F. In our example, it will be:

$$\min\{-200, -160\} = -200 \Rightarrow \text{column } x_1$$

In the next step, we must determine which one of the temporary basic variables will become non-basic. The row containing this variable is the pivot row. For this purpose, we must denote rows with positive elements in the pivot column. They are potential pivot rows. Among the potential rows, we will select the one that has the smallest value of the quotient of the appropriate element in the column "Solution" and of the element in the pivot column. In our example, it is:

$$\min\left\{\frac{40}{1}, \frac{240}{8}\right\} = \min\{40, 30\} = 30 \Rightarrow \text{row } x_4$$

The pivot element is the element at the intersection of the pivot row and the pivot column. After selecting the pivot element, we generate the new table in the following way:

- All elements in the pivot column are equal to zero, except the element which was the pivot element. The value of that element is 1.

- The new values of the elements in the pivot row are equal to the old values divided by the pivot number.
- The other elements are determined in the following way. Suppose that element a_{ij} is a pivot number, and that we want to determine the new value of the element a_{rs}. We should also note elements a_{is} and a_{rj}, that are the elements at the intersection of the pivot row and the column s, and at the intersection of the row r and the pivot column. The element at the intersection of row r and column s will have the new value

$$a_{rs} - \frac{a_{is} \cdot a_{rj}}{a_{ij}}.$$

By using these rules, we obtain the new simplex table. The next table for our example is given in Table 1.3 below. The new solution is $x_1 = 30$, $x_2 = 0$, $x_3 = 10$, $x_4 = 0$. For this solution, the objective function value is 6000.

Since we have one negative value in row F, this solution is not optimal. It can be improved, and, for that purpose, we should make the new iteration. We perform iterations until the optimal solution is reached.

The pivot column will be x_2 as only this variable has positive values in row F. The pivot row is obtained in the following way:

$$\min\left\{\frac{10}{0.5}, \frac{30}{0.5}\right\} = \min\{20, 60\} = 20 \Rightarrow \text{pivot row is } x_3$$

The pivot element is at the intersection of row x_3 and column x_2.

The new values for the elements of the table should be determined in the same way as in the previous step. The new table is shown in Table 1.4. The new solution is $x_1 = 20$, $x_2 = 20$, $x_3 = 0$, $x_4 = 0$, and $x_5 = 0$. The objective function value for this solution is 7200.

We note that there are no longer any negative elements in F row. Consequently, we conclude that this solution is the optimal solution.

Table 1.3 Simplex table with the second feasible solution

	F	x_1	x_2	x_3	x_4	Solution
F	1	0	−60	0	25	6000
x_3	0	0	0.5	1	−0.125	10
x_1	0	1	0.5	0	0.125	30

Table 1.4 Simplex table with the final solution of the problem

	F	x_1	x_2	x_3	x_4	Solution
F	1	0	0	120	15	7200
x_2	0	0	1	2	−0.25	20
x_1	0	1	0	−1	0.25	20

There are many software packages for solving linear programs. We are going to present how previous examples could be solved by Excel Solver, LPSolve IDE, or LINDO.

In the Excel sheet, we choose the cells for the decision variables, as well as one cell to calculate the objective function and to present the constraints. Figure 1.3 shows the appearance of the Excel sheet for the previous example. Let us suppose that the cells for the decision variables are A2 and B2. In cell B4, we calculate the objective function value in the following way: "=20*A2+15*B2". In this example, we have two constraints. The parts of the constraints to the left of the sign "≤" should be calculated in the cells A7 and A8. In cell A7, we should write "=A2+B2", and, in cell A8, Cell A8: "=8*A2+5*B2". The values given at the right side of the sign are just given as the cells: C7 and C8.

When the sheet is prepared, we start the Solver (from the pallet data). In the window of the Solver, we shall:

- enter the cell of the objective function,
- choose whether we shall maximize or minimize the objective function,
- enter constraints, and
- choose the solving method.

Figure 1.4 shows how the Solver looks after entering the values related to our example. By pressing the button Solve, we obtain the solution given in Figure 1.5. We notice that $x_1 = 20$, $x_2 = 20$, and $F = 7200$.

In the LPSolve IDE we write the objective function and the constraints in a similar way as we write on paper. Figure 1.6 shows the implementation of our linear program. When insertion of the objective function and

	A	B	C	D	E
1	x1	x2			
2					
3					
4	F =				
5					
6					
7	0	<=	1500		
8	0	<=	10000		
9					
10					

Figure 1.3 Data prepared in Excel for solving the problem

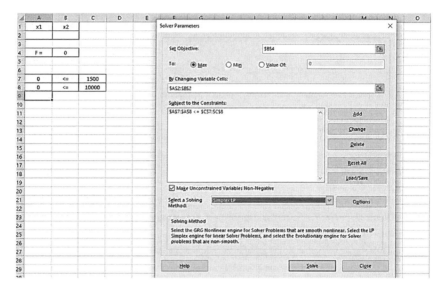

Figure 1.4 Completed Excel Solver

constraints is complete, we press the button Solve. The solution is given in Figure 1.7.

In a similar way, we can solve the linear program in LINDO. Figure 1.8 shows the implementation of our example. In the first row, we write the criteria of optimization and the objective function. After that, we write the set of constraints. Before we start with the constraints, we have to set "Subject to", and, at the end of the set of constraints, we have to write "End". The solution of this Solver is given in Figure 1.9.

	A	B	C	D	E
1	x1	x2			
2	20	20			
3					
4	F =	7200			
5					
6					
7	40	<=	40		
8	240	<=	240		
9					
10					

Figure 1.5 Solution obtained in Excel

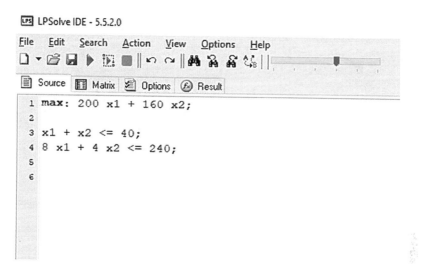

Figure 1.6 Linear program inserted in LPSolve IDE

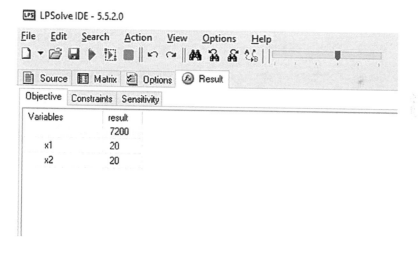

Figure 1.7 Solution obtained for the problem in LPSolve IDE

Example 1.2

The 2000 tons of cargo must be unloaded within 8 hours. Two cranes with capacities equal to 160 t/h and 200 t/h, respectively, can be used for unloading, with average cost per hour for the first crane of $100/h, and for the second, $150/h. Determine the optimal working times of the cranes to minimize the total cost of the cranes used.

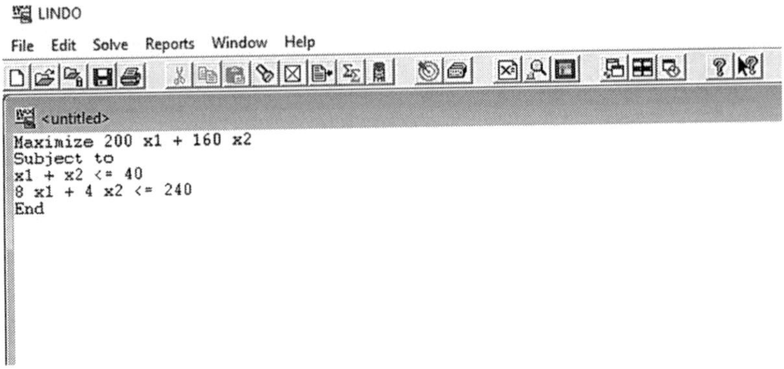

Figure 1.8 The linear program inserted in LINDO

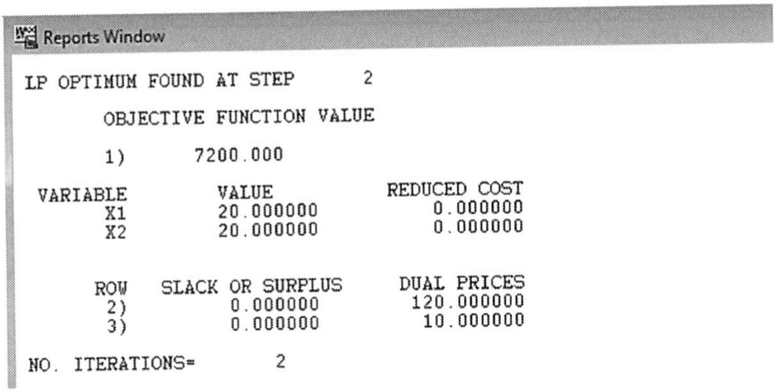

Figure 1.9 Solution of the problem obtained from LINDO

SOLUTION:

Let us denote variables x_1 and x_2 in the following way:

x_1 – working time for the first crane (in hours)
x_2 – working time for the second crane (in hours)

The linear program reads:
Minimize

$$F = 100 \cdot x_1 + 150 \cdot x_2$$

subject to:

$$160 \cdot x_1 + 200 \cdot x_2 \geq 2000$$

$$x_1 \leq 8$$

$$x_2 \leq 8$$

$$x_1 \geq 0, x_2 \geq 0$$

Since our problem has two decision variables, we present the problem graphically (Figure 1.10). By replacing "$\leq$" and "$\geq$" with "$=$", we create the equation for each constraint. We obtain the feasible solution space of the considered problem by taking into account all the constraints (Figure 1.10).

Figure 1.11 shows the solution space of the considered problem. To discover the optimal solution, we have to discover the feasible solution that has the smallest objective function value. In our case, the optimal solution is at the intersection of the following equations:

$$x_1 = 8 \text{ and } 160 \cdot x_1 + 200 \cdot x_2 = 2000$$

By solving the system of equations, we obtain the following: $x_1 = 8$ and $x_2 = 3.6$. The objective function value for this solution is equal to $1340.

The linear program from the previous example can be written as the set of equations in the following way:

Minimize

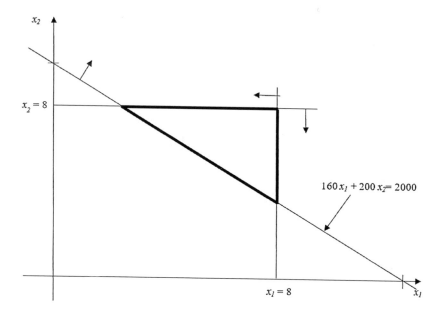

Figure 1.10 The feasible solutions space

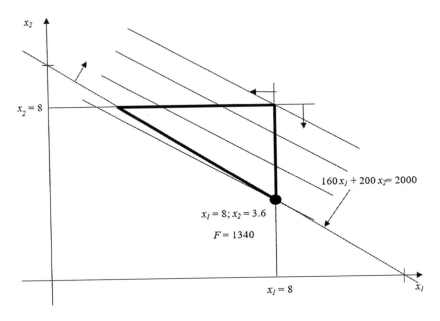

Figure 1.11 The solution of the problem

$$F = 100 \cdot x_1 + 150 \cdot x_2$$

subject to:

$$160 \cdot x_1 + 200 \cdot x_2 - x_3 = 2000$$

$$x_1 + x_4 = 8$$

$$x_2 + x_5 = 8$$

$$x_1 \geq 0, x_2 \geq 0, x_3 \geq 0, x_4 \geq 0, x_5 \geq 0$$

Variable x_3 is a surplus variable, and x_4 and x_5 are slack variables. Since we do not have a slack variable as the first constraint, we should add a new artificial variable in order to obtain the first feasible solution, i.e. to fill the simplex table. Let us denote this variable as R. The final value of R must be zero. Because of that, we add this variable to the objective function with a large positive coefficient. In general, we denote this coefficient as M. Now, our model has the following form:

Minimize

$$F = 100 \cdot x_1 + 150 \cdot x_2 + M \cdot R$$

subject to:

$$160 \cdot x_1 + 200 \cdot x_2 - x_3 + R = 2000$$

$$x_1 + x_4 = 8$$

$$x_2 + x_5 = 8$$

$$x_1 \geq 0, x_2 \geq 0, x_3 \geq 0, x_4 \geq 0, x_5 \geq 0$$

In this example we will take for M the value of 10000. Also, we will rewrite the objective function in a way that moves decision variables to the left-hand side of the sign "=". Now, we have the following equations:

$$F - 100 \cdot x_1 - 150 \cdot x_2 - 10000 \cdot R = 0$$

$$160 \cdot x_1 + 200 \cdot x_2 - x_3 + R = 2000$$

$$x_1 + x_4 = 8$$

$$x_2 + x_5 = 8$$

We fill the simplex table by taking into account the equations obtained. In this table, we write the coefficients of the decision variables. Since we have four equations, the table has four rows (Table 1.5).

We note that the column R does not have an appropriate form (all elements must be equal to zero, except the element at the intersection with row R; this element must be 1). Because of that, we have to make new values for row F in the following way: the elements in row R should be multiplied by the value for M (in our case, 10000). The value obtained should be added to the appropriate value of the element in row F. For example, the new elements in row F, are:

for the column $F : 1 + 10000 \cdot 0 = 1$

for the column $x_1 : -100 + 10000 \cdot 160 = 1599900$

for the column $x_2 : -150 + 10000 \cdot 200 = 1999850$

$$\vdots$$

for the column solution: $0 + 10000 \cdot 200 = 2000000$

Table 1.5 Initial simplex table

	F	x_1	x_2	x_3	x_4	x_5	R	Solution
F	1	−100	−150	0	0	0	−10000	0
R	0	160	200	−1	0	0	1	2000
x_4	0	1	0	0	1	0	0	8
x_5	0	0	1	0	0	1	0	8

Table 1.6 Simplex table with the first feasible solution

	F	x_1	x_2	x_3	x_4	x_5	R	Solution
F	1	1599900	1999850	−10000	0	0	0	20000000
R	0	160	200	−1	0	0	1	2000
x_4	0	1	0	0	1	0	0	8
x_5	0	0	1	0	0	1	0	8

The new simplex table has the form shown in Table 1.6.

We have to select the pivot column. For that purpose, we are going to select the decision variable that has the largest positive value in row F:

$$\max\{1599900,1999850\} = 1999850 \Rightarrow \text{column } x_2$$

In the next step, we determine the pivot row. Rows with positive elements in the pivot column are potential pivot rows. Among the potential pivot rows, we will select the one that has the smallest value of the quotient right-hand sides, with the element in the pivot column:

$$\min\left\{\frac{2000}{200}, \frac{8}{1}\right\} = \min\{10,8\} = 8 \Rightarrow \text{row } x_5$$

The new simplex table is given in Table 1.7. Since we still have positive values in row F, we should make a new iteration.

The pivot column will be x_1 since only this variable has a positive value in row F. The pivot row is obtained in the following way:

$$\min\left\{\frac{400}{160}, \frac{8}{1}\right\} = \min\{2.5,8\} = 2.5 \Rightarrow \text{pivot row is } R$$

The pivot number is at the intersection of row R and column x_1.

The new values for the elements of the table should be determined in the same way as in the previous step. The new simplex table is given in Table 1.8. We note again that the solution is not optimal as variable x_5 has positive values in row F. This time, the pivot element is at the intersection of the column x_5 and the row x_4.

Table 1.7 Simplex table with the second feasible solution

	F	x_1	x_2	x_3	x_4	x_5	R	Solution
F	1	1599900	0	−10000	0	−1999850	0	4001200
R	0	160	0	−1	0	−200	1	400
x_4	0	1	0	0	1	0	0	8
x_2	0	0	1	0	0	1	0	8

The solution obtained in the new table (Table 1.9) is the optimal solution as the values of all decision variables in row F are negative or zero. From the table, we note the following:

$F = 1340$,

$x_1 = 8$,

$x_2 = 3.6$, and

$x_4 = 4.4$.

Let us show how this problem can be solved by Excel Solver, LPSolve IDE, and LINDO.

Figure 1.12 shows the appearance of the Excel sheet. Cells A2 and B2 will be used for the values of the decision variables x_1 and x_2. Cell B4

Table 1.8 Simplex table with the third feasible solution

	F	x_1	x_2	x_3	x_4	x_5	R	Solution
F	1	0	0	−0.625	0	25	−9999.38	1450
x_1	0	1	0	−0.00625	0	−1.25	0.00625	2.5
x_4	0	0	0	0.00625	1	1.25	−0.00625	5.5
x_2	0	0	1	1	0	1	0	8

Table 1.9 Final simplex table

	F	x_1	x_2	x_3	x_4	x_5	R	Solution
F	1	0	0	−0.75	−20	0	−9999.25	1340
x_1	0	1	0	0	1	0	0	8
x_4	0	0	0	0.005	0.8	1	−0.005	4.4
x_2	0	0	1	0.995	−0.8	0	0.005	3.6

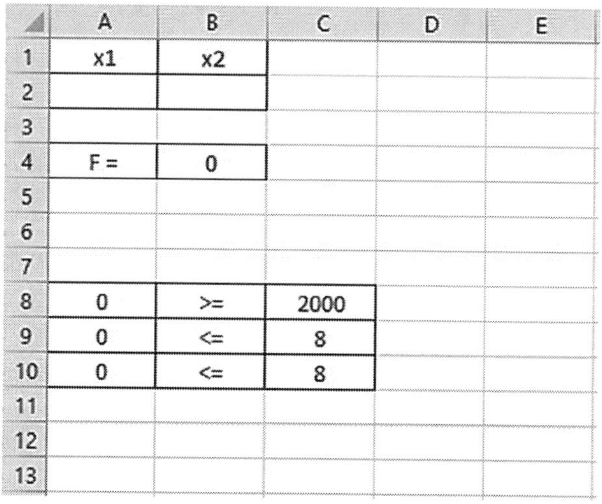

⧄	A	B	C	D	E
1	x1	x2			
2					
3					
4	F =	0			
5					
6					
7					
8	0	>=	2000		
9	0	<=	8		
10	0	<=	8		
11					
12					
13					

Figure 1.12 The appearance of the Excel sheet

will be used for the objective function value. We use cells A8, A9, A10, C8, C9, and C10 for the constraints. Let us explain how cells B4, A8, A9, and A10 should be completed. In the cell, we write the function that calculates the value of the objective function $F = 100 \cdot x_1 + 150 \cdot x_2$. In our sheet, we use cells A2 and B2 for the x_1 and x_2 values, respectively. Our function for cell B4 is: "=100*A2+150*B2". In a similar way, we can write the function for cell A8. In this cell, we calculate a part of the first constraint to the left of the sign "$\geq$", i.e., we calculate: $160 \cdot x_1 + 200 \cdot x_2$. This function can be written in the following way: "=160*A2+200*B2". To the left of the sign "$\leq$" in the second constraint is just x_2. The function for the cell A9 is written in the following way: "= A2". In a similar way, we write the function for the cell A10: "= B2".

When the sheet is completed, we start the Solver (from the pallet data). In the window of the Solver, we should:

- enter the cell of the objective function,
- choose whether we maximize or minimize the objective function,
- enter constraints, and
- choose the solving method.

Figure 1.13 shows what Solver looks like after entering the values related to our example. After using the button Solve, we obtain the

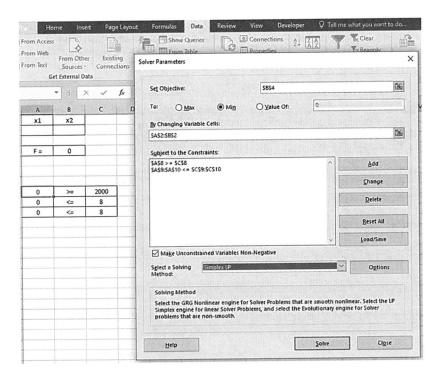

Figure 1.13 Excel Solver

	A	B	C	D	E
1	x1	x2			
2	8	3.6			
3					
4	F =	1340			
5					
6					
7					
8	2000	>=	2000		
9	8	<=	8		
10	3.6	<=	8		
11					
12					

Figure 1.14 The solution obtained

solution given in Figure 1.14. The solution is: $x_1 = 8$, $x_2 = 3.6$, and $F = 1340$.

In the LPSolve IDE, we write the objective function and the constraints in a similar way as we write on paper. Figure 1.15 shows the implementation of our linear program. When insertion of the objective function and constraints is completed, we press the button Solve. The solution is given in Figure 1.16.

In a similar way, we can solve the linear program in LINDO. Figure 1.17 shows the implementation of our example. In the first row, we write the criteria for optimization and the objective function. After

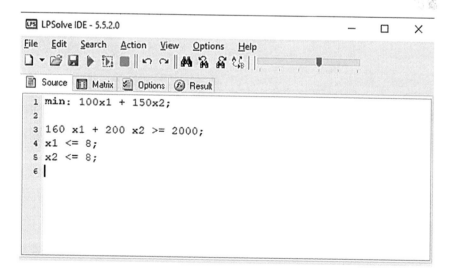

Figure 1.15 Linear program implementation in the LPSolve IDE

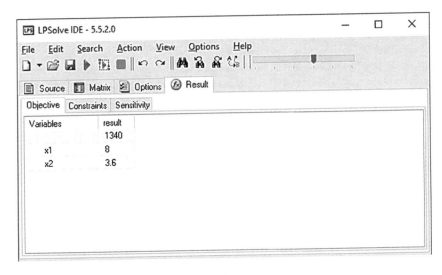

Figure 1.16 The solution obtained in the LPSolve IDE

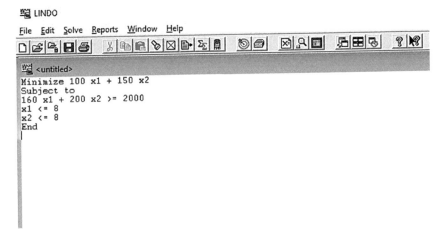

Figure 1.17 Linear program in Lindo

that, we write the set of constraints. Before we start with the constraints, we have to set "Subject to", and, at the end of the set of constraints, we have to write "End".

1.3 INTEGER AND MIXED-INTEGER PROGRAMMING

Analysts frequently realize that some or all of the variables in the formulated linear program must be *integers*. This means that some or all variables

take exclusively integer values. In order to make the formulated problem easier, analysts often allow these variables to take fractional values. For example, an analyst knows that the number of first-class passengers must be in the range between 30 and 40. A linear program could produce the "optimal solution" that tells us that the number of first-class passengers equals 37.8. In this case, we can neglect the fractional part, and we can decide to preserve 37 (or 38) seats for the first-class passengers. In this way, we are making a small numerical error, but we are capable of solving the problem easily. In some other situations, it is not possible for the analyst to behave in this way. Imagine that you have to decide about the alignment of a new highway. You must choose one out of numerous generated alternatives. This is a kind of "yes/no" ("1/0") decisions: "Yes", if the alternative is chosen, "No", if otherwise. In other words, we can introduce binary variables into the analysis. The variable has a value of 1 if the i-th alternative is chosen, and a value of 0 otherwise, i.e.:

$$x_i = \begin{cases} 1 & \text{if the } i\text{-th alternative is chosen} \\ 0 & \text{otherwise} \end{cases}$$

The value 0.7, for example, of the variable means nothing to us. We are not able to decide about the best highway alignment if the variables take fractional values.

There are various logical constraints that should be taken into account when handling variables that take exclusively integer values. For example, in a situation where the decision maker has to choose, at most, one alternative among n available alternatives, the constraint reads:

$$\sum_{i=1}^{n} x_i \leq 1$$

In some situations, at least one alternative must be chosen among n alternatives. This constraint reads:

$$\sum_{i=1}^{n} x_i \geq 1$$

When we solve problems similar to the highway alignment problem, we work exclusively with integer variables. These kinds of problems are known as *integer programs*, and the corresponding area is known as *integer programming*. Integer programs usually describe the problems in which one, or more, alternatives must be selected from a finite set of generated alternatives. There are also problems in which some variables can take only integer values, whereas some other variables can take fractional values. These problems are known as *mixed-integer programs*. It is much harder to solve integer programming problems than linear programming problems.

The following is an integer programming model:

Maximize

$$F = \sum_{j=1}^{n} c_j x_j$$

subject to:

$$\sum_{j=1}^{n} a_{ij} x_j \leq b_j \quad \forall\, i = 1,2,\ldots,m$$

$$x_j \geq 0 \quad \text{and integer } \forall\, j = 1,2,\ldots,n$$

There are numerous software systems that solve linear, integer, and mixed-integer linear programs (CPLEX, Excel and Quattro Pro Solvers, FortMP, LAMPS, LINDO, LINGO, MILP88, MINTO, MIPIII, MPSIII, OML, and OSL).

1.3.1 Branch-and-bound method

The branch-and-bound method has been one of the most-used techniques for solving nondeterministic polynomial-time (NP)-hard optimization problems for years. The method was proposed by Ailsa Land and Alison Doig in 1960. The name of the method was suggested by Little et al. (1963). This method solves integer problems by considering their continuous version. In the first step, the problem identified is solved as a continuous model.

Let us note, for example, an integer-constrained variable x_i. Let us denote, by x_i^*, its optimum continuous value, which is fractional. Obviously, there are no feasible integer solutions within the following interval:

$$\lfloor x_i^* \rfloor < x_i < \lfloor x_i^* \rfloor + 1$$

Therefore, a feasible integer value of x_i must satisfy one of the following two conditions:

$$x_i \leq \lfloor x_i^* \rfloor$$

or

$$x_i \geq \lfloor x_i^* \rfloor + 1$$

When constraint $x_i \leq \lfloor x_i^* \rfloor$ and constraint $x_i \geq \lfloor x_i^* \rfloor + 1$ are imposed on the original solution space, two mutually exclusive problems are generated

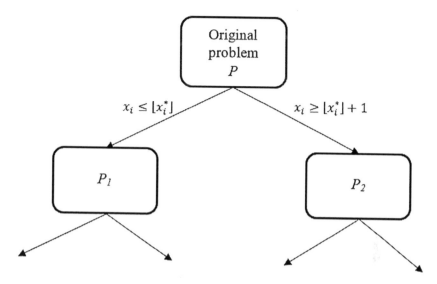

Figure 1.18 The original problem, P, is partitioned into two sub-problems P_1 and P_2

(Figure 1.18). In this way, an original problem P is partitioned into two sub-problems, P_1 and P_2. The term "partitioned" is most often replaced by the term "branched".

In the next step, we solve the sub-problems P_1 and P_2. These sub-problems are solved as linear programs. If the optimal solution for the sub-problem obtained is feasible, with respect to the integer problem, it is stored and has the status of the best-known solution. In this case, there is no need to further branch this sub-problem. In the opposite case, the sub-problem is again branched into two sub-problems (Figure 1.18). In the same way, as we did for the variable x_i, we, once more, impose the integer conditions on one of the integer variables that, at present, has a fractional optimal value. Every time we discover a feasible integer solution better than the best-known solution, it becomes a new best-known solution.

The process of branching continues in the search for an optimal solution to the analyzed problem. This process is stopped in the following cases: (a) each sub-problem ends with an integer solution; (b) the analyst becomes aware that a better solution cannot be obtained.

When does an analyst become aware that a better solution cannot be obtained? The branch-and-bound technique uses the concept of bounding. Let us imagine that the continuous optimum solution of a sub-problem is worse that the best-known integer solution. In this case, there is no need to further investigate the sub-problem. The branch-and-bound technique uses the objective function value of the best-known integer solution as a bound to reject poorer sub-problems.

Example 1.3

A company has to transport 180 pallets to a customer. Pallets could be transported in 40-foot or 20-foot containers. The company has five 40-foot containers and five 20-foot containers. The transportation cost for one 40-foot container equals $1000 and, for one 20-foot container equals $600. Taking into consideration the dimensions and weights of each pallet and the dimensions and capacity of the containers, it is decided that one 40-foot container can be loaded with 50 pallets, and one 20-foot container can be loaded with 22 pallets. Determine the number of 40-foot containers and the number of 20-foot containers that should be transported to minimize the total transportation cost.

SOLUTION:

Let us denote by x_1 the number of 40-foot containers that should be transported, and by x_2 the number of 20-foot containers that should be transported to the customer. The mathematical formulation of the problem reads:

Minimize

$$F = 1000x_1 + 600x_2$$

subject to:

$$50x_1 + 22x_2 \geq 180$$

$$x_1 \leq 5$$

$$x_2 \leq 5$$

$$x_1 \geq 0, x_2 \geq 0 \text{ and integer}$$

We solve this program by the branch-and-bound method. In the first step, we solve the previous program with continuous variables (x_1 and x_2). Let us denote this program by P. The program P reads:

Minimize

$$F = 1000x_1 + 600x_2$$

subject to:

$$50x_1 + 22x_2 \geq 180$$

$$x_1 \leq 5$$

$$x_2 \leq 5$$

$$x_1 \geq 0 \quad x_2 \geq 0$$

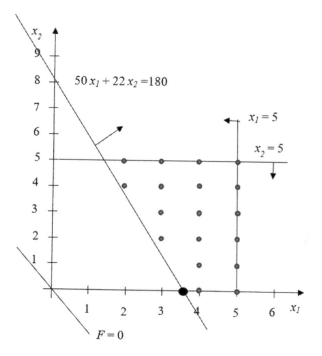

$50 x_1 + 22 x_2 = 180$

$x_1 = 5$

$x_2 = 5$

$F = 0$

Figure 1.19 Graphical representation of the linear program P

The graphical representation of this problem is shown in Figure 1.19. The optimal solution of this program reads: $x_1 = 3.6$, $x_2 = 0$, and $F = 3600$. Since the variable x_1 is not an integer, we make two new programs. Let us denote them by P1 and P2. Program P1 consists of the objective function and the constraints of the program P and one new constraint, $x_1 \leq 3$. Program P2 consists of the objective function and the constraints of the program P and one new constraint, $x_1 \geq 4$.

Program P1 reads:
Minimize

$$F = 1000 x_1 + 600 x_2$$

subject to:

$$50 x_1 + 22 x_2 \geq 180$$

$$x_1 \leq 5$$

$$x_2 \leq 5$$

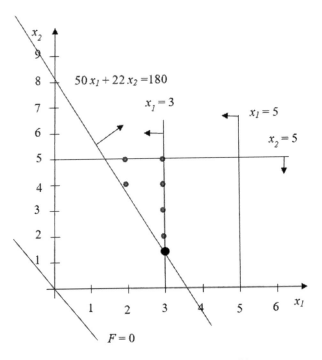

Figure 1.20 Graphical representation of the linear problem PI

$$x_1 \leq 3$$

$$x_1 \geq 0 \; x_2 \geq 0$$

Figure 1.20 shows the graphical representation of the program P1. By solving this problem, we obtain the following solution: $x_1 = 3$, $x_2 = 1.364$, $F = 3818.18$. We can note that x_2 does not have an integer value. Consequently, we form two new programs, P11 and P12. Program P11 is generated by adding the constraint $x_2 \leq 1$ to the program P1. The program P12 is created by adding the constraint $x_2 \geq 2$ to the program P1.

The program P2 reads:
Minimize

$$F = 1000x_1 + 600x_2$$

subject to:

$$50x_1 + 22x_2 \geq 180$$

$$x_1 \leq 5$$

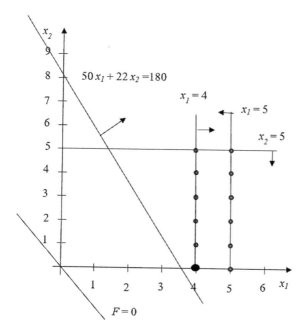

$50\,x_1 + 22\,x_2 = 180$

$x_1 = 4$

$x_1 = 5$

$x_2 = 5$

$F = 0$

Figure 1.21 Graphical representation of the linear problem P2

$x_2 \leq 5$

$x_1 \geq 4$

$x_1 \geq 0 \ x_2 \geq 0$

The graphical representation of this program is given in Figure 1.21. By solving this program, we obtain the following solution: $x_1 = 4$, $x_2 = 0$, and $F = 4000$. The variables have integer values in this solution. Since we do not yet have a solution with integer values of the variables, this solution should be saved as the solution of the initial problem. Additionally, since both variables are integers, we do not have to make new programs.

Let us return to branch P1 and check the programs P11 and P12. The program P11 reads:
Minimize

$$F = 1000x_1 + 500x_2$$

subject to:

$$50x_1 + 22x_2 \geq 180$$

$$x_1 \leq 5$$

$$x_2 \leq 5$$

$$x_1 \leq 3$$

$$x_2 \leq 1$$

$$x_1 \geq 0 \quad x_2 \geq 0$$

The graphical representation of this program is given in Figure 1.22. We note that this program does not have a feasible solution.

Let us analyze the program P12. This program has the following form:

Minimize

$$F = 1000x_1 + 500x_2$$

subject to:

$$50x_1 + 22x_2 \geq 180$$

$$x_1 \leq 5$$

$$x_2 \leq 5$$

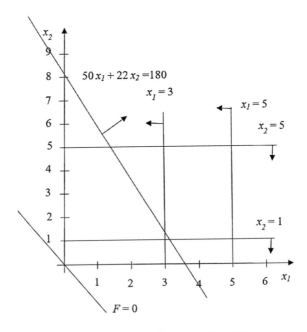

Figure 1.22 Graphical representation of the linear problem P11

$$x_1 \le 3$$

$$x_2 \ge 2$$

$$x_1 \ge 0 \ x_2 \ge 0$$

The graphical solution is given in Figure 1.23. The solution of this program is $x_1 = 2.72$, $x_2 = 2$, and $F = 3920$. The variable x_1 in this solution does not have an integer value and this program should be branched on two new programs, P121 and P122, by adding the new constraints $x_1 \le 2$ and $x_1 \ge 3$ to the program P12.
　The program P121 is:
　Minimize

$$F = 1000 x_1 + 500 x_2$$

subject to:

$$50 x_1 + 22 x_2 \ge 180$$

$$x_1 \le 5$$

$$x_2 \le 5$$

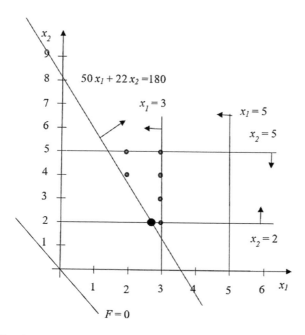

Figure 1.23 Graphical representation of the linear problem P12

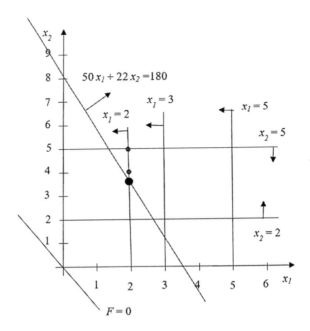

Figure 1.24 Graphical representation of the linear problem P121

$$x_1 \leq 3$$

$$x_2 \geq 2$$

$$x_1 \leq 2$$

$$x_1 \geq 0 \quad x_2 \geq 0$$

The graphical solution of this program is given in Figure 1.24. The solution of this program is $x_1 = 2$, $x_2 = 3.636$, and $F = 4181.82$. We note that x_2 does not have an integer value. We make a new branch since the objective function of this program ($F = 4181.82$) is greater than the objective function value of the previously found integer solution ($F = 4000$), so that this branch will be bounded.
   Solution reads:

$$x_1 = 2$$

$$x_2 = 3.636$$

$$F = 4181.82$$

As 4181.82 > 4000, this branch should be bounded.

Let us check whether it is possible to find a better solution by solving program P122. This program reads:

Minimize

$$F = 1000x_1 + 500x_2$$

subject to:

$$50x_1 + 22x_2 \geq 180$$

$$x_1 \leq 5$$

$$x_2 \leq 5$$

$$x_1 \leq 3$$

$$x_2 \geq 2$$

$$x_1 \geq 3$$

$$x_1 \geq 0 \quad x_2 \geq 0$$

The graphical representation of this problem is given in Figure 1.25. The solution of this program is: $x_1 = 3$, $x_2 = 2$, and $F = 4200$. We note

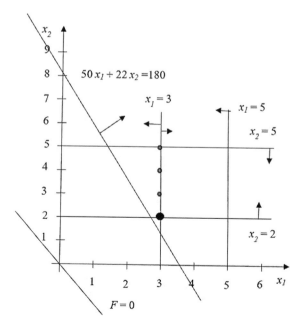

Figure 1.25 Graphical representation of the linear problem P122

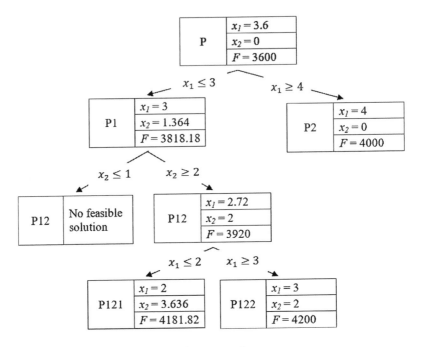

Figure 1.26 The overview of the branching procedure

that both variables have integer values. Since the objective function value is larger than the objective function value of the previously discovered integer solution (F = 4000), we do not accept the new solution.

Since we do not have unsolved programs, we stop with the algorithm. We conclude that the solution of the initial problem is $x_1 = 4$, $x_2 = 0$, and $F = 4000$. The overview of the branching procedure is shown in Figure 1.26.

The branch-and-bound method can be carried out using various software. Let us demonstrate how Excel Solver, LPSolver IDE, and LINDO can be used to solve the example in question.

Figure 1.27 shows the input data prepared for starting the Excel Solver. The functions on the cells B4, A7, A8, and A9 are as follows:

- cell B4: "=1000*A2 + 600*B2"
- cell A7: "=50*A2 + 22*B2"
- cell A8: "=A2"
- cell A9: "=B2".

Figure 1.28 shows how input data should be inserted into Excel Solver. There is a small difference in the method of data insertion into the linear program and the integer program. For the integer programs, we have to add one more constraint in order to make integer variables.

	A	B	C	D	E
1	x1	x2			
2					
3					
4	F =	0			
5					
6					
7	0	>=	180		
8	0	<=	5		
9	0	<=	5		
10					
11					
12					

Figure 1.27 The prepared input data

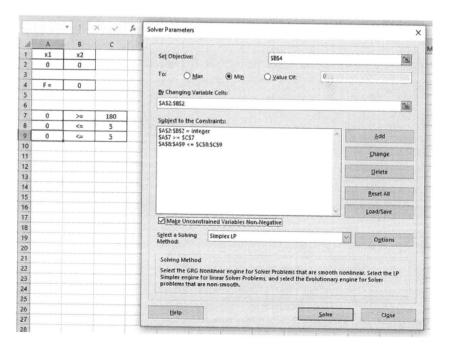

Figure 1.28 The Excel Solver for integer programming

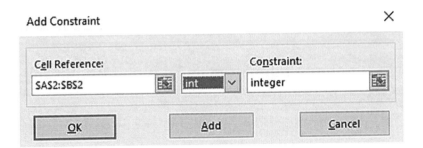

Figure 1.29 Constraint that variables are integers

This step can be done in the way shown in Figure 1.29. The final solution of this example is given in Figure 1.30.

Figure 1.31 shows how the problem in question can be solved by LPSolve IDE software. The only difference between carrying out the linear and integer programs in LPSolve IDE is that one more constraint is added. In that constraint, all integer variables should be defined. In the example in question, the constraint is shown in line 7 of Figure 1.31.

The solution of this example in LPSolve IDE is given in Figure 1.32.

Let us demonstrate how the same problem can be solved in LINDO software. As we can see from Figure 1.33, the process is almost the same as for linear programming. Again, we must define variables as integers. This must be done as the end of the constraint. We define that a variable is an integer by writing "gin" in front of that variable (see Figure 1.33. We write "gin" in front of x_1 and x_2). In our example we define that both variables are integers. The solution obtained is given in Figure 1.34.

	A	B	C	D	E
1	x1	x2			
2	4	0			
3					
4	F =	4000			
5					
6					
7	200	>=	180		
8	4	<=	5		
9	0	<=	5		
10					
11					
12					

Figure 1.30 The solution obtained from the Excel Solver

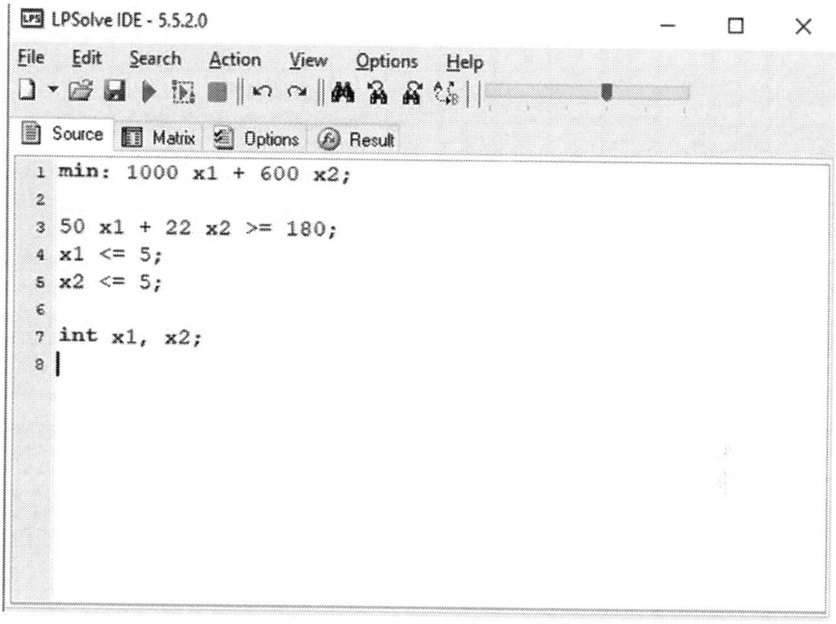

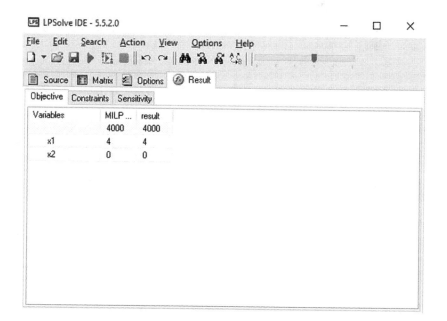

Figure 1.31 Integer program inserted in LPSolve IDE

Figure 1.32 The solution of the problem obtained from LPSolve IDE

LINDO

File Edit Solve Reports Window Help

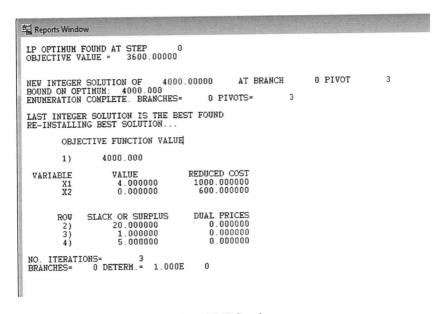

```
<untitled>
Minimize 1000 x1 + 600 x2
Subject to
50 x1 + 22 x2 >= 180
x1 <= 5
x2 <= 5
End
gin x1
gin x2
```

Figure 1.33 The Integer program inserted in LINDO software

```
Reports Window
LP OPTIMUM FOUND AT STEP      0
OBJECTIVE VALUE =   3600.00000

NEW INTEGER SOLUTION OF     4000.00000     AT BRANCH      0 PIVOT      3
BOUND ON OPTIMUM:  4000.000
ENUMERATION COMPLETE. BRANCHES=     0 PIVOTS=      3

LAST INTEGER SOLUTION IS THE BEST FOUND
RE-INSTALLING BEST SOLUTION...

        OBJECTIVE FUNCTION VALUE

    1)    4000.000

  VARIABLE        VALUE        REDUCED COST
       X1        4.000000       1000.000000
       X2        0.000000        600.000000


      ROW    SLACK OR SURPLUS    DUAL PRICES
       2)       20.000000         0.000000
       3)        1.000000         0.000000
       4)        5.000000         0.000000

NO. ITERATIONS=      3
BRANCHES=     0 DETERM.= 1.000E   0
```

Figure 1.34 The solution obtained from LINDO software

Example 1.4

The liberalization of airline tariffs has led to intensive competition among air carriers. Under such conditions, an air carrier logically wants to sell the seats available in a way that maximizes profit. The liberalization of airline tariffs has also resulted in a large number of different tariffs existing on the same flight. Passengers paying lower tariffs (as a rule, making private trips) often reserve seats before passengers paying higher tariffs (business class passengers who decide to travel several days or hours before the flight), which is why a certain number of passengers, who are prepared to pay a higher tariff, cannot find a vacant seat on the flight they want. The simplest reservation system is often called distinct fare class inventories, indicating separate seat inventories for each fare class. Once a seat is assigned to a fare class inventory, it may be booked only in that fare class, or else remain unsold. In the case of a nested reservation system, the high-fare request will not be rejected as long as any seats are available in lower fare classes. Let us consider the airline seat inventory control problem for a direct, nonstop flight. The capacity of an aircraft (the number of seats in the aircraft) equals 150. Let us assume that passengers are offered two tariff classes: $130, and $100. We assume that we are able to predict exactly the total number of requests for the two different passenger tariff classes. We expect 40 passenger requests for first class, and 120 passenger requests for second class. We decide to sell at least 15 seats to the passengers paying lower tariffs. We have to determine the total numbers of seats which must be sold in the different passenger tariff classes in order to achieve the maximum airline revenue.

SOLUTION:

Since we wish to determine the total numbers of seats sold in each of the different passenger tariff classes, the variables of the model can be defined as:

x_1 – the total number of seats planned to be sold in the first-class passenger tariff

x_2 – the total number of seats planned to be sold in the second-class passenger tariff

Since each seat from the first class tariff sells for $130, the total revenue from selling x_1 seats is $130 \cdot x_1$. In the same way, the total airline revenue from x_2 seats in second class is equal to $100 \cdot x_2$. The total airline revenue equals the sum of the two revenues, i.e.: $130 \cdot x_1 + 100 \cdot x_2$.

From the problem formulation, we conclude that there are specific restrictions on seat sales and on demand. The seat sales restrictions may be expressed verbally in the following way:

- Total number of seats sold in both classes together must be less than or equal to the aircraft capacity.
- Total number of seats sold in any class must be less than or equal to the total number of passenger requests.
- Total number of seats sold in second class must be at least 15.
- Total number of seats sold in first class and in second class cannot be less than zero (non-negativity constraint).

The following is the mathematical model for airline revenue management problem:

Maximize

$$F = 130x_1 + 100x_2$$

subject to

$$x_1 + x_2 \leq 150$$

$$x_1 \leq 40$$

$$x_2 \leq 120$$

$$x_2 \geq 15$$

$$x_i \geq 0 \quad \text{and integer} \quad i = 1,2$$

The implementation of this integer program in LPSolve IDE software is given in Figure 1.35. By solving this problem, we obtain the following solution: $x_1 = 40$, $x_2 = 110$. The objective function value equals \$16200.

Example 1.5

Many departing passengers walk significant distances in airport terminal buildings between check-in desks and gates (the word "gate" is used in the literature to describe aircraft stands at the airport terminals, as well as "off-pier stands" at the apron). Simultaneously, the distances walked by arriving passengers, between gates and the baggage claim area, can be also considerable. In addition, many transit passengers are expected to walk significant distances between specific

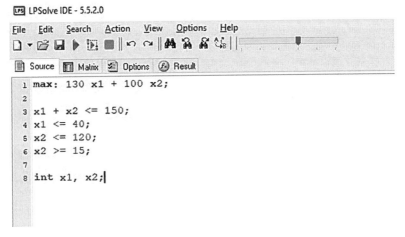

Figure 1.35 The implementation of the integer program in question in LPSolve IDE software

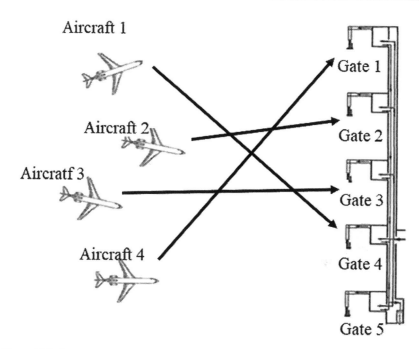

Aircraft 1

Gate 1

Aircraft 2

Gate 2

Aircratf 3

Gate 3

Aircraft 4

Gate 4

Gate 5

Figure 1.36 Graphical representation of the solution of the gate assignment problem

gates when changing planes (Figure 1.36). The total passenger walking distance within an airport terminal building may vary, depending on the specific assignment of aircraft to particular parking positions.

The standard *Gate Assignment Problem* can be defined in the following way. For a given set of parking positions and a given set of aircraft that can use any of these parking positions, find a parking position assignment for aircraft that will minimize the total walking distance of all passengers arriving, transiting and departing by aircraft parked at the set of parking positions. The decision maker must assign aircraft to available gates and must determine the start and end time of serving aircraft at the gate to which it has been assigned.

Airport gates are one of the greatest congestion points of the air transportation system. The total daily number of aircraft operations at large airports could be more than 1000, whereas the total number of gates is frequently more than 100.

Let us consider an example where we have to assign 4 aircraft to 5 available gates. Departure and arrival walking distances are given in Table 1.10.

The number of departing arriving passengers are given in Table 1.11.

SOLUTIONS:

For each pair of aircraft i and gates j we have to determine the total walking distance d_{ij}. According to data given in

Table 1.10 Walking distances for departure and arrival passengers

Gate	Walking distance for departure passengers between check-in desks and gate (m)	Walking distance for arrival passengers between gate and the baggage claim area (m)
1	320	250
2	405	360
3	290	300
4	150	320
5	500	275

Table 1.11 Numbers of departing and arriving passengers

Aircraft	Number of departing passengers	Number of arriving passengers
1	156	145
2	85	76
3	98	110
4	170	186

Table 1.12 Total walking distance (m) of all passengers for every aircraft-gate pair

d_{ij}	Gate 1	Gate 2	Gate 3	Gate 4	Gate 5
Aircraft 1	86170	115380	88740	69800	117875
Aircraft 2	46200	61785	47450	37070	63400
Aircraft 3	58860	79290	61420	49900	79250
Aircraft 4	100900	135810	105100	85020	136150

Table 1.10 and Table 1.11, these distances are calculated in the following way:

$$d_{11} = 156 \cdot 320 + 145 \cdot 250 = 86170$$

$$d_{12} = 156 \cdot 405 + 145 \cdot 360 = 115380$$

$$\vdots$$

$$d_{45} = 170 \cdot 500 + 186 \cdot 275 = 136150$$

Table 1.12 shows the total walking distance of all passengers for every aircraft–gate pair.

The Gate Assignment Problem is formulated in the following way:
Minimize

$$F = \sum_{i=1}^{4} \sum_{j=1}^{5} d_{ij} x_{ij}$$

subject to:

$$\sum_{j=1}^{5} x_{ij} = 1 \qquad \forall i = 1,2,3,4$$

$$\sum_{i=1}^{4} x_{ij} \leq 1 \quad \forall j = 1,2,3,4,5$$

$$x_{ij} \in \{0,1\} \qquad \forall i = 1,\dots,4; j = 1,\dots,5$$

The objective function represents the total walking distance. This objective function should be minimized. The first constraint guarantees that each aircraft will be served on one of the gates. A maximum of one aircraft can be served at each gate (as there are fewer aircraft than gates, this constraint has a sign $\leq$). This is defined by the second constraint. The third constraint defines the decision variables x_{ij} as being binary.

The mathematical formulation now reads:

Minimize

$$F = 86170x_{11} + 115380x_{12} + 88740x_{13} + \cdots + 136150x_{45}$$

subject to:

$$x_{11} + x_{12} + x_{13} + x_{14} + x_{15} = 1$$

$$x_{21} + x_{22} + x_{23} + x_{24} + x_{25} = 1$$

$$x_{31} + x_{32} + x_{33} + x_{34} + x_{35} = 1$$

$$x_{41} + x_{42} + x_{43} + x_{44} + x_{45} = 1$$

$$x_{11} + x_{21} + x_{31} + x_{41} \leq 1$$

$$x_{12} + x_{22} + x_{32} + x_{42} \leq 1$$

$$x_{13} + x_{23} + x_{33} + x_{43} \leq 1$$

$$x_{14} + x_{24} + x_{34} + x_{44} \leq 1$$

$$x_{15} + x_{25} + x_{35} + x_{45} \leq 1$$

$$x_{ij} \in \{0,1\} \qquad \forall i = 1,\dots,4; j = 1,\dots,5$$

By solving this problem, we obtained the solution where x_{14}, x_{22}, x_{33} and x_{41} equal one, and all other variables are zero. This means that

Table 1.13 Characteristics of trips

Trip	Departure terminal	Departure time	Arrival terminal	Arrival time
1	a	7:00	b	8:00
2	b	7:30	c	8:15
3	b	8:05	d	8:50
4	d	9:00	a	10:00
5	c	9:10	a	9:55
6	c	9:20	d	10:05

gate 4 should be assigned to aircraft 1, gate 2 to aircraft 2, gate 3 to aircraft 3, and gate 1 to aircraft 4. The graphical representation of this solution is given in Figure 1.36. The total walking distance for this solution for all the passengers equals 293,905 m.

Example 1.6

(Modified from Ceder (2015))
    There are two bus terminals a and b. There are six planned bus trips (Table 1.13 and Figure 1.37). The buses must be assigned to the set of planned timetabled trips in such a way as to minimize the total number

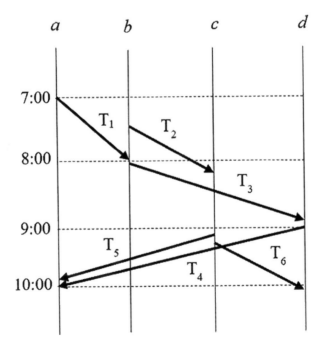

Figure 1.37 Graphical representation of the bus scheduling problem

Table 1.14 Travel times (min) of empty buses between the terminals

d_{ij}	a	b	c	d
a	0	50	40	50
b	50	0	35	40
c	40	35	0	35
d	50	40	35	0

of buses needed. The travel times of empty buses between terminals (d_{ij}) are given in Table 1.14.

SOLUTION:

Since buses have to perform 6 trips, we conclude that the number of buses necessary is six or less. We will need six buses if it is impossible to perform two trips by one bus. Our problem reads:

Maximize

$$F = \sum_{i \in T} \sum_{j \in T} x_{ij}$$

subject to:

$$\sum_{j \in T} x_{ij} \le 1 \quad \forall i \in T$$

$$\sum_{i \in T} x_{ij} \le 1 \quad \forall j \in T$$

$$x_{ij} \le k_{ij} \quad \forall i, j \in T$$

$$x_{ij} \in \{0,1\} \quad \forall i, j \in T$$

where:

T – set of trips

x_{ij} – decision variable that is equal to 1 if trips i and j should be joined, where the trip i is a predecessor of trip j, otherwise is equal to 0

k_{ij} – coefficient that takes values 1 or 0, according to the following expression:

$$k_{ij} = \begin{cases} 1, & \text{if } b_i + d_{ij} \le a_i \\ 0, & \text{otherwise} \end{cases}.$$

The objective function represents the sum of the joined trips. The first constraint guarantees that each trip may be joined with a maximum of one successor trip. The second constraint defines that each trip may be joined with a maximum of one predecessor trip. If trips i and j cannot be joined, the decision variable x_{ij} is equal to 0 (third constraint). The fourth constraint defines decision variables x_{ij} as binary.

Figures 1.38 and 1.39 show how this problem can be solved in Excel. The following are the functions at the cells B22, from B17 to G17 and from H11 to H16:

For cell B22: "=SUM(B11:G16)"
For cell B17: "=SUM(B11:B16)"
...
For cell G17: "=SUM(G11:G16)"
For cell H11: "=SUM(B11:G11)"
...
For cell H16: "=SUM(B16:G16)".

By solving the problem, we obtained the following solution: $x_{13} = 1$, $x_{34} = 1$, $x_{25} = 1$, with all other variables equaling zero. From this solution, we see that the first bus performs trips 1, 3, and 4. The second bus services trips 2 and 5. It is obvious that we need one more bus for trip 6. Consequently, the total number of buses is equal to three.

	A	B	C	D	E	F	G	H	I	J	K
1	kij	1	2	3	4	5	6				
2	1	0	0	1	1	1	1				
3	2	0	0	0	1	1	1				
4	3	0	0	0	1	0	0				
5	4	0	0	0	0	0	0				
6	5	0	0	0	0	0	0				
7	6	0	0	0	0	0	0				
8											
9											
10	xij	1	2	3	4	5	6				
11	1							0	<=	1	
12	2							0	<=	1	
13	3							0	<=	1	
14	4							0	<=	1	
15	5							0	<=	1	
16	6							0	<=	1	
17		0	0	0	0	0	0				
18		<=	<=	<=	<=	<=	<=				
19		1	1	1	1	1	1				
20											
21											
22	F =	0									
23											

Figure 1.38 The Excel spreadsheet

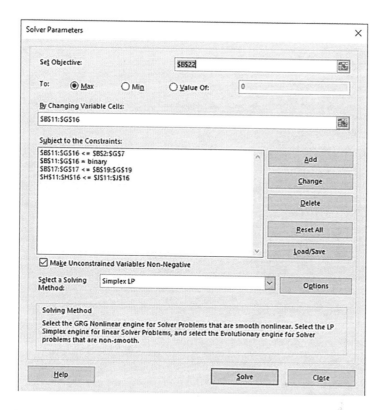

Figure 1.39 The completed Excel Solver

Example 1.7

There are three sections on a freeway. Capacities of the sections are given in Table 1.15. At each section, there is one on-ramp (Figure 1.40). The demands for on-ramps are given in Table 1.16. Before an on-ramp, the total traffic flow is $y = 4500$ vehicles/hour. Determine the maximal number of vehicles per hour that can enter the freeway from the on-ramps in the following two cases:

a) from each on-ramp, at least half of the on-ramp demand vehicles must enter per hour.

Table 1.15 Capacity of each section of freeway

Section	Section capacity (vehicles/h)
1	5000
2	5700
3	6300

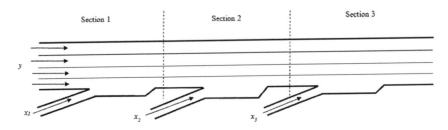

Figure 1.40 Three sections on a freeway

Table 1.16 Demand at on-ramps

Ramp	Demand (vehicles/h)
1	600
2	700
3	800

b) from each on-ramp, the same percent of vehicles per hour must enter, taking into consideration on-ramp demand.

SOLUTION:

Let us denote by x_1, x_2, and x_3 the integer decision variables, which represent the number of vehicles per hour that can enter the freeway from the on-ramps 1, 2, and 3, respectively. The objective function can be written in the following way: $F = x_1 + x_2 + x_3$. This objective function, that represents the total number of vehicles per hour that enter from the on-ramps, should be maximized. Taking into account the capacities of the sections, we write the following constraints:

$$x_1 + y \leq 5000$$

$$x_1 + x_2 + y \leq 5700$$

$$x_1 + x_2 + x_3 + y \leq 6300$$

In the same way, we define three constraints that take care of on-ramp demands. Decision variables x_1, x_2, and x_3 must be smaller or equal to on-ramp demand, i.e.:

$$x_1 \leq 600$$

$$x_2 \leq 700$$

$$x_3 \leq 800$$

Let us define the constraints in the cases a and b.

a) In this case, the variables x_1, x_2, and x_3 must be larger than or equal to half of the on-ramp demands, i.e.:

$$x_1 \geq 300$$

$$x_2 \geq 350$$

$$x_3 \geq 400$$

The integer program reads:
Maximize

$$F = x_1 + x_2 + x_3$$

subject to:

$$x_1 \leq 500$$

$$x_1 + x_2 \leq 1200$$

$$x_1 + x_2 + x_3 \leq 1800$$

$$x_1 \leq 600$$

$$x_2 \leq 700$$

$$x_3 \leq 800$$

$$x_1 \geq 300$$

$$x_2 \geq 350$$

$$x_3 \geq 400$$

$$x_i \geq 0 \text{ and integer} \quad i = 1,2,3$$

The solution of this problem is $x_1 = 500$, $x_2 = 700$, $x_3 = 600$. The objective function value equal 1800.

b) In this case, there are requirements that, from each on-ramp, the same percent of vehicles must enter, taking into account on-ramp demands. Bearing this in mind, we write:

$$\frac{x_1}{600} = \frac{x_2}{700} = \frac{x_3}{800} = p$$

where p is the continuous decision variable. Multiplying this variable by 100, we can obtain the percent of on-ramp demands (in vehicles per hour) that can enter the freeway. From the previous relation, we write the following three constraints:

$$x_1 - 600p = 0$$

$$x_2 - 700p = 0$$

$$x_3 - 800p = 0$$

The whole model reads:
Maximize

$$F = x_1 + x_2 + x_3$$

subject to:

$$x_1 + y \leq 5000$$

$$x_1 + x_2 + y \leq 5700$$

$$x_1 + x_2 + x_3 + y \leq 6300$$

$$x_1 \leq 600$$

$$x_2 \leq 700$$

$$x_3 \leq 800$$

$$x_1 - 600p = 0$$

$$x_2 - 700p = 0$$

$$x_3 - 800p = 0$$

$$x_i \geq 0 \text{ and integer} \quad i = 1,2,3$$

$$p \geq 0$$

The solution of this problem is $x_1 = 498$, $x_2 = 581$, $x_3 = 664$, $p = 0.83$. The objective function value equals 1743.
We notice that in this example we had integer variables, and one continuous variable. Generally speaking, when some variables can be only integer values, and some others can be fractional values, such a problem is known as a *Mixed-Integer Program*. In transportation

engineering, mixed-integer programs are much more common than linear or integer programs. The next two examples that we will consider represent mixed-integer programs.

Example 1.8

There are 7 vehicles in the depot. All vehicles have the same capacity of 200 units. There are 25 nodes that need to be served. Characteristics of the nodes are given in Table 1.17. Every node should belong to just one route. Service at every node must begin within the specified time window. The depot is denoted as zero node. Determine a set of vehicles' routes that has the minimum total length.

Table 1.17 Node characteristics

Node number	Coordinate X	Coordinate Y	Capacity	Time window		Service time
0	40	50	0	0	1236	0
1	40	66	20	170	225	90
2	20	85	40	475	528	90
3	33	35	10	16	80	90
4	28	35	10	1001	1066	90
5	44	5	20	286	347	90
6	40	15	40	35	87	90
7	48	30	10	632	693	90
8	47	40	10	12	77	90
9	95	30	30	387	456	90
10	90	35	10	203	260	90
11	85	35	30	47	124	90
12	70	58	20	458	523	90
13	63	58	10	737	802	90
14	65	85	40	475	518	90
15	60	85	30	561	622	90
16	55	80	10	743	820	90
17	45	30	10	734	777	90
18	42	10	40	186	257	90
19	0	45	20	567	624	90
20	20	55	10	449	504	90
21	25	52	40	169	224	90
22	30	50	10	10	73	90
23	22	75	30	30	92	90
24	42	65	10	15	67	90
25	45	70	30	825	870	90

SOLUTION:

Our problem is known as the vehicle routing problem with time windows (VRPTW). The mathematical formulations of this problem is given in the following way (Desrochers et al., 1988; Cordeau et al., 2007):

Minimize

$$F = \sum_{k \in K} \sum_{(i,j) \in A} c_{ij} \cdot x_{ij}^{k}$$

subject to:

$$\sum_{k \in K} \sum_{j \in \delta^{+}(i)} x_{ij}^{k} = 1 \qquad \forall i \in N$$

$$\sum_{j \in \delta^{+}(0)} x_{0j}^{k} = 1 \qquad \forall\, k \in K$$

$$\sum_{i \in \delta^{-}(j)} x_{ij}^{k} - \sum_{i \in \delta^{+}(j)} x_{ji}^{k} = 0 \qquad \forall k \in K, j \in N$$

$$\sum_{i \in \delta^{-}(n+1)} x_{i,n+1}^{k} = 1 \qquad \forall k \in K$$

$$w_{j}^{k} \geq w_{i}^{k} + s_{i} + t_{ij} - M \cdot \left(1 - x_{ij}^{k}\right) \quad \forall k \in K, (i,j) \in A$$

$$a_{i} \leq w_{i}^{k} \leq b_{i} \qquad \forall k \in K, i \in V$$

$$\sum_{i \in N} q_{i} \cdot \sum_{j \in \delta^{+}(i)} x_{ij}^{k} \leq Q \qquad \forall k \in K$$

$$x_{ij}^{k} \in \{0,1\} \qquad \forall k \in K, (i,j) \in A$$

where:

x_{ij}^{k} – binary variable that the takes value 1 if vehicle k, after visiting node i, goes to node j

w_{i}^{k} – decision variable that denotes the beginning of the service of node i by vehicle k

Q – vehicle capacity

c_{ij} – transportation cost from node i to node j (or the length of the path from node i to node j)

a_{i} – earliest time point by which service can start at node i

b_{i} – latest time point by which service can start at node i

s_{i} – duration of the service at node i

q_{i} – demand of node i

K – set of vehicles

$$\delta^+(i) = \{j : (i,j) \in A\}$$

$$\delta^-(j) = \{i : (i,j) \in A\}$$

The objective function corresponds to the total transportation costs, that need to be minimized. The first constraint ensures that all customers will be served. The second and third constraints guarantee that all the vehicles depart and come back, respectively, to the depot. The third constraint is the conservation flow constraint. The fifth and sixth constraints take care of the time windows. The seventh constraint takes into account the available vehicle capacity. The eight constraint defines decision variable x_{ij}^k as a binary variable.

By solving the mixed-integer program generated for the problem in question, we generated the seven routes, that have a total length of 549.07 distance units. The routes could be determined from the values of x variables. Variables x, that have values equal to1, are given in Table 1.18. By taking into account these variables, we obtained the seven routes that are given in Table 1.19.

The routes obtained are shown in the Figure 1.41.

Example 1.9

In one day, 15 ships should arrive at the port. Their lengths (in m), arrival, and service times are given in Table 1.20. The length of the quay is 1000 m. Service of all ships must be finished within 1920 min

Table 1.18 Variables x which are equal to 1

Variable	Value	Variable	Value	Variable	Value
$x_{0,8}^1$	1	$x_{0,3}^4$	1	$x_{7,17}^6$	1
$x_{8,26}^1$	1	$x_{3,4}^4$	1	$x_{17,26}^6$	1
$x_{0,23}^2$	1	$x_{4,26}^4$	1	$x_{18,5}^6$	1
$x_{2,26}^2$	1	$x_{0,22}^5$	1	$x_{0,24}^7$	1
$x_{23,2}^2$	1	$x_{19,26}^5$	1	$x_{1,14}^7$	1
$x_{0,11}^3$	1	$x_{20,19}^5$	1	$x_{14,15}^7$	1
$x_{9,12}^3$	1	$x_{21,20}^5$	1	$x_{15,16}^7$	1
$x_{10,9}^3$	1	$x_{22,21}^5$	1	$x_{16,25}^7$	1
$x_{11,10}^3$	1	$x_{0,6}^6$	1	$x_{24,1}^7$	1
$x_{12,13}^3$	1	$x_{5,7}^6$	1	$x_{25,26}^7$	1
$x_{13,26}^3$	1	$x_{6,18}^6$	1		

Table 1.19 The optimal set of routes

Route 1	0 – 8 –0
Route 2	0 – 23 – 2 – 0
Route 3	0 – 11 – 10 – 9 – 12 – 13 – 0
Route 4	0 – 3 – 4 – 0
Route 5	0 – 22 – 21 – 20 – 19 – 0
Route 6	0 – 6 – 18 – 5 – 7 – 17 – 0
Route 7	0 – 24 – 1 – 14 – 15 – 16 – 25 – 0

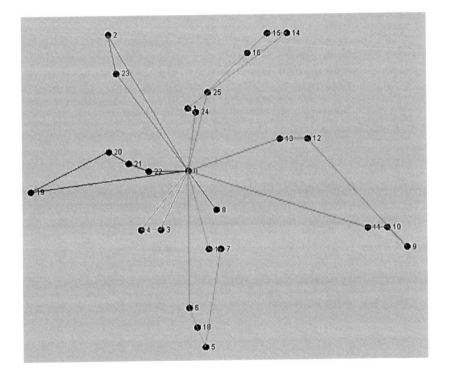

Figure 1.41 The optimal vehicle routes

of the beginning of the day. Determine the plan of ship services in a way that minimizes the total time that ships will spend at the port.

SOLUTION:

This problem is known in the literature as the berth allocation problem. When the length of the quay is given, instead of the fixed number of berths, the problem is known as the continuous berth allocation problem (as in the problem in question). In such a case, the ships' service plan is usually represented by a time–space diagram. One example of such a diagram is given in Figure 1.42.

Table 1.20 Characteristics of the ships

Ship	Length of ship (m)	Arrival time (min)	Service time (min)
1	150	15	300
2	200	60	240
3	250	90	360
4	100	210	180
5	300	255	360
6	150	360	480
7	200	410	180
8	250	480	540
9	130	590	300
10	180	750	240
11	200	1000	420
12	150	1215	360
13	180	1270	120
14	250	1370	540
15	200	1410	240

The mathematical formulation reads (Lee et al. 2010):
Minimize

$$F = \sum_{i=1}^{N} w_i \cdot \left(c_i - a_i \right)$$

subject to:

$$u_j - u_i - p_i - \left(\sigma_{ij} - 1 \right) \cdot T \geq 0 \qquad \forall 1 \leq i,j \leq N, i \neq j$$

$$v_j - v_i - s_i - \left(\delta_{ij} - 1 \right) \cdot S \geq 0 \qquad \forall 1 \leq i,j \leq N, i \neq j$$

$$\sigma_{ij} + \sigma_{ji} + \delta_{ij} + \delta_{ji} \geq 1 \qquad \forall 1 \leq i,j \leq N, i \neq j$$

$$\sigma_{ij} + \sigma_{ji} \leq 1 \qquad \forall 1 \leq i,j \leq N, i \neq j$$

$$\delta_{ij} + \delta_{ji} \leq 1 \qquad \forall 1 \leq i,j \leq N, i \neq j$$

$$p_i + u_i = c_i \qquad \forall 1 \leq i \leq N$$

$$a_i \leq u_i \leq \left(T - p_i \right) \qquad u_i \in \mathfrak{R}^+, \forall 1 \leq i \leq N$$

$$0 \leq v_i \leq \left(S - s_i \right) \qquad v_i \in \mathfrak{R}^+, \forall 1 \leq i \leq N$$

$$\sigma_{ij} \in \{0,1\}, \delta_{ij} \in \{0,1\} \qquad \forall 1 \leq i,j \leq N, i \neq j$$

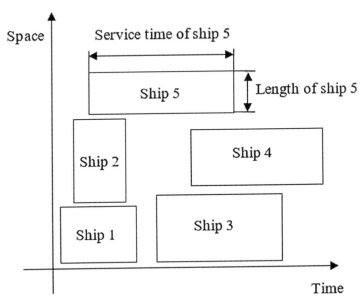

Figure 1.42 The example of the time–space diagram for continuous berth allocation problem

where:
  S – the length of the continuous berth
  T – the length of the planning horizon
  N – the total number of incoming ships,
  p_i – the service time for ship i, $1 \le i \le N$
  s_i – the length of ship i, $1 \le i \le N$
  a_i – arrival time of ship i, $1 \le i \le N$

and decision variables:
  w_i – the weight assigned for ship i, $1 \le i \le N$
  u_i – the mooring time of ship i, $1 \le i \le N$
  v_i – the starting berth position occupied by ship i, $1 \le i \le N$
  c_i –the departure time of ship i, $1 \le i \le N$

$$\sigma_{ij} = \begin{cases} 1, & \text{if ship } i \text{ is completely on the left of ship } j \text{ in the time-space diagram} \\ 0, & \text{otherwise} \end{cases}$$

$$\delta_{ij} = \begin{cases} 1, & \text{if ship } i \text{ is completely below ship } j \text{ in the time-space diagram} \\ 0, & \text{otherwise} \end{cases}$$

The objective function is the sum of the weighted turnaround times for all incoming ships. The first five constraints guarantee that the ships do not overlap in the time–space diagram. The completion time for

Table 1.21 The solution obtained for the continuous berth
allocation problem

	v_i	u_i	c_i
Ship 1	0	15	315
Ship 2	250	60	300
Ship 3	750	90	450
Ship 4	150	210	390
Ship 5	450	255	615
Ship 6	300	360	840
Ship 7	100	410	590
Ship 8	750	480	1020
Ship 9	0	590	890
Ship 10	450	750	990
Ship 11	0	1000	1420
Ship 12	200	1215	1575
Ship 13	350	1270	1390
Ship 14	530	1370	1910
Ship 15	780	1410	1650

each ship (c_i) is calculated according to the sixth constraint. The feasible domain of the decision variables is defined by the seventh, eighth, and ninth constraints.

In our example, we do not have values for the weights of ships (the weight represents the importance that the decision maker gives to the specific ship). We assume that the weight of each ship is equal to 1. By solving the mixed-integer program based on previous mathematical formulations, we obtain the solution given in the Table 1.21. The table contains the values of the v_i, u_i and c_i decision variables. From these values, we can conclude where the ships should be located on the quay, when the service of each ship starts, and when the service of each ship is finished. The graphical representation of this solution is given in Figure 1.43.

1.4 DYNAMIC PROGRAMMING

In traffic engineering, we often encounter problems for which successful solutions need to be made by a series of decisions over time. For example, city administrations face capital budgeting problems. The city administration often needs to allocate the available budget to several different transport infrastructure projects. At the end of each year, a decision is made for the next year. Such and similar problems are most often solved in stages, where the stages are connected through recursive computation. Dynamic programming solves problem in stages. In this way, the whole problem is broken down into smaller problems.

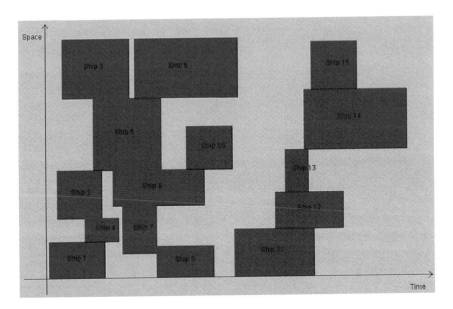

Figure 1.43 The graphical representation of the solution of the continuous berth allocation problem

Let us clarify the basic principles of dynamic programming by analyzing the following example. International freight is usually shipped in inter-
modal containers *via* multiple modes of transportation (rail, ship, truck).
There is no handling of the freight itself when freight changes its mode of
transportation. We consider shipping of one 20-foot container that should
be shipped from the supplier (node 0) to the client (node 9). The company
could send the container by truck to the railway station at node 1, to the
railway station at node 2, or to the railway station at node 3 (Figure 1.44).
From any of these railway stations, the container could be sent by rail to
ports (node 4, node 5, or node 6). From any of these ports, the container
could be sent by container ship to port 7 or port 8. Finally, from port 7 or
port 8, the container could be sent by truck to the final destination (node 9).

The transportation costs along specific links are given in Table 1.22.

Let us consider, for example, container route $(0, 2, 4, 8, 9)$.
The total transportation costs of this container route are equal to
$500 + 400 + 1,600 + 800 = 3,300$. The total transportation costs in the case of
the container route $(0, 3, 5, 7, 9)$ are equal to $600 + 350 + 1550 + 700 = 3,200$.
Obviously, route $(0, 2, 4, 8, 9)$ is cheaper than route $(0, 3, 5, 7, 9)$. By pro-
ceeding in this way, we can examine all possible routes in the network, calculate their costs, and finally discover the cheapest route. We could perform
exhaustive enumerations, i.e., we could compute the total cost for each of
the possible container routes. The optimal solution would be the container
route that has the lowest transportation costs. In the case of medium- and

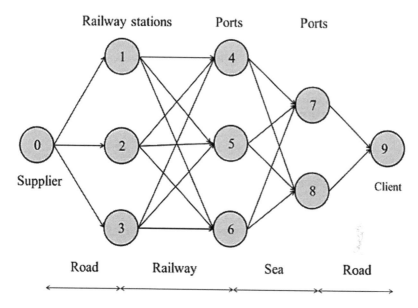

Figure 1.44 The transportation costs along specific links given in Table 1.22

Table 1.22 Transportation costs

Link	Cost ($)	Link	Cost ($)
(0, 1)	450	(3, 5)	350
(0, 2)	500	(3, 6)	300
(0, 3)	600	(4, 7)	1,500
(1, 4)	250	(4, 8)	1,600
(1, 5)	300	(5, 7)	1,550
(1, 6)	350	(5, 8)	1,700
(2, 4)	400	(6, 7)	1,450
(2, 5)	250	(6, 8)	1,850
(2, 6)	300	(7, 9)	700
(3, 4)	550	(8, 9)	800

large-sized transportation networks, exhaustive enumeration would lead to computational inefficiency. In addition, exhaustive enumeration does not use any information on the routes previously examined to remove future, poorer routes.

Our task is to discover the cheapest shipping route for the container. The cheapest shipping route problem could be solved in stages. Dynamic programming is a mathematical procedure that solves problems in stages (Figure 1.45). Dynamic programming was introduced in 1953 by Richard Ernest Bellman (1920–1984), an American applied mathematician. In each

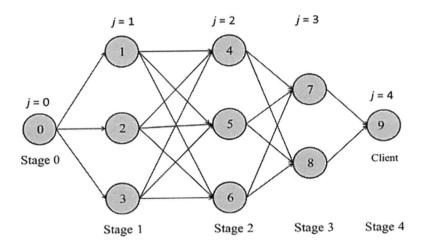

Figure 1.45 Network model of the cheapest shipping route problem

stage, the analyst considers one optimizing variable. The computations at the various stages are linked in such a way that the optimal solution of the problem in question is obtained when the final stage is reached.

Each point, where the container is being transferred from one transportation mode to the other, represents a stage. The initial stage $(j = 0)$ represents the supplier. The first stage $(j = 1)$ refers to the container transfer from the truck to the train. It can happen at node 1, node 2, or node 3. The second stage $(j = 2)$ refers to the container transfer from the train to the container ship. This operation could happen at node 4, node 5, or node 6. The third stage $(j = 3)$ refers to the container transfer from the container ship to the truck. This could occur at node 7 or node 8. Finally, the last stage $(j = 4)$ represents the client.

Let us introduce the following notation:

x_1 – the total cost in stage 1
x_2 – the total cost in stage 1 and stage 2
x_3 – the total cost in stage 1, stage 2, and stage 3
x_4 – the total cost in stage 1, stage 2, stage 3, and stage 4

The lengths of the links connecting the nodes of consecutive stages represent the transportation cost of the link. The lengths of the links that connect stage 0 with the nodes in stage 1 are indicated in Figure 1.46.

Let us introduce the following notation:

$f_j(x_j)$ – the length of the shortest path that leads to node x_j at stage j.

The following is satisfied in the initial stage (supplier): $f_0(0) = 0$. We are now ready to calculate transportation costs stage by stage.

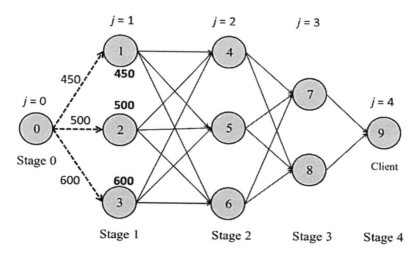

Figure 1.46 Calculating transportation costs of the first stage

Stage 1 (container transfer from the truck to the train):

The length of the shortest path that leads to node x_1 at stage 1 is:

$$f_1(x_1) = f_0(x_0) + C_1(0, x_1)$$

where $C_1(0, x_1)$ are the transportation costs of link $(0, x_1)$. We have the following:

$$f_1(1) = 0 + 450 = 450$$

$$f_1(2) = 0 + 500 = 500$$

$$f_1(3) = 0 + 600 = 600$$

The lengths of the shortest paths that lead to nodes 1, 2, and 3 are indicated in Figure 1.46 in bold letters.

Stage 2 (container transfer from the train to the ship):

We now calculate the lengths of the shortest paths to nodes 4, 5, and 6. All these nodes have three incoming links from the stage 1. Consequently, we write the following:

$$\begin{pmatrix} \text{length of shortest} \\ \text{path to node } x_2 \end{pmatrix} = \min_{\substack{\text{feasible} \\ \text{links} \\ (x_1, x_2)}} \left\{ \begin{pmatrix} \text{length of shortest} \\ \text{path to node } x_1 \end{pmatrix} + \begin{pmatrix} \text{length of link} \\ (x_1, x_2) \end{pmatrix} \right\}$$

In other words:

$$f_2(x_2) = \min_{\substack{\text{feasible} \\ \text{links} \\ (x_1, x_2)}} \{f_1(x_1) + C_2(x_1, x_2)\}$$

i.e.:

$$f_2(4) = \min(450 + 250, \ 500 + 400, \ 600 + 550)$$
$$= \min(700, \ 900, \ 1150) = 700$$

$$f_2(5) = \min(450 + 300, \ 500 + 250, \ 600 + 350)$$
$$= \min(750, \ 750, \ 950) = 750$$

$$f_2(6) = \min(450 + 350, \ 500 + 300, \ 600 + 300)$$
$$= \min(800, \ 800, \ 900) = 800$$

Figure 1.47 illustrates the calculation of the transportation costs of the second stage.

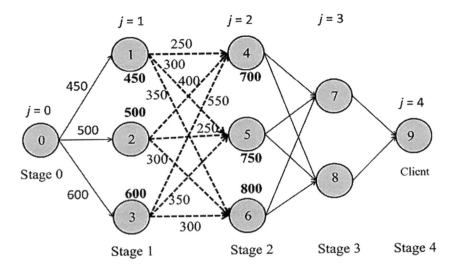

Figure 1.47 Calculating transportation costs of the second stage

Stage 3 (container transfer from the container ship to the truck):

The length of the shortest path $f_3(x_3)$ to node x_3 equals:

$$f_3(x_3) = \min_{\substack{\text{feasible} \\ \text{links} \\ (x_2, x_3)}} \{f_2(x_2) + C_3(x_2, x_3)\}$$

$$f_3(7) = \min(700 + 1,500, \ \ 750 + 1,550, \ 800 + 1,450)$$

$$= \min(2,200, \ 2,300, \ \ 2,250) = 2,200$$

$$f_3(8) = \min(700 + 1,600, \ \ 750 + 1,700, \ 800 + 1,850)$$

$$= \min(2,300, \ 2,450, \ 2,650) = 2,300$$

Figure 1.48 illustrates calculation of the transportation costs of the third stage.

Stage 4 (delivering container to the client):

The length of the shortest path $f_3(x_3)$ to node x_3 equals:

$$f_4(x_4) = \min_{\substack{\text{feasible} \\ \text{links} \\ (x_2, x_3)}} \{f_3(x_3) + C_4(x_3, x_4)\}$$

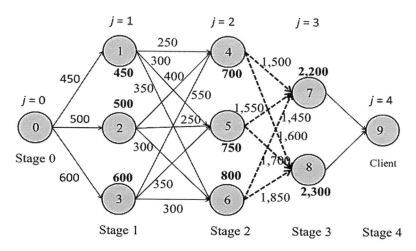

Figure 1.48 Calculating transportation costs of the third stage

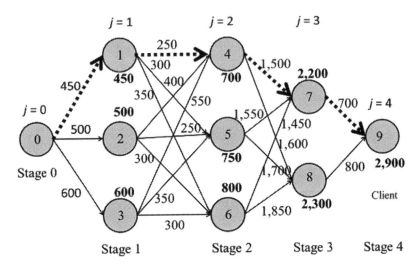

Figure 1.49 The cheapest shipping route for the container

i.e.:

$$f_4(9) = \min(2,200 + 700, \ 2,300 + 800) = \min(2,900, \ 3,100) = 2,900$$

The length of the shortest (cheapest) path from the supplier to the client is:

$$(0, 1, 4, 7, 9)$$

The optimal path is shown in Figure 1.49.

1.4.1 Forward and backward recursive equation

If we analyze the equations that we used to solve the container shipment problem, we see that they are recursive, since knowing $f_0(x_0)$, we compute $f_j(x_j) f_0(x_0) = 0$ for stage j from $f_{j-1}(x_{j-1})$ at stage $j-1$. The dynamic programming recursive equations read:

$$f_0(x_0) = 0$$

$$f_j(x_j) = \min_{\substack{\text{feasible proposals } p_j}} \{f_{j-1}(x_{j-1}) + C_j(p_j)\} \qquad j = 1, \dots, n$$

The computations are carried out from the first stage to the last stage. This type of computation is called a forward procedure. Graphical representation of the forward pass is given in Figure 1.50.

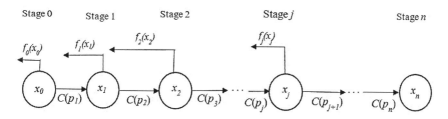

Stage 0   Stage 1   Stage 2   Stage j Stage n

Figure 1.50 **Graphical representation of the forward procedure of dynamic programing**

There are also backward procedures, where computations start from the last stage and then go backward toward the first stage. The equations for the backward procedure can be written in the following way:

$$f_n(x_n) = 0$$

$$f_j(x_j) = \min_{\text{feasible proposals } p_j} \left\{ f_{j+1}(x_{j+1}) + C_j(p_j) \right\} \qquad j = 1,\ldots,n$$

Example 1.10

A public transit operator serves 7 bus routes with 89 buses. Depot and bus lines served by the depot are shown in Figure 1.51.

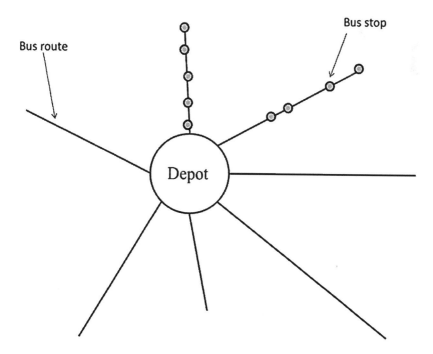

Figure 1.51 **Depot and bus routes served by the depot**

Characteristics of the routes are shown in Table 1.23. The capacity of one C_b bus is $C_b = 100$.

where:

λ_i [Pass./h] – passenger arrival rate at route i

q_i^{max}[Pass./h] – maximum load section at route i

T_i [h] – bus turnaround time at route i (an elapsed interval of time in which a bus can pass line i in both directions)

N_i – planned number of buses at route i

C_i [Pass./h] – the offered capacity at route i

α_i – load factor at route i

$N_{i,min}$ –- minimal number of buses necessary

b_i^{max} – maximum allowed headway at route i

A certain number of the planned buses could, from time to time, be out of operation due to various technical reasons. A smaller number of buses in operation than is planned could also be caused by driver absenteeism. Suppose that 15 buses are out of operation. Assign the remaining 74 buses on the routes in a way to minimize the total waiting time of passengers. The time interval over which we consider the problem is $T^o = 1$ h.

SOLUTION:

The time interval over which we consider the problem is $T^o = 1$ h. Let us note the i-th bus route (Figure 1.52).

We denote by q_i^{max} the number of carried passengers per hour on the maximum load section of bus route i. We also denote by M_i the assigned number of buses on route i. The service frequency f_i represents the number of vehicles per time unit past a specific point in the same direction. The frequency equals:

$$f_i = \frac{M_i}{T_i} \left[\frac{veh}{h} \right]$$

The frequency f_i is expressed as the number of vehicles per hour. Headway in public transportation operations represents the time interval between vehicles past a specific point. Headways are expressed in minutes. As the frequency represents the number of vehicles per time unit past a specific point in the same direction, the frequency is the inverse of the headway, i.e.:

$$f_i = \frac{1}{b_i}$$

It has been shown, and widely accepted, that, in the case of regular bus arrivals, the average waiting time per passenger, w_i, represents one half of a headway, i.e.:

$$w_i = \frac{1}{2} \cdot b_i = \frac{1}{2} \frac{1}{f_i} = \frac{T_i}{2M_i}$$

Table 1.23 Bus route characteristics

Bus route	λ_i [Pass./h]	q_i^{max} [Pass./h]	T_i [h]	N_i	f_i [Departures/h]	h_i [h]	C_i [seats/h]	Load factor α_i	$N_{i,min}$	$h_i^{max}[h]$
1	1500	300	1	5	5	0.2	500	0.6	3	0.25
2	7000	1500	1.5	30	30	0.05	2000	0.75	23	0.1
3	3500	800	1	10	10	0.1	500	0.8	8	0.2
4	2000	400	1	5	5	0.2	500	0.8	4	0.2
5	6000	1400	1	20	20	0.05	2000	0.7	14	0.167
6	3000	600	1.5	10	15	0.1	1000	0.6	9	0.2
7	1000	200	1	4	4	0.25	400	0.5	2	0.333

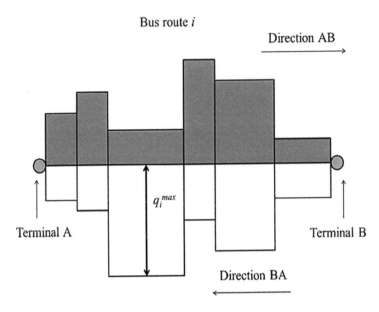

Figure 1.52 The number of passengers carried per hour on different load sections of bus route *i*

The total passenger waiting time T_w on all 7 bus routes during the time interval T^o in question equals:

$$T_w = \sum_{i=1}^{7} \lambda_i \cdot T^o \cdot \frac{T_i}{2 \cdot M_i}$$

If we wish to transport all passengers on all seven bus routes, the following inequality must be satisfied:

$$q_i^{max} \cdot T_i \leq C_b \cdot M_i \qquad \forall i = 1,2,\ldots,7$$

i.e.:

$$M_i \geq \frac{q_i^{max} \cdot T_i}{C_b} \qquad \forall i = 1,2,\ldots,7$$

For every bus line *i*, we can also introduce the maximum allowed headway, h_i^{max}. The maximum allowed headway h_i^{max} on bus route *i* is one of the basic indicators of the level of service offered to the passengers. We will assign buses to the bus routes in such a way that no one bus route has headway greater than the maximum allowed headway. In a case of existing maximum allowed headways h_i^{max} for every bus route *i*, the following inequality must be satisfied:

$$\frac{T_i}{M_i} \le b_i^{max}$$

We try to assign available buses to bus routes in such a way as to minimize the total passenger waiting time on all bus routes. Since there is a shortage of buses, the assigned number of buses M_i at any bus route must be less than or equal to the planned number of buses, N_i. Mathematical formulation of our problem reads:

Minimize

$$T_w = \sum_{i=1}^{7} \lambda_i \cdot T^o \cdot \frac{T_i}{2 \cdot M_i}$$

subject to:

$$max \left\{ \frac{T_i}{b_i^{max}}, \frac{q_i^o \cdot T_i}{C_b} \right\} \le M_i \le N_i \quad i = 1, 2, \ldots, 7$$

$$\sum_{i=1}^{7} M_i = 74$$

$$M_i \ge 0 \quad \text{and} \quad \text{integer} \quad i = 1, 2, \ldots, 7$$

The objective function, which represents the total waiting time of all passengers, should be minimized. First constraint defines how many buses we should assign to each route. The total number of available buses is 74. This is given by second constraint. Decision variables M_i are defined as integers by third constraint.

By taking into account first constraint, we calculate the following possible number of buses at specific routes:

$$4 \le M_1 \le 5$$

$$23 \le M_2 \le 30$$

$$8 \le M_3 \le 10$$

$$M_4 = 5$$

$$14 \le M_5 \le 20$$

$$9 \le M_6 \le 15$$

$$3 \le M_7 \le 4$$

We decide that stages $i = 1, 2, \ldots, 7$ in our example are bus routes. We denote by 1, 2, ...,7 the first, second, ..., seventh stages (bus routes).

State x_i at stage i represents the total number of buses assigned to the first, second, ..., and the j-th bus route, i.e.:

x_1 – number of buses assigned to the first route
x_2 – number of buses assigned to the first and second routes
x_3 – number of buses assigned to the first, second, and third routes
$\vdots$
x_7 – number of buses assigned to the first, second, third, fourth, fifth, sixth, and the seventh routes

In other words:

$$x_i = M_1 + M_2 + \cdots + M_i, \qquad i = 1, 2, ..., 7$$

The values of the $x_1, x_2, x_3, x_4 x_5, x_6, x_7$ must be within the following intervals:

x_1:

$$4 \leq x_1 \leq \min\{5, 74 - (23 + 8 + 5 + 14 + 9 + 3)\} = \min\{5, 12\}$$

$$4 \leq x_1 \leq 5$$

x_2:

$$4 + 23 \leq x_2 \leq \min\{5 + 30, 74 - (8 + 5 + 14 + 9 + 3)\} = \min\{35, 35\}$$

$$27 \leq x_2 \leq 35$$

x_3:

$$27 + 8 \leq x_3 \leq \min\{35 + 10, 74 - (5 + 14 + 9 + 3)\} = \min\{45, 43\}$$

$$35 \leq x_3 \leq 43$$

x_4:

$$35 + 5 \leq x_4 \leq \min\{43 + 5, 74 - (14 + 9 + 3)\} = \min\{48, 48\}$$

$$40 \leq x_4 \leq 48$$

x_5:

$$40 + 14 \leq x_5 \leq \min\{48 + 20, 74 - (9 + 3)\} = \min\{68, 62\}$$

$$54 \leq x_5 \leq 62$$

x_6:

$$54 + 9 \leq x_6 \leq \min\{62 + 15, 74 - 3\} = \min\{77, 71\}$$

$$63 \leq x_6 \leq 71$$

The x_7 must be equal to 74, i.e.:

$$x_7 = 74$$

Let us calculate the total waiting time if we assign buses 4 and 5 to bus route 1. The waiting times are:

$$f_1(4) = \lambda_1 \cdot T^o \frac{T_1}{2 \cdot M_1} = 1500 \cdot 1 \cdot \frac{1}{2 \cdot 4} = 187.5$$

$$f_1(5) = \lambda_1 \cdot T^o \frac{T_1}{2 \cdot M_1} = 1500 \cdot 1 \cdot \frac{1}{2 \cdot 5} = 150$$

These values are shown in Table 1.24. From the table, we can conclude that optimal values for M_1, if x_1 are 4 and 5, are 4 and 5, respectively.

The optimal values of M_2, for different values of x_2, are given in Table 1.25.

Let us clarify the performed calculations. The total waiting time, if we assign 27 buses to bus routes 1 and 2, is:

$$f_2(27) = \lambda_2 \cdot T^o \frac{T_2}{2 \cdot M_2} + f_1(4) = 7000 \cdot 1 \cdot \frac{1.5}{2 \cdot 23} = 228.26 + 187.5 = 415.76$$

If we assign 28 buses to bus routes 1 and 2, the total waiting time equals:

$$f_2(28) = \min_{M_2=23,24} \left\{ \lambda_2 \cdot T^o \frac{T_2}{2 \cdot M_2} + f_1(28 - M_2) \right\}$$

$$= \min_{M_2=23,24} \left\{ 7000 \cdot 1 \cdot \frac{1.5}{2 \cdot M_2} + f_1(28 - M_2) \right\}$$

$$f_2(28) = \min_{M_2=23,24} \left\{ \frac{5250}{M_2} + f_1(28 - M_2) \right\}$$

$$= \min \begin{cases} 228.26 + f_1(5) = 378.26 & for\ M_2 = 23 \\ 218.75 + f_1(4) = 406.25 & for\ M_2 = 24 \end{cases} = 378.26 \Rightarrow M_2^* = 23$$

The waiting time of 378.26 hours is obtained for $M_2^* = 23$.

Table 1.24 Optimal values of M_1 for different values of x_1

x_1	$M_1 = 4$	$M_1 = 5$	$f_1(x_1)$	M_1^*
4	187.5	–	187.5	4
5	–	150	150	5

Table 1.25 Optimal values of M_2 for different values of x_2

x_2	\multicolumn{8}{c	}{M_2}	$f_2(x_2)$	M_2^*
	23	24	25	26	27	28	29	30		
27	415.76	–	–	–	–	–	–	–	415.76	23
28	378.26	406.25	–	–	–	–	–	–	378.26	23
29	–	368.75	397.5	–	–	–	–	–	368.75	24
30	–	–	360	389.42	–	–	–	–	360	25
31	–	–	–	351.92	381.94	–	–	–	351.92	26
32	–	–	–	–	344.44	375	–	–	344.44	27
33	–	–	–	–	–	337.5	368.53	–	337.5	28
34	–	–	–	–	–	–	331.03	362.5	331.03	29
35	–	–	–	–	–	–	–	325	325	30

If we assign 29 buses to bus routes 1 and 2, the total waiting time equals:

$$f_2(29) = \min_{M_2=24,25} \left\{ \frac{5250}{M_2} + f_1(29 - M_2) \right\}$$

$$= min \begin{cases} 218.75 + f_1(5) = 368.75 & for\ M_2 = 24 \\ 210 + f_1(4) = 397.5 & for\ M_2 = 25 \end{cases}$$

$$= 368.75 \Rightarrow M_2^* = 24$$

The waiting time of 368.75 h is obtained for $M_2^* = 24$.

We calculated waiting times for $x_2 = 30$, $x_2 = 31$, $x_2 = 32$, $x_2 = 33$, $x_2 = 34$, and $x_2 = 35$ in the same way.

The results obtained are given in Table 1.25. For example, we see from Table 1.25 that, for $x_2 = 27$, the value of M_2 could only be equal to 23. Consequently, $f_2(x_2)$ equals 415.5 and M_2^* is 23. In the case where $x_2 = 28$, the M_2 can take values 23 and 24. The minimum of 378.26 and 406.25 is 378.26. In other words, since $f_2(x_2) = \min\{378.26, 406.25\} = 378.26$, the optimal value for M_2^* is 23. We fill the table for the other values of x_2 in the same way.

Table 1.26 Optimal values of M_3 for different values of x_3

x_3	$M_3 = 8$	$M_3 = 9$	$M_3 = 10$	$f_3(x_3)$	M_3^*
35	634.511	–	–	634.511	8
36	597.011	610.205	–	597.011	8
37	587.5	572.705	590.761	572.705	9
38	578.75	563.194	553.261	553.261	10
39	570.673	554.444	543.75	543.75	10
40	563.194	546.368	535	535	10
41	556.25	538.889	526.923	526.923	10
42	549.784	531.944	519.444	519.444	10
43	543.75	525.479	512.5	512.5	10

Table 1.27 The optimal values of M_4 for different values of x_4

x_4	$M_4 = 5$	$f_4(x_4)$	M_4^*
40	834.511	834.25	5
41	797.011	796.75	5
42	772.705	772.44	5
43	753.261	753	5
44	743.75	743.75	5
45	735	735	5
46	726.923	726.92	5
47	719.444	719.44	5
48	712.5	712.5	5

Table 1.28 The optimal values of M_5 for different values of x_5

	M_5								
x_5	14	15	16	17	18	19	20	$f_5(x_5)$	M_5^*
54	1048.54	–	–	–	–	–	–	1048.54	14
55	1011.04	1034.25	–	–	–	–	–	1011.04	14
56	986.73	996.75	1021.75	–	–	–	–	986.73	14
57	967.29	972.44	984.25	1010.72	–	–	–	967.29	14
58	958.04	953.00	959.94	973.22	1000.92	–	–	953.00	15
59	949.29	943.75	940.50	948.91	963.42	992.14	–	940.50	16
60	941.21	935.00	931.25	929.47	939.11	954.64	984.25	929.47	17
61	933.73	926.92	922.50	920.22	919.67	930.33	946.75	919.67	18
62	926.79	919.44	914.42	911.47	910.42	910.89	922.44	910.42	18

The optimal values of M_3, for different values of x_3, are given in Table 1.26.

The optimal values of M_4, for different values of x_4, are given in Table 1.27.

The optimal values of M_5, for different values of x_5, are given in Table 1.28.

The optimal values of M_6, for different values of x_6, are given in Table 1.29.

The optimal values of M_7, for different values of x_7, are given in Table 1.30.

We can conclude from Table 1.30 that $M_7^* = 4$. When we assign 4 buses to bus route 7, it leaves $74 - 4 = 70$ buses for other routes. We see from Table 1.30 that, for $x_6 = 70$, M_6^* equals 13 ($M_6^* = 13$). When we assign 13 buses to route 6, it leaves $70 - 13 = 57$ buses for the first 5 routes. We make conclusions about the optimal number of buses assigned to specific bus routes from the remaining tables. The optimal numbers of buses that should be assigned to specific bus routes are given in Table 1.31.

The graphical representation of the optimal solution is given in Figure 1.53.

Table 1.29 The optimal values of M_6 for different values of x_6

x_6	9	10	11	12	13	14	15	$f_6(x_6)$	$M_6{}^*$
				M_6					
63	1298.54							1298.54	9
64	1261.04	1273.54	-	-	-	-	-	1261.04	9
65	1236.73	1236.04	1253.08	-	-	-	-	1236.04	10
66	1217.29	1211.73	1215.58	1236.04	-	-	-	1211.73	10
67	1203.00	1192.29	1191.27	1198.54	1221.61	-	-	1191.27	11
68	1190.50	1178.00	1171.83	1174.23	1184.11	1209.25	-	1171.83	11
69	1179.47	1165.50	1157.55	1171.83	1159.80	1171.75	1198.54	1157.55	11
70	1169.67	1154.47	1145.05	1140.50	1140.36	1147.44	1161.04	1140.36	13
71	1160.42	1144.67	1134.02	1128.00	1126.08	1128.00	1136.73	1126.08	13

Table 1.30 The optimal values of M_7, for different values of x_7

x_7	$M_7 = 3$	$M_7 = 4$	$f_7(x_7)$	$M_7{}^*$
74	1292.74	1265.36	1265.36	4

Table 1.31 The optimal numbers of buses that should be assigned to specific bus routes

Bus route	Number of buses to be assigned to the bus route
1	$M_1{}^* = 5$
2	$M_2{}^* = 23$
3	$M_3{}^* = 10$
4	$M_4{}^* = 5$
5	$M_5{}^* = 14$
6	$M_6{}^* = 13$
7	$M_7{}^* = 4$

PROBLEMS

1.1. A port operator should load at least 2000 t of goods to the ship in 8 h. Two cranes should be used for loading. The productivity of the first crane is 200 t/h, and of the second one 170 t/h. The cost of using the first crane is $40/h and the cost of using the second crane is $30/h. The cranes can work simultaneously. Determine the optimal solution (the amount of goods loaded by each crane) in such a way as to minimize the total unloading cost (make and solve the linear program).

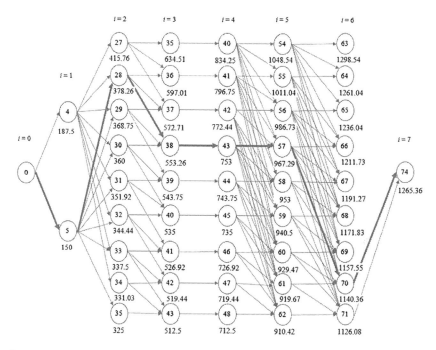

Figure 1.53 Optimal solution of the bus allocation problem

1.2. Four supermarkets (S1, S2, S3 and S4) need one type of goods. They have the following demands: 3, 5, 8, and 6 tons of goods. There are three warehouses (W1, W2 and W3) that supply the supermarkets. The warehouses have the following quantity of goods: 7, 12, and 9 tons of goods. The unit transportation costs from each warehouse to each supermarket are given in Table 1.32. Determine the optimal plan of transportation to minimize the total transportation costs.

1.3. There are 4 ships and 5 berths. Ships 1 and 2 could be served at berths 1, 2, and 3. Ships 3 and 4 could be served at any berth. One berth can serve just one ship and one ship can be served at only one berth. Ships' service times (in hours) are given in Table 1.33. Create an integer problem to minimize the total waiting time.

1.4. Two distribution centers should be opened in the transportation network given in Figure 1.54. Distribution centers could be located at

Table 1.32 Transportation costs ($/t)

	S1	S2	S3	S4
W1	30	15	25	20
W2	25	23	18	30
W3	20	20	27	25

Table 1.33 Ship service time at the berths (in h)

	Berth 1	Berth 2	Berth 3	Berth 4	Berth 5
Ship 1	10	12	13	–	–
Ship 2	9	10	11	–	–
Ship 3	7	9	10	12	14
Ship 4	8	10	12	15	17

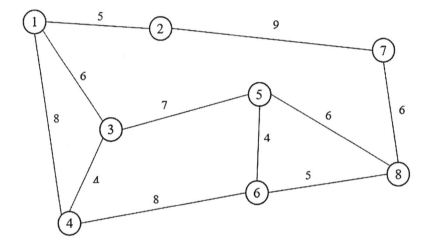

Figure 1.54 Transportation network

any node of the network. The nodes need the following number of deliveries per day: node 1 – 10, node 2 – 8, node 3 – 15, node 4 – 12, node 5 – 18, node 6 – 5, node 7 – 15, and node 8 – 20.

Determine the nodes where the distribution should be located and allocate nodes to distribution centers in the way to minimize the total transportation cost.

1.5. One hundred pallets should be loaded onto trucks within 8 h. Three forklifts could be used for loading. The following are the loading times of the forklifts: 5 min per pallet (for the first forklift), 8 min per pallet (for the second forklift) and 10 min per pallet (for the third forklift). The fixed cost of using a forklift is the same for each forklift and is equal to $50 (this cost should be taken into account if the forklift works). The variable costs for forklifts 1, 2, and 3 are $4, $3, and $2 per pallet, respectively. Determine the number of pallets loaded by each forklift to minimize the total cost. Write the integer linear program, the goal of which is to minimize the total cost.

BIBLIOGRAPHY

Bazaraa, M. S., Jarvis, J. J., and Sherali, H. D., *Linear Programming and Network Flows*, 4th ed., John Wiley and Sons, Inc., Hoboken, NJ, 2010.

Bellman, R., The Theory of Dynamic Programming, *Bulletin of the American Mathematical Society*, 60, 503–516, 1954.

Bellman, R. (1957), *Dynamic Programming*, Princeton University Press. Dover paperback edition, 2003.

Bellman, R. and Dreyfus, S., *Applied Dynamic Programming*, The RAND Corporation, Santa Monica, CA, 1962.

Bertsimas, D. and Tsitsiklis, J. N., *Introduction to Linear Optimization*, Athena Scientific, Nashua, NH, 1997.

Ceder, A., *Public Transit Planning and Operation: Modeling, Practice and Behavior*, 2nd ed., CRC Press, Taylor & Francis Group, 2015.

Chen, D.-S., Batson, R. G., and Dang, Y., *Applied Integer Programming: Modeling and Solution*, John Wiley and Sons, Hoboken, NJ, 2010.

Cordeau, J. F., Laporte, G., Savelsbergh, M. W. P., and Vigo, D., Vehicle Routing. In C. Barnhart and G. Laporte (Eds.), *Transportation*, Elsevier, Amsterdam, 2007.

Dantzig, G. B., *Linear Programming and Extensions*, Princeton University Press, Princeton, NJ, 1963.

Dantzig, G. B. and Thapa, M. N., *Linear Programming 1: Introduction*, Springer Verlag, New York, 1997.

Dantzig, G. B. and Thapa, M. N., *Linear Programming 2: Theory and Extensions*, Springer Verlag, New York, 2003.

Desrochers, M., Lenstra, J. K., Savelsbergh, M. W. P. and Soumis, F., Vehicle Routing with Time Windows: Optimization and Approximation. In: B. L. Golden and A. A. Assad (Eds.), *Vehicle Routing: Methods and Studies*, North-Holland, Amsterdam, 65–84, 1988.

Gass, S. I., *Linear Programming: Methods and Applications*, 5th ed., McGraw-Hill, New York, 1985.

Hillier, F. and Lieberman, G., *Introduction to Operations Research*, McGraw - Hill Publishing Company, New York, 1990.

Karlof, J. K., *Integer Programming: Theory and Practice*, CRC Press, Boca Raton, FL, 2006.

Kaufmann, A., *Graphs, Dynamic Programming, and Finite Games*, Academic Press, New York-London, 1967.

Lee, D. H., Chen, J. H., and Cao, J. X., The Continuous Berth Allocation Problem: A Greedy Randomized Adaptive Search Solution, *Transportation Research Part E*, 46, 1017–1029, 2010.

Little, J. D. C., Murty, K. G., Sweeney, D.W., and Karel, C., An Algorithm for the Travelling-Salesman Problem, *Operations Research*, 11, 6, 972–989, 1963.

Luenberger, D. and Yinyu, Y., *Introduction to Linear and Nonlinear Programming*, 3rd ed., International Series in Operations Research and Management Science, Volume 116. Springer, New York, 2008.

Nemhauser, G. L. and Wolsey, L. A., *Integer and Combinatorial Optimization*, Wiley Interscience, New York, 1988.

Potts, R. B. and Oliver, R. M., *Flows in Transportation Networks*, Academic Press, New York, 1972.

Schrijver, A., *Theory of Linear and Integer Programming*, John Wiley and Sons, New York, 1998.

Sierksma, G. and Zwols, Y., *Linear and Integer Optimization: Theory and Practice*, 3rd ed., CRC Press, Boca Raton, FL, 2015.

Teodorović, D. and Rallis, T., A Model for Assigning Vehicles to Scheduled Routes When There Is a Shortage of Vehicles, *Transportation Planning and Technology*, 12, 135–150, 1988.

Teodorović, D. and Vukadinović, K., Logistics Engineering Tool Chest. In G. Don Taylor (ed.), *Logistics Engineering Handbook*, 3-1- 3-30, CRC Press, Boca Raton-London-New York, 2008.

Taha, H., *Operations Research*, MacMillan Publishing Co., Inc., New York, 1982.

Winston, W., *Operations Research*, Duxbury Press, Belmont, CA, 1994.

Wolsey, L. A., *Integer Programming*, John Wiley and Sons, New York, 1998.

Chapter 2

Optimal paths

2.1 TRANSPORTATION NETWORKS BASICS

We represent transportation networks by geometrical figures (Figures 2.1, 2.2, 2.3, 2.4, 2.5, and 2.6). Some points (nodes) in these figures are mutually connected by lines (links). In the transportation networks, the nodes are usually intersections of lines. The nodes are intersections of streets in urban areas, roads in sub-urban areas, and roads/highways in inter-urban areas. The nodes could also be bus, tram, and trolleybus stations, road freight terminals, etc. The links are streets and roads/highways converging to and diverging from the corresponding intersections and stations/terminals. In rail transportation, the nodes are junctions along the lines and the stations/terminals, and the links are the rail lines/tracks connecting them. In air transportation, the nodes are airports and intersections of airways, while the links are airways.

We denote transportation networks in a similar way, as *graphs*. The notation $G = (N, A)$ of the transportation network refers to a set of *nodes* (or *vertices*), N, and a set of links (or *arcs*, *edges*, or *branches*), A, that connect pairs of nodes. We denote by (i, j) a link that connects node $i \in N$ to node $j \in N$. We generally designate one or more numerical characteristics to every link $(i, j) \in A$. Most often, these numerical characteristics represent "travel cost", c_{ij}, or "link capacity", u_{ij}. Similarly, numerical characteristics $b(i)$ are assigned to every node $i \in N$. The numerical characteristic could represent, for example, the total number of passengers generated at node i, or the yearly number of aircraft landings at node i. In this book, we use the terms "network" and "graph" interchangeably.

The network is called an *oriented network* (or a *directed network*) when all links in the network are oriented. In an oriented network, link (i, j) goes from node i to node j. In the opposite case, where none of the branches are oriented, the network is said to be *non-oriented*. When certain branches in the network are oriented and some are non-oriented, the network is called a *mixed* network.

Figures 2.1, 2.2, and 2.3 provide examples of oriented, non-oriented, and mixed networks, respectively.

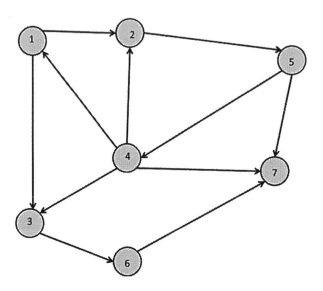

Figure 2.1 Oriented network

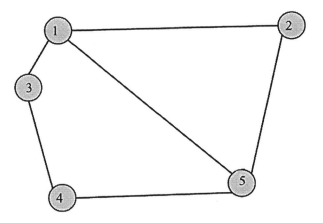

Figure 2.2 Non-oriented network

The *indegree of a node* in an oriented network signifies the number of head endpoints adjacent to a node. In the same way, the *outdegree of a node* in an oriented network is defined as the number of tail endpoints adjacent to a node. The indegree of node 5 in a network shown in Figure 2.1 equals 1. The outdegree of the same node equals 2. The *degree of a node* in a non-oriented network represents the number of links incident to the node. For example, the degree of node 1 in a non-oriented network shown in Figure 2.2 equals 3. The *path* going from node i to node j is a sequence of all links and all nodes that should be passed when moving from node i to node j.

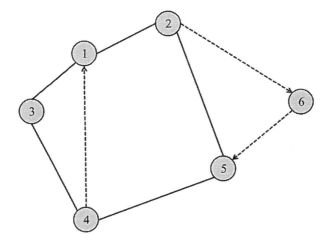

Figure 2.3 Mixed network

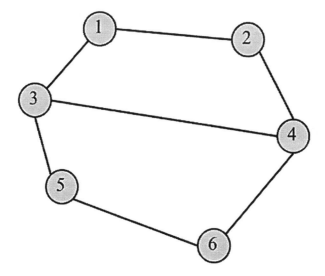

Figure 2.4 Non-oriented connected network

A path could be defined by a list of nodes or by a list of links that should be passed. The sequence (3, 4, 5) denotes the path that we follow when traveling from node 3 through node 4 to node 5 (Figure 2.2). This path could also be denoted as ((3, 4), (4, 5)), meaning that we traverse links (3, 4) and (4, 5) when traveling from node 3 to node 5. A *cycle* is a path, of which the origin and destination nodes overlap. The path (1, 5, 4, 3, 1) is a cycle as node 1 is the initial and the final node of the path (Figure 2.2). A path is said to be *simple*, when all links appear only once in the path. A path is *elementary*

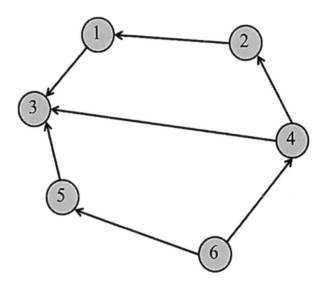

Figure 2.5 Oriented connected network

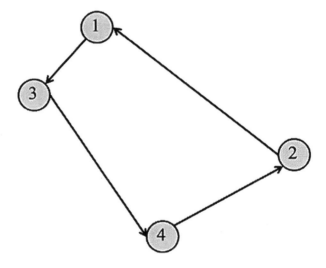

Figure 2.6 Strongly connected oriented network

when all nodes appear only once in the path. Node i is *connected* to node j if there is a path that leads from node i to node j. A non-oriented network is connected if there is a path between every pair of nodes $i, j \in N$. An oriented network is connected if a corresponding non-oriented network (the network that is created if orientation is removed from the oriented network) is connected. An oriented network that has paths between all pairs of nodes is called a *strongly connected* oriented network. A non-oriented connected network is shown in Figure 2.4.

Let us arbitrarily orient all links of the network shown in Figure 2.4. We generated the oriented network shown in Figure 2.5. The network shown in Figure 2.5 is an oriented connected network, given that a corresponding non-oriented network (Figure 2.4) is connected. There are no paths between certain pairs of nodes in the network shown in Figure 2.5 (there is no path between node 3 and node 4, for example, and no path between node 5 and node 6, etc.). Figure 2.6 shows one strongly connected oriented network.

2.2 FINDING THE SHORTEST PATH IN A TRANSPORTATION NETWORK

When traveling through the network, we are faced with the problem of finding the paths that are "optimal". In other words, the paths that we are looking for must possess optimal properties. The problems of finding optimal paths in transportation networks are known as *shortest path problems*. Depending on the context of the problem considered, the "shortest path" could be the shortest path, the fastest path, the most reliable path, the path with the greatest capacity, etc. Network links are characterized by length. "Link" "length" could represent distance, travel time, travel cost, link reliability, etc. Link lengths are mostly treated as deterministic quantities.

Link lengths are treated as random variables in some problems. Most frequently, these link lengths represent travel times. There are random variations in travel times, caused by weather conditions, randomness in traffic flows, traffic accidents, and other factors. In these cases, the *shortest paths in a probabilistic network* should be determined. In some cases, when searching for the optimal path, we simultaneously try to optimize two or more objectives. For example, when searching for the best path, we could try to simultaneously optimize travel time, as well as travel costs. In such cases we are dealing with multicriteria *shortest path problems*.

We use the expression "*shortest path*" to denote the optimal path. Depending on the context of the problem considered, the following variants of the shortest path problem could appear:

- Shortest path between two specified nodes
- Shortest paths from a given node to all other nodes
- k shortest paths from a given node to all other nodes
- Shortest paths between all pairs of nodes
- k shortest paths between all pairs of nodes
- Shortest path between two specified nodes that must pass through some pre-specified nodes
- Shortest path between two specified nodes that must pass through some pre-specified links

2.2.1 Dijkstra's algorithm

Dijkstra (1959) developed one of the most efficient algorithms for finding shortest paths from one node to all other nodes in a network. This algorithm assumes that all the lengths $d(i, j)$ of all links in the network $G = (N, A)$ are non-negative.

We denote, by a, the node for which we want to discover the shortest paths to all other nodes in the network. During the process of discovering these shortest paths, each node can be in one of two possible states: in an open state, if the node is denoted by a temporary label, or in a closed state, if it is denoted by a permanent label. In the case of a permanent label, we are not sure whether the discovered path is the shortest path. Dijkstra's algorithm gradually changes temporary labels into permanent labels. The initial distances between any two nodes in the network are defined as follows. The distance from node a to node a is zero. The distance between two nodes is equal to ∞ if there is no link between these two nodes. If there is a link that connects two nodes, the distance between these nodes is equal to the length of the link that connects them.

If there are several links that connect two nodes, the distance between these nodes is equal to the length of the shortest link that connects them. Each node i in the network is denoted by the following two labels:

d_{ai}	the length of the shortest known path from node a to node i discovered so far,
q_i	the predecessor node to node i on the shortest path from node a to node i discovered so far

We denote, by c, the last node to be given a permanent label. We also denote, by +, the predecessor node to node a. Dijkstra's algorithm is as follows:

Step 1:	The process starts from node a. Set $d_{aa} = 0$, $q_a = +$, $d_{ai} = \infty$ for $i \neq a$, and $q_i = -$ for $i \neq a$. The only node which is in a closed state is node a. Therefore, we write $c = a$.
Step 2:	Examine all links (c, i) which exit from the last node that is in a closed state (node c). If node i is also in a closed state, we move on to the next node. If node i is in an open state and $d_{ac} + d(c, i) < d_{ai}$, then set: $d_{ai} = d_{ac} + d(c, i)$ and $q_i = c$.
Step 3:	Compare the values d_{ai} with all nodes that are in an open state and choose the node with the smallest d_{ai} value. Let this be node j. Node j passes from an open to a closed state.
Step 4:	Node j is in a closed state. If all nodes are in a closed state, stop. If there are still some open state nodes, return to Step 2 and continue.

Example 2.1

By using Dijkstra's algorithm, discover the shortest paths from node 0 to all other nodes in the transportation network shown in Figure 2.7

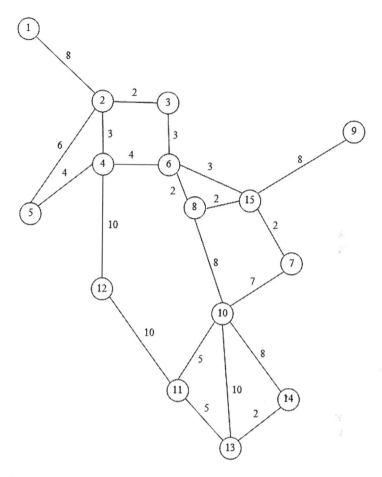

Figure 2.7 Network in which the shortest paths from node 1 to all other nodes should be determined (Mandl's network)

(Mandl's network). The numbers next to the network links shown in Figure 2.7 indicate the link lengths.

SOLUTION:

The length of the shortest path from node 1 to node 1 is equal to zero $(d_{11}=0)$. We denote by the symbol + the predecessor node to node 1 $(q_1 = +)$. The lengths of all other shortest paths from node 1 to all other nodes $i \neq 1$ are, for the present, unexamined, so, for all other nodes, $i \neq 1$, we set $d_{1i}=\infty$. Given that predecessor nodes to nodes $i \neq 1$ are unidentified, we set $q_i =-$ for all $i \neq 1$. The single node that is now in a closed state is node 1. Therefore, $c=1$. To the node 1 symbol we put the label $(0, +)$, and add the symbol ′ that denotes that node 1 is in a closed state. This completes the first step of the algorithm.

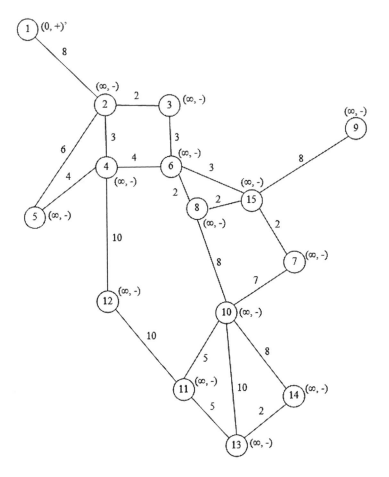

Figure 2.8 Network after the first algorithmic step

After the first algorithmic step, the transportation network is shown in Figure 2.8.

ITERATION 1:

Let us return to the second step of the algorithm. The last node that switched from an open to a closed state was node 1. This means that $c = 1$. Let us examine all links that leave node 1 and go towards nodes that are in an open state. There is only one such link (link (1, 2)).

Because: $d_{11} + d(1,2) = 0 + 8 = 8 < d_{12} = \infty$, it follows:

$$d_{12} = d_{11} + d(1,2) = 8 \text{ and } q_2 = 1.$$

Table 2.1 Matrix of shortest distances from node 1

Iteration	Nodes													
	2	3	4	5	6	7	8	9	10	11	12	13	14	15
1	**8**	∞	∞	∞	∞	∞	∞	∞	∞	∞	∞	∞	∞	∞
2		**10**	**11**	14	∞	∞	∞	∞	∞	∞	∞	∞	∞	∞
3			**11**	14	13	∞	∞	∞	∞	∞	∞	∞	∞	∞
4				14	**13**	∞	∞	∞	∞	∞	21	∞	∞	∞
5				**14**		∞	15	∞	∞	∞	21	∞	∞	16
6						∞	**15**	∞	∞	∞	21	∞	∞	16
7						∞		23	∞	∞	21	∞	∞	**16**
8						**18**		24	23	∞	21	∞	∞	
9								24	23	∞	**21**	∞	∞	
10								24	23	**31**		∞	∞	
11								**24**		28		33	31	
12										28		33	31	
13												33	**31**	
14												**33**		

From the still-open nodes, we choose the node with the smallest shortest distance, i.e.:

$$\min\{8,\infty,\infty,\infty,\infty,\infty,\infty,\infty,\infty,\infty,\infty,\infty,\infty,\infty\} = 8$$

Node 2 passes to the closed state (see the first row of Table 2.1 and Figure 2.9).

ITERATION 2:

There are three links leaving node 2 (links (2, 3), (2, 4), and (2, 5)). We can write the following, after examining the lengths of all the links leaving node 2 that are in a closed state:

Because: $d_{12} + d(2,3) = 8 + 2 = 10 < d_{13} = \infty$, it follows:
$d_{13} = d_{12} + d(2,3) = 10$ and
$q_2 = 2$.

Because: $d_{12} + d(2,4) = 8 + 3 = 11 < d_{14} = \infty$, it follows:
$d_{14} = d_{12} + d(2,4) = 11$ and
$q_4 = 2$.

Because: $d_{12} + d(2,5) = 8 + 6 = 14 < d_{15} = \infty$, it follows:
$d_{15} = d_{12} + d(2,5) = 14$ and
$q_4 = 2$.

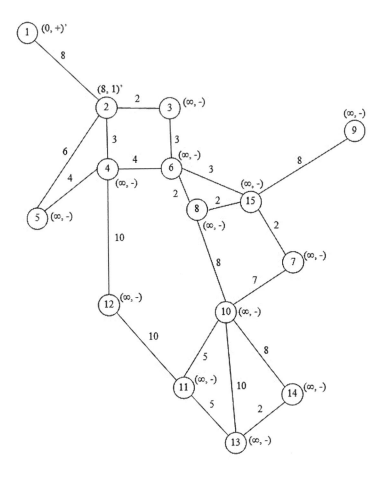

Figure 2.9 Network after the first iteration

To discover the next node that should be closed, we are looking for the lowest value among values given in the second row of Table 2.1. Also, we can see these values from Figure 2.10.

$$\min\{10,11,14,\infty,\infty,\infty,\infty,\infty,\infty,\infty,\infty,\infty,\infty\} = 10$$

We conclude that node 3 passes to a closed state.

ITERATION 3:

In the third iteration, we are trying to discover shorter distances from 1 to the still-open nodes (4, 5, 6, 7, 8, 9, 10, 11, 12, 13, 14, and 15) by using the paths *via* node 3, etc. We have one link from node 3, namely (3,6).

Because: $d_{13} + d(3,6) = 10 + 3 = 13 < d_{16} = \infty$, it follows:

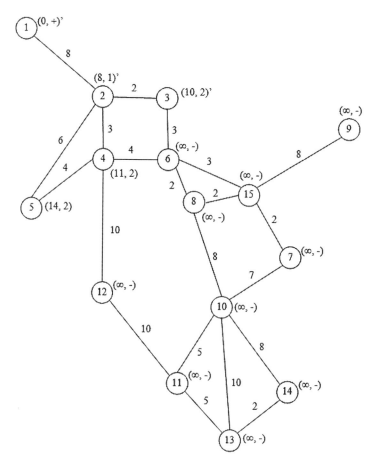

Figure 2.10 Network after the second iteration

$$d_{16} = d_{13} + d(3,6) = 13 \text{ and}$$
$$q_4 = 3.$$

To discover the next node to be closed, we are looking for the lowest value among values given in the second row of Table 2.1, i.e.:

$$\min\{11, 14, 13, \infty, \infty, \infty, \infty, \infty, \infty, \infty, \infty, \infty\} = 11$$

The node that passes to the closed state is 3.

The process continues until all nodes become closed. Table 2.1 shows the "evolution" of discovering the shortest distances from node 1 to all the other nodes. The finally discovered shortest distances are denoted by bold numbers (Table 2.1). From the matrix, we can also read the iteration when these distances were discovered.

Table 2.2 Matrix of predecessor nodes

Iteration	Nodes													
	2	3	4	5	6	7	8	9	10	11	12	13	14	15
1	**1**	—	—	—	—	—	—	—	—	—	—	—	—	—
2		**2**	2	2	—	—	—	—	—	—	—	—	—	—
3			**2**	2	3	—	—	—	—	—	—	—	—	—
4				2	3	—					4	—	—	—
5			**2**			—	6	—	—	—	4	—	—	6
6						—	6	—	—	—	4	—	—	6
7							—	—	8	—	4	—	—	6
8							**15**	15	8	—	4	—	—	
9								15	8	—	4	—	—	
10								15	8	12		—	—	
11								**15**		10		10	10	
12											10	10	10	
13												10	10	
14												10		

In the same way, the matrix shown in Table 2.2 shows the "evolution" of discovering the predecessors. The final solution is shown in Figure 2.11.

2.2.2 Shortest paths between all pairs of nodes

Let our problem be to find the shortest path between all nodes in transportation network $G=(N, A)$. Floyd's algorithm (1962) is one of the best-known algorithms that can be used to find the shortest paths between all pairs of nodes. We denote all nodes of the network by positive whole numbers: $1, 2, ...,n$.

We now introduce D_0, which is the beginning matrix of the shortest path lengths, and Q_0, which is the predecessor node matrix. We denote, by d_{ij}^k, the length of the shortest path from node i to node j, which is found in the k-th passage through the algorithm, and, by q_{ij}^k, the immediate predecessor node of node j on the shortest path from node I, which was also discovered on the k-th passage. Elements d_{ij}^0 of matrix D_0 are defined in the following manner:

If a branch exists between node i and node j, the length of the shortest path d_{ij}^0 between these nodes equals length $d(i, j)$ of branch (i, j), which connects them. Should there be several branches between node i and node j, the length of the shortest path, d_{ij}^0, must equal the length of the shortest branch, i.e.:

$$d_{ij}^0 = \min\{d_1(i,j), d_2(i,j),...,d_m(i,j)\} \tag{3.3}$$

where m is the number of branches between node i and node j.

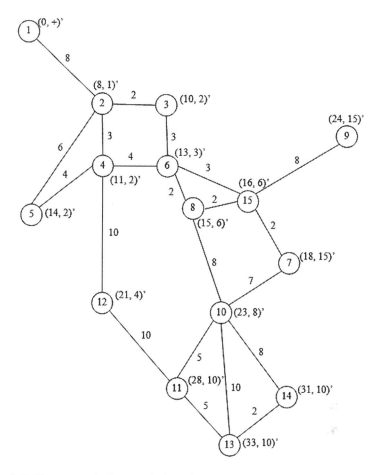

Figure 2.11 Shortest paths from node 1 to all other nodes

It is clear that $d_{ij}^0 = 0$ when $i=j$. In the case where there is no branch between node i and node j, we have no information at the beginning concerning the length of the shortest path between these two nodes, so we treat them as though they were infinitely far from each other, i.e. that the following is true for such pairs of nodes:

$$d_{ij}^0 = \infty \tag{3.4}$$

Elements q_{ij}^0 of the predecessor matrix Q_0 are defined as follows:

First, we assume that $q_{ij}^0 = i$, for $i \neq j$, i.e. that for every pair of nodes (i, j), for $i \neq j$, the immediate predecessor of node j on the shortest path leading from node i to node j is actually node i. Since we have defined the elements of matrixes D_0 and Q_0, we can now examine Floyd's algorithm, which contains the following steps:

Step 1:   let $k = 1$.

Step 2:   we calculate elements q_{ij}^k of the shortest path length matrix D_k found after the k-th passage through the algorithm, using the following equation:

$$q_{ij}^k = \min\left\{d_{ij}^{k-1}, d_{ik}^{k-1} + d_{kj}^{k-1}\right\}.$$

Step 3:   elements of the predecessor node matrix Q_k found after the k-th passage through the algorithm are calculated as follows:

$$q_{ij}^k = \begin{cases} q_{kj}^{k-1} & \text{for} \quad d_{ij}^k \neq d_{ij}^{k-1} \\ q_{ij}^{k-1} & \text{otherwise} \end{cases}$$

Step 4:   If $k = n$, the algorithm is finished. If $k < n$, increase k by 1, i.e. $k = k + 1$, and return to Step 2.

Let us now look at the algorithm in a little more detail. In Step 2, each time we go through the algorithm, we are checking as to whether a shorter path exists between nodes i and j other than the path we already know about, which was established during one of the earlier passages through the algorithm. If we establish that $d_{ij}^k \neq d_{ij}^k$, i.e. if we establish, during the k-th passage through the algorithm, that the length of the shortest path d_{ij}^k between nodes i and j is less than the length of the shortest path d_{ij}^{k-1} known before the k-th passage, we have to change the immediate predecessor node to node j.

Since the length of the new shortest path is:

$$d_{ij}^k = d_{ik}^{k-1} + d_{kj}^{k-1} \tag{3.5}$$

it is clear that, in this case, node k is the new immediate predecessor node to j, and therefore:

$$q_{ij}^k = q_{kj}^{k-1} \tag{3.6}$$

This is actually carried out in the third algorithmic step. It is also clear that the immediate predecessor node to node i does not change if, at the end of Step 2, we have established that no other new, shorter path exists. This means that:

$$q_{ij}^k = q_{ij}^{k-1} \text{ when } d_{ij}^k = d_{ij}^{k-1} \tag{3.7}$$

When we go through the algorithm n times (n is the number of nodes in the transportation network), elements d_{ij}^{k-1} of the final matrix D_n will constitute the shortest path's lengths between pairs of nodes (i, j), and elements q_{ij}^n of matrix Q_n will enable us in to identify all of the nodes which are on the shortest path, from node i to node j.

Example 2.2

Determine the shortest paths between all pairs of nodes on transportation network $T(N, A)$ shown in Figure 2.12. Link lengths are shown in Figure 2.12.

SOLUTION:

The starting matrix D_0 is shown in Table 2.3.

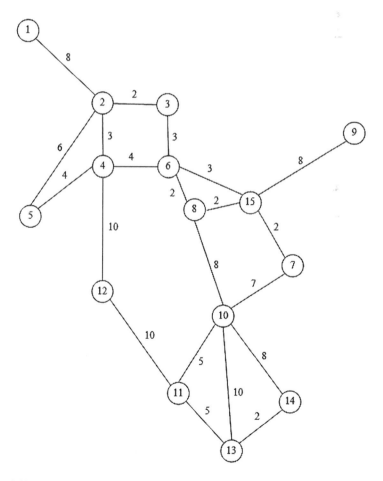

Figure 2.12 Network in which the shortest paths between all pairs of nodes should be determined

Table 2.3 Initial matrix of distances (D_0)

	1	2	3	4	5	6	7	8	9	10	11	12	13	14	15
1	0	8	∞	∞	∞	∞	∞	∞	∞	∞	∞	∞	∞	∞	∞
2	8	0	2	3	6	∞	∞	∞	∞	∞	∞	∞	∞	∞	∞
3	∞	2	0	∞	∞	3	∞	∞	∞	∞	∞	∞	∞	∞	∞
4	∞	3	∞	0	4	4	∞	∞	∞	∞	∞	10	∞	∞	∞
5	∞	6	∞	4	0	∞	∞	∞	∞	∞	∞	∞	∞	∞	∞
6	∞	∞	3	4	∞	0	∞	2	∞	∞	∞	∞	∞	∞	3
7	∞	∞	∞	∞	∞	∞	0	∞	∞	7	∞	∞	∞	∞	2
8	∞	∞	∞	∞	∞	2	∞	0	∞	8	∞	∞	∞	∞	2
9	∞	∞	∞	∞	∞	∞	∞	∞	0	∞	∞	∞	∞	∞	8
10	∞	∞	∞	∞	∞	∞	7	8	∞	0	5	∞	10	8	∞
11	∞	∞	∞	∞	∞	∞	∞	∞	∞	5	0	10	5	∞	∞

All elements of the main diagonal of the matrix D_0 are equal to zero ($d_{ij}^{k-1} = 0$ for $i=j$). Let us note element $d_{1,2}^0$ of the matrix D_0. This element is equal to 8, as the length of the link that connects node 1 and node 2 equals 8. Element $d_{3,1}^0$ is equal to ∞, since there is no link leading node 3 to node 1 (there is only a link oriented from node 1 to node 3). Element $d_{5,1}^0$ of the matrix D_0 is equal to ∞, since there is no link connecting node 5 and node 1.

The initial matrix Q_0 is shown in Table 2.4.

In the beginning, we consider node i as a predecessor node to node j on the shortest path leading from node i to node j (for $i \neq j$). For example, the following relationship is satisfied:

$$q_{2,1}^0 = q_{2,3}^0 = q_{2,4}^0 = q_{2,5}^0 = 2$$

Table 2.4 Initial matrix of predecessor nodes (Q_0)

	1	2	3	4	5	6	7	8	9	10	11	12	13	14	15
1	1	1	1	1	1	1	1	1	1	1	1	1	1	1	1
2	2	2	2	2	2	2	2	2	2	2	2	2	2	2	2
3	3	3	3	3	3	3	3	3	3	3	3	3	3	3	3
4	4	4	4	4	4	4	4	4	4	4	4	4	4	4	4
5	5	5	5	5	5	5	5	5	5	5	5	5	5	5	5
6	6	6	6	6	6	6	6	6	6	6	6	6	6	6	6
7	7	7	7	7	7	7	7	7	7	7	7	7	7	7	7
8	8	8	8	8	8	8	8	8	8	8	8	8	8	8	8
9	9	9	9	9	9	9	9	9	9	9	9	9	9	9	9
10	10	10	10	10	10	10	10	10	10	10	10	10	10	10	10
11	11	11	11	11	11	11	11	11	11	11	11	11	11	11	11
12	12	12	12	12	12	12	12	12	12	12	12	12	12	12	12
13	13	13	13	13	13	13	13	13	13	13	13	13	13	13	13
14	14	14	14	14	14	14	14	14	14	14	14	14	14	14	14
15	15	15	15	15	15	15	15	15	15	15	15	15	15	15	15

Let us start with a first algorithmic step ($k = 1$). Let us calculate the elements of matrix D_1. The following relationships are satisfied:

$$d_{1,2}^1 = \min\left\{d_{1,2}^0, d_{1,1}^0 + d_{1,2}^0\right\} = \min\left\{8, 0 + 8\right\} = 8$$

$$d_{1,3}^1 = \min\left\{d_{1,3}^0, d_{1,1}^0 + d_{1,3}^0\right\} = \min\left\{\infty, 0 + \infty\right\} = \infty$$

$$\vdots$$

$$d_{1,15}^1 = \min\left\{d_{1,15}^0, d_{1,1}^0 + d_{1,15}^0\right\} = \min\left\{\infty, 0 + \infty\right\} = \infty$$

$$d_{2,1}^1 = \min\left\{d_{2,1}^0, d_{2,1}^0 + d_{1,1}^0\right\} = \min\left\{8, 8 + 0\right\} = 8$$

$$d_{2,3}^1 = \min\left\{d_{2,3}^0, d_{2,1}^0 + d_{1,3}^0\right\} = \min\left\{2, 8 + \infty\right\} = 2$$

....

....

$$d_{15,14}^1 = \min\left\{d_{15,14}^0, d_{15,1}^0 + d_{1,14}^0\right\} = \min\left\{\infty, \infty + \infty\right\} = \infty$$

After the first step, the matrix of distances and the matrix of predecessor nodes remain the same, i.e.:

$$D_1 = D_0$$

$$Q_1 = Q_0$$

We return to step 2. For $k = 2$, we obtain the following:

$$d_{1,2}^2 = \min\left\{d_{1,2}^1, d_{1,2}^1 + d_{2,2}^1\right\} = \min\left\{8, 8 + 0\right\} = 8$$

$$d_{1,3}^2 = \min\left\{d_{1,3}^1, d_{1,2}^1 + d_{2,3}^1\right\} = \min\left\{\infty, 8 + 2\right\} = 10$$

Since $d_{1,3}^2 \neq d_{1,3}^1$, we obtain:

$$q_{1,3}^2 = q_{2,3}^1 = 2$$

$$d_{1,4}^2 = \min\left\{d_{1,4}^1, d_{1,2}^1 + d_{2,4}^1\right\} = \min\left\{\infty, 8 + 3\right\} = 11$$

Since $d_{1,4}^2 \neq d_{1,4}^1$, we obtain:

$$q_{1,4}^2 = q_{2,4}^1 = 2$$

$$d_{1,5}^2 = \min\left\{d_{1,5}^1, d_{1,2}^1 + d_{2,5}^1\right\} = \min\left\{\infty, 8+6\right\} = 14$$

Since $d_{1,5}^2 \neq d_{1,5}^1$, we obtain:

$$q_{1,5}^2 = q_{2,5}^1 = 2$$

$$\vdots$$

$$d_{15,14}^2 = \min\left\{d_{15,14}^1, d_{15,2}^1 + d_{2,14}^1\right\} = \min\left\{\infty, \infty+\infty\right\} = \infty$$

The matrix of distances after the second step (D_2) is given in Table 2.5. The matrix of predecessor nodes after the second step (Q_2) is given in Table 2.6.

The matrix of distances after the last step (D_{15}), and the matrix of predecessor nodes after the last step (Q_{15}) are given in Table 2.7 and Table 2.8, respectively.

From the matrices D_{15} and Q_{15}, we obtain the complete information about the shortest paths between all pairs of nodes. So, for example, the length of the shortest path from node 12 to node 14 is equal to 17. The predecessor node to node 14 is node 13, since $q_{12,14} = 13$. The predecessor node to node 13 on the shortest path from node 12 is node 11, since $q_{12,13} = 11$. Since $q_{12,11} = 12$, we conclude that the shortest path from node 12 to node 14 reads (12, 11, 13, 14).

Table 2.5 The matrix of distances after the second step (D_2)

	1	2	3	4	5	6	7	8	9	10	11	12	13	14	15
1	0	8	10	11	14	∞	∞	∞	∞	∞	∞	∞	∞	∞	∞
2	8	0	2	3	6	∞	∞	∞	∞	∞	∞	∞	∞	∞	∞
3	10	2	0	5	8	3	∞	∞	∞	∞	∞	∞	∞	∞	∞
4	11	3	5	0	4	4	∞	∞	∞	∞	∞	10	∞	∞	∞
5	14	6	8	4	0	∞	∞	∞	∞	∞	∞	∞	∞	∞	∞
6	∞	∞	3	4	∞	0	∞	2	∞	∞	∞	∞	∞	∞	3
7	∞	∞	∞	∞	∞	∞	0	∞	∞	7	∞	∞	∞	∞	2
8	∞	∞	∞	∞	∞	2	∞	0	∞	8	∞	∞	∞	∞	2
9	∞	∞	∞	∞	∞	∞	∞	∞	0	∞	∞	∞	∞	∞	8
10	∞	∞	∞	∞	∞	∞	7	8	∞	0	5	∞	10	8	∞
11	∞	∞	∞	∞	∞	∞	∞	∞	∞	5	0	10	5	∞	∞
12	∞	∞	∞	10	∞	∞	∞	∞	∞	∞	10	0	∞	∞	∞
13	∞	∞	∞	∞	∞	∞	∞	∞	∞	10	5	∞	0	2	∞
14	∞	∞	∞	∞	∞	∞	∞	∞	∞	8	∞	∞	2	0	∞
15	∞	∞	∞	∞	∞	3	2	2	8	∞	∞	∞	∞	∞	0

Table 2.6 The matrix of predecessor nodes after the second step (Q_2)

	1	2	3	4	5	6	7	8	9	10	11	12	13	14	15
1	1	1	2	2	2	1	1	1	1	1	1	1	1	1	1
2	2	2	2	2	2	2	2	2	2	2	2	2	2	2	2
3	2	3	3	2	2	3	3	3	3	3	3	3	3	3	3
4	2	4	2	4	4	4	4	4	4	4	4	4	4	4	4
5	2	5	2	5	5	5	5	5	5	5	5	5	5	5	5
6	6	6	6	6	6	6	6	6	6	6	6	6	6	6	6
7	7	7	7	7	7	7	7	7	7	7	7	7	7	7	7
8	8	8	8	8	8	8	8	8	8	8	8	8	8	8	8
9	9	9	9	9	9	9	9	9	9	9	9	9	9	9	9
10	10	10	10	10	10	10	10	10	10	10	10	10	10	10	10
11	11	11	11	11	11	11	11	11	11	11	11	11	11	11	11
12	12	12	12	12	12	12	12	12	12	12	12	12	12	12	12
13	13	13	13	13	13	13	13	13	13	13	13	13	13	13	13
14	14	14	14	14	14	14	14	14	14	14	14	14	14	14	14
15	15	15	15	15	15	15	15	15	15	15	15	15	15	15	15

Table 2.7 Matrix of distances after thelast step (D_{15})

	1	2	3	4	5	6	7	8	9	10	11	12	13	14	15
1	0	8	10	11	14	13	18	15	24	23	28	21	33	31	16
2	8	0	2	3	6	5	10	7	16	15	20	13	25	23	8
3	10	2	0	5	8	3	8	5	14	13	18	15	23	21	6
4	11	3	5	0	4	4	9	6	15	14	19	10	24	22	7
5	14	6	8	4	0	8	13	10	19	18	23	14	28	26	11
6	13	5	3	4	8	0	5	2	11	10	15	14	20	18	3
7	18	10	8	9	13	5	0	4	10	7	12	19	17	15	2
8	15	7	5	6	10	2	4	0	10	8	13	16	18	16	2
9	24	16	14	15	19	11	10	10	0	17	22	25	27	25	8
10	23	15	13	14	18	10	7	8	17	0	5	15	10	8	9
11	28	20	18	19	23	15	12	13	22	5	0	10	5	7	14
12	21	13	15	10	14	14	19	16	25	15	10	0	15	17	17
13	33	25	23	24	28	20	17	18	27	10	5	15	0	2	19
14	31	23	21	22	26	18	15	16	25	8	7	17	2	0	17
15	16	8	6	7	11	3	2	2	8	9	14	17	19	17	0

PROBLEMS

2.1. A transportation network is shown in Figure 2.13. The nodes in the network are cities. The distances between the cities are indicated in the figure. Determine the shortest distances from city 1 to all the other cities, using Dijkstra's algorithm.

Table 2.8 Matrix of nodes predecessor nodes after the last step (Q_{15})

	1	2	3	4	5	6	7	8	9	10	11	12	13	14	15
1	1	1	2	2	2	2	3	15	6	15	8	10	4	10	10	6
2	2	2	2	2	2	3	15	6	15	8	10	4	10	10	6
3	2	3	3	2	2	3	15	6	15	8	10	4	10	10	6
4	2	4	2	4	4	4	15	6	15	8	10	4	10	10	6
5	2	5	2	5	5	4	15	6	15	8	10	4	10	10	6
6	2	3	6	6	4	6	15	6	15	8	10	4	10	10	6
7	2	3	6	6	4	15	7	15	15	7	10	4	10	10	7
8	2	3	6	6	4	8	15	8	15	8	10	4	10	10	8
9	2	3	6	6	4	15	15	15	9	7	10	4	10	10	9
10	2	3	6	6	4	8	10	10	15	10	10	11	10	10	7
11	2	3	6	6	4	8	10	10	15	11	11	11	11	13	7
12	2	4	2	12	4	4	15	6	15	11	12	12	11	13	6
13	2	3	6	6	4	8	10	10	15	13	13	11	13	13	7
14	2	3	6	6	4	8	10	10	15	14	13	11	14	14	7
15	2	3	6	6	4	15	15	15	15	7	10	4	10	10	15

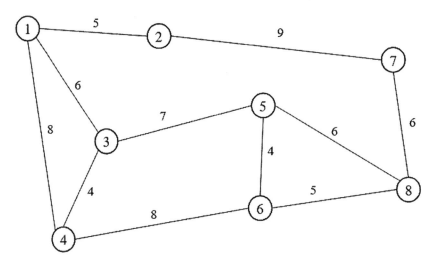

Figure 2.13 Transportation network in which the shortest distances between city 1 and all the other cities should be determined, using Dijkstra's algorithm

2.2. Figure 2.14 shows a transportation network. The nodes in the network are cities. The distances between the cities are indicated in the figure. Determine the shortest distances from city 4 to all the other cities, using Dijkstra's algorithm.

2.3. Determine the shortest distances between all pairs of nodes in the network given in Figure 2.15.

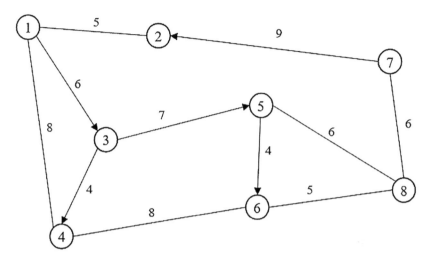

Figure 2.14 Transportation network in which the shortest distances between city 4 and all the other cities should be determined, using Dijkstra's algorithm

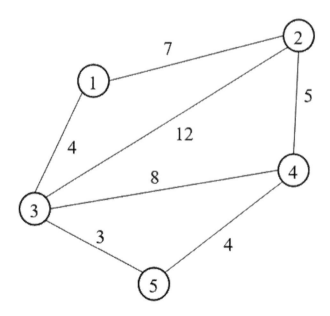

Figure 2.15 Transportation network in which the shortest distances between all pairs of nodes should be determined by Floyd's algorithm

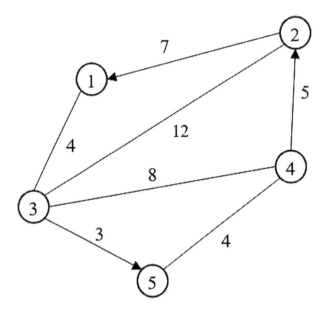

Figure 2.16 Transportation network in which the shortest distances between all pairs of nodes should be determined by Floyd's algorithm

2.4. Determine the shortest distances between all pairs of nodes in the network given in Figure 2.16.

BIBLIOGRAPHY

Busacker, R. G. and Saaty, T. L., *Finite Graphs and Networks: An Introduction with Applications*, McGraw–Hill, New York, 1965.

Dantzig, G. B., On the Shortest Route Problem through a Network, *Management Science*, 6, 187–190, 1960.

Deo, N. and Pang, C., Shortest Path Algorithms: Taxonomy and Annotation, *Networks*, 14, 275–323, 1984.

Dial, R., Glover, F., Karney, D., and Klingman, D., A Computational Analysis of Alternative Algorithms and Labeling Techniques for Finding Shortest Path Trees, *Networks*, 9, 215–248, 1979.

Dijkstra, E. W., A Note on Two Problems in Connection with Graphs, *Numerische Mathematik*, 1, 269–271, 1959.

Floyd, R. W., Algorithm 97–Shortest Path, *Communications of ACM*, 5, 344–348, 1962.

Ford, L. R., Jr. and Fulkerson, D. R., *Flows in Networks*, Princeton University Press, Princeton, NJ, 1962.

Gallo, G. S. and Pallottino, S., Shortest Path Algorithms, *Annals of Operations Research*, 7, 3–79, 1988.

Glover, F., Glover, R., and Klingman, D., Computational Study of an Improved Shortest Path Algorithm, *Networks*, 14, 25–36, 1984.

Glover, F., Klingman, D., Phillips, N., and Schneider, R. F., New Polynomial Shortest Path Algorithms and Their Computational Attributes, *Management Science*, 31, 1106–1128, 1985.

Golden, B., Shortest Path Algorithms: A Comparison, *Operations Research*, 24, 1164–1168, 1976.

Hillier, F. and Lieberman, G., *Introduction to Operations Research*, McGraw–Hill Publishing Company, New York, 1990.

Larson, R. and Odoni, A., *Urban Operations Research*, Prentice Hall, Englewood Cliffs, NJ, 1981.

Minty, G. J., A Comment on the Shortest Route Problem, *Operations Research*, 5, 724, 1957.

Newell, G. F., *Traffic Flow on Transportation Networks*, MIT Press, Cambridge, MA, 1980.

Pallottino, S., Shortest-path Methods: Complexity, Inter-Relations and New Propositions, *Networks*, 14, 257–267, 1984.

Potts, R. B. and Oliver, R. M., *Flows in Transportation Networks*, Academic Press, New York, 1972.

Sheffi, Y., *Urban Transportation Networks*, Prentice Hall, Englewood Cliffs, NJ, 1985.

Steenbrink, P. A., *Optimization of Transport Networks*, John Wiley & Sons, Inc, New York, 1974.

Taha, H., *Operations Research*, MacMillan Publishing Co., Inc., New York, 1982.

Teodorović, D., and Janić, M., *Transportation Engineering: Theory, Practice and Modeling*, Elsevier, New York, 2016.

Winston, W., *Operations Research*, Duxbury Press, Belmont, CA, 1994.

Chapter 3

Multi-attribute decision-making

3.1 MULTI-ATTRIBUTE DECISION-MAKING (MADM) METHODS

Let us imagine, for example, that a public transit operator would like to buy new buses. The engineers responsible for the acquisition have to rank the bus models under consideration (the set of alternatives), and finally to decide on one model (one alternative) from the set of potential alternatives. The criteria that they think about when choosing a new bus model can be price, approximate future maintenance cost, fuel consumption, depreciation, safety, comfort, etc.

Traffic engineers and planners take into account the characteristics of the bus rapid transit, light rail, and metro systems when making a decision about the most suitable transportation mode for new suburbs. They could analyze, for example, construction time, capital costs, operating costs, line capacity, speeds, travel times, future headways, etc.

Government, industry, and/or traffic authorities regularly have to evaluate a set of transportation projects ("alternatives"). The ranking of the alternatives is typically carried out according to a number of criteria ("attributes") that, as a rule, are mutually conflicting.

Multi-attribute decision-making methods (MADM) take into account various types of criteria with different units. The number of alternatives in the real-life problems tackled with MADM has been predetermined. The MADM methods can be used to identify a single preferred alternative, to rank the alternatives, or to make the distinction between acceptable and unacceptable alternatives.

3.1.1 The common characteristics of MADM problems

Practically all MADM problems have the following common characteristics:

- There is a need to rank the two or more alternatives and to select the best one.
- There are multiple criteria (attributes) that the decision maker uses to rank the alternatives.

- The criteria are mutually conflicting.
- The attributes are measured in different units.

Frequently the term "alternative" is used to describe a transportation project. The terms "option", "policy", "action", and "candidate" are also used instead of "alternative" in the literature. The alternatives are typically ranked according to a number of attributes. The number of attributes depends on the nature of the problem considered. The attributes being considered also have different units of measurement. For example, the attributes and the corresponding units of measurements could be price ($); fuel consumption (miles per gallon); waiting time (minutes); comfort (non-numerical way (words)), etc.

We denote the total number of alternatives (transportation projects,) by m, and the total number of attributes or criteria, according to which the considered alternatives are compared, by n. In the decision matrix D, x_{ij} is the performance of alternative A_i with respect to attribute X_j.:

$$
D = \begin{array}{c}
\quad \\
A_1 \\
A_2 \\
\vdots \\
A_m
\end{array}
\begin{array}{|cccc|}
X_1 & X_2 & \cdots & X_n \\
\hline
x_{11} & x_{12} & \cdots & x_{1n} \\
x_{21} & x_{22} & \cdots & x_{2n} \\
\vdots & \vdots & & \vdots \\
x_{m1} & x_{m2} & \cdots & x_{mn}
\end{array}
$$

Essentially, all MADMs need information about the relative importance ("weight") of each attribute. Weights could be assigned by the analyst (decisionmaker), or they could be calculated by a variety of methods.

There are benefit attributes and cost attributes. In the case of benefit attributes, the higher values are desirable (profit, revenue, fuel efficiency,...). In the case of cost attributes, the smaller values are preferable (direct operation cost, passenger waiting time at hub,...).

3.1.2 Normalization of the attribute values

Outcomes for alternatives with respect to individual criteria are expressed in different units (i.e. on different scales). Most MADMs require that these values be expressed on the same scale. Therefore, prior to applying any of the MADMs, the normalization of the attribute values is carried out.

In the case of vector normalization, the normalized values r_{ij} are calculated as:

$$
r_{ij} = \frac{x_{ij}}{\sqrt{\sum_{k=1}^{m} x_{kj}^2}} \qquad i = 1, 2, \ldots, m \qquad j = 1, 2, \ldots, n
$$

After performing vector normalization, all columns, i.e. all attributes, have the equal unit value of the vector.

Example 3.1

The decision matrix D reads:

$$D = \begin{vmatrix} 1.5 & 20 & 27,000 & 70 & 9 & 3 \\ 2.5 & 90 & 20,000 & 70 & 21 & 8 \\ 7 & 120 & 60,000 & 100 & 104 & 5 \end{vmatrix}$$

The matrix D has 3 rows (3 alternatives) and 7 columns (7 attributes). We calculate normalized values r_{ij} as:

$$r_{ij} = \frac{x_{ij}}{\sqrt{\sum_{k=1}^{m} x_{kj}^2}} \qquad i = 1, 2, \ldots, m \qquad j = 1, 2, \ldots, n$$

For example, r_{23} equals:

$$r_{21} = \frac{x_{21}}{\sqrt{\sum_{k=1}^{3} x_{k1}^2}} = \frac{2.5}{\sqrt{1.5^2 + 2.5^2 + 7^2}} = 0.33$$

The normalized values r_{ij} are given in matrix R:

$$R = \begin{vmatrix} 0.20 & 0.13 & 0.39 & 0.50 & 0.08 & 0.30 \\ 0.33 & 0.59 & 0.29 & 0.50 & 0.20 & 0.81 \\ 0.92 & 0.79 & 0.87 & 0.71 & 0.98 & 0.51 \end{vmatrix}$$

The linear scale transformation, in the case of benefit attributes, assumes division of the outcome of a certain attribute by the attribute's maximum value, i.e.:

$$r_{ij} = \frac{x_{ij}}{x_j^*} \qquad i = 1, 2, \ldots, m \qquad j = 1, 2, \ldots, n$$

Where x_j^* is the j-that tribute maximum value.

The linear scale transformation, in the case of cost attributes, is computed as:

$$r_{ij} = 1 - \frac{x_{ij}}{x_j^*} \qquad i = 1, 2, \ldots, m \qquad j = 1, 2, \ldots, n$$

Example 3.2

The decision matrix D reads:

$$D = \begin{vmatrix} 1.5 & 20 \\ 2.5 & 90 \\ 7 & 120 \end{vmatrix}$$

The matrix D has 3 rows (3 alternatives) and 2 columns (2 attributes). Both attributes are benefit attributes. We perform linear scale transformation. We calculate the normalized values r_{ij} as:

$$r_{ij} = \frac{x_{ij}}{x_j^*} \qquad i = 1, 2, \ldots, m \qquad j = 1, 2, \ldots, n$$

For example, r_{31} equals:

$$r_{21} = \frac{x_{21}}{x_1^*} = \frac{2.5}{7} = 0.36$$

The normalized values r_{ij} are (matrix R).

$$R = \begin{vmatrix} 0.21 & 0.17 \\ 0.36 & 0.75 \\ 1 & 1 \end{vmatrix}$$

If attributes were cost attributes, we would calculate the normalized values r_{ij} as:

$$r_{ij} = 1 - \frac{x_{ij}}{x_j^*} \qquad i = 1, 2, \ldots, m \qquad j = 1, 2, \ldots, n$$

and the normalized values r_{ij} would be (matrix R):

$$R = \begin{vmatrix} 0.79 & 0.83 \\ 0.64 & 0.25 \\ 0 & 0 \end{vmatrix}$$

3.1.3 Attribute weights

In the case of the cardinal weights of the attributes, numerical values ("importance") are assigned to each attribute. All weights must be numerical values greater than or equal to zero, and smaller than or equal to one. The following relation must be satisfied:

$$\sum_{j=1}^{n} w_j = 1$$

In other words, the total sum of all the weights must be equal to one. Attributes could be also organized into a simple rank order. In this case, we list the most important attribute first and the least important attribute last. The number of attributes (criteria) used for ranking of alternatives is equal to n. Analysts usually assign 1 to the most important attribute, and n to the least important attribute. The attribute weights are calculated as follows:

$$w_k = \frac{\dfrac{1}{r_k}}{\displaystyle\sum_{j=1}^{n} \dfrac{1}{r_j}}$$

where:
   r_k–rank of the k-th attribute

Example 3.3

The analyst ranks alternative airport locations according to the following criteria:

   X_1– Total construction cost
   X_2– Distance from downtown
   X_3– Connectivity with highway and railway networks

Let us assume that the rank order of the criteria is the following:

   X_3 – Connectivity with highway and railway networks
   X_1– Total construction cost
   X_2– Distance from downtown

The ranks are:

   $r_3 = 1$

   $r_1 = 2$

   $r_2 = 3$

The corresponding criterion weights respectively equal:

$$w_1 = \frac{\dfrac{1}{r_1}}{\dfrac{1}{r_1} + \dfrac{1}{r_2} + \dfrac{1}{r_3}} = \frac{\dfrac{1}{2}}{\dfrac{1}{2} + \dfrac{1}{3} + \dfrac{1}{1}} = 0.272$$

$$w_2 = \cfrac{\cfrac{1}{r_2}}{\cfrac{1}{r_1} + \cfrac{1}{r_2} + \cfrac{1}{r_3}} = \cfrac{\cfrac{1}{3}}{\cfrac{1}{2} + \cfrac{1}{3} + \cfrac{1}{1}} = 0.181$$

$$w_3 = \cfrac{\cfrac{1}{r_3}}{\cfrac{1}{r_1} + \cfrac{1}{r_2} + \cfrac{1}{r_3}} = \cfrac{\cfrac{1}{1}}{\cfrac{1}{2} + \cfrac{1}{3} + \cfrac{1}{1}} = 0.547$$

When we determine weights from the ranks, the sum of all the weights must also be equal to one, i.e.:

$$w_1 + w_2 + w_3 = 0.272 + 0.181 + 0.547 = 1$$

3.1.4 Dominance

The dominance method is based on an idea of a dominated alternative. An alternative is dominated if there is a new alternative which exceeds it in one or more attributes and equals it in all other attributes.

The dominance method contains the following steps:

Step 1: Compare the first two alternatives. If one alternative is dominated by the other, reject the dominated one.

Step 2: Compare the winner (the remaining alternative) with the third alternative. Reject the dominated alternative and retain the winner. Compare the winner (the remaining alternative) with the fourth alternative, etc.

Step 3: The nondominated set of alternatives is determined after a total of $(m-1)$ comparisons, where m is the total number of alternatives.

Example 3.4

The characteristics of the bus rapid transit, light rail, and metro systems are given in Table 3.1.
where:

      Alternative A_1– Bus rapid transit
      Alternative A_2– Light rail
      Alternative A_3– Metro
      X_1– Construction time (years)

Table 3.1 Comparison of bus rapid transit, light rail, and metro by the dominance method

	X_1	X_2	X_3	X_4	X_5	X_6
A_1	1.5	20	27,000	70	9	3
A_2	2.5	90	20,000	70	21	8
A_3	7	120	60,000	100	104	5

X_2 – Minimum headway (sec)
X_3 – Line capacity (passengers/direction/hour)
X_4 – Maximum speed (kph)
X_5 – Capital costs (monetary units)
X_6 – Operating costs (monetary units)

Compare bus rapid transit, light rail, and metro by the dominance method

SOLUTION:

We first compare the alternative A_1 (Bus Rapid Transit) with the alternative A_2 (Light Rail). These two alternatives have the same maximum speeds. In all other attributes, the alternative A_1 (Bus Rapid Transit) exceeds the alternative A_2 (Light Rail). We conclude that the alternative A_2 (Light Rail) is dominated. We reject this alternative. The winner is alternative A_1 (Bus Rapid Transit). We compare now the alternative A_1 (Bus Rapid Transit) with the alternative A_3 (Metro). The alternative A_1 (Bus Rapid Transit) exceeds the alternative A_3 (Metro) in the following attributes: construction time, minimum headway, capital costs, and operating costs. On the other hand, the alternative A_3 (Metro) outperforms the alternative A_1 (Bus Rapid Transit) in line capacity, and maximum speed.

We conclude that the dominated set is composed of the alternative A_1 (Bus Rapid Transit) and the alternative A_3 (Metro).

3.1.5 Maximin

The maximin method is based on the principle that the total accomplishment of an alternative is influenced by its weakest attribute. This method requires measuring attributes on a common scale. The maximin method contains the following steps:

Step 1: Transform the values x_{ij} in a common interval.
Step 2: Determine the weakest attribute for each alternative.
Step 3: Select as the best alternative, A^*, the alternative that has the highest value on the weakest attribute.

Example 3.5

The characteristics of the bus rapid transit, light rail, and metro are given in Table 3.2.
where:

Alternative A_1– Bus rapid transit
Alternative A_2– Light rail
Alternative A_3– Metro
X_1– Construction time (years)
X_2– Minimum headway (s)
X_3– Line capacity (passengers/direction/hour)

Table 3.2 Comparison of bus rapid transit, light rail, and metro by the maximin method

	X_1	X_2	X_3	X_4	X_5	X_6
A_1	1.5	20	27,000	70	9	3
A_2	2.5	90	20,000	70	21	8
A_3	7	120	60,000	100	104	5

X_4– Maximum speed (kph)
X_5– Capital costs (monetary units)
X_6– Operating costs (monetary units)

Compare bus rapid transit, light rail, and metro by the Maximin method.

SOLUTION:

We denote by x_{ij} the values that alternatives A_i $(i = 1,2,3)$ take with respect to a particular attribute (criterion) X_j $(i = 1,2,3,4,5,6)$. Our criteria are measured in different units. Therefore, we transform the values x_{ij} to a common interval. We use the following transformation for the construction time, minimum headway, capital costs, and operating costs:

$$a_{ij} = \frac{x_j^{min}}{x_{ij}}$$

where:
a_{ij}–normalized value of x_{ij}
x_j^{min}–minimal value of the attribute X_j among all the alternatives

We use the following transformation for the line capacity and maximum speed:

$$a_{ij} = \frac{x_{ij}}{x_j^{max}}$$

where:
a_{ij}– normalized value of x_{ij}
x_j^{max}– maximal value of the attribute X_j among all the alternatives

The normalized values a_{ij} are given in Table 3.3.
The values of the weakest attribute for the alternatives are:

A_1:   0.45

Table 3.3 Maximin method: The normalized a_{ij} values

	X_1	X_2	X_3	X_4	X_5	X_6
A_1	1	1	0.45	0.7	1	1
A_2	0.6	0.22	0.33	0.7	0.43	0.37
A_3	0.21	0.17	1	1	0.09	0.6

$$A_2: \quad 0.22$$

$$A_3: \quad 0.09$$

The highest value of the weakest attributes equals $\max(0.45, 0.22, 0.09) = 0.45$. This value corresponds to the alternative A_1. We conclude that the best alternative A^*, according to the maximin method, is the alternative A_1.

Example 3.6

The minimax concept is similar to the maximin concept. In some cases, we use the minimax concept to discover the best alternative. Let us consider rural areas denoted by A, B, C, D, and E (Figure 3.1). A joint fire-fighting brigade is to be designed for these five areas and the optimal location of the fire-fighting brigade must be determined. The optimal location must minimize the greatest distance between potential fire locations and the fire-fighting brigade station. The fire station can be in only one of the five areas.

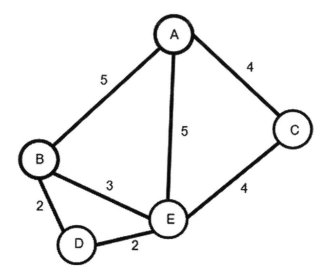

Figure 3.1 Transportation network for which the location of the fire station must be determined

Link lengths are shown in Figure 3.1. We find the distances of the shortest paths between all pairs of nodes. These distances are shown in the matrix $[d_{ij}]$: Remark by the author: I do not see matrix!!!!

$$[d_{ij}] = \begin{array}{c} \\ A \\ B \\ C \\ D \\ E \end{array} \begin{array}{ccccc} A & B & C & D & E \\ \begin{bmatrix} 0 & 5 & 4 & 7 & 5 \\ 5 & 0 & 7 & 2 & 3 \\ 4 & 7 & 0 & 6 & 4 \\ 7 & 2 & 6 & 0 & 2 \\ 5 & 3 & 4 & 2 & 0 \end{bmatrix} \end{array}$$

The node that has the minimum value of the maximum elements of its row is the optimal location for the fire station. In our case, node E is the optimal location for the fire station.

3.1.6 Maximax

The maximax method is based on the principle that the total performance of an alternative is influenced by its best attribute value. This method requires measuring attributes on a common scale. The maximax method contains the following steps:

Step 1: Transform the values x_{ij} in the common interval.
Step 2: Determine the best attribute value for each alternative.
Step 3: Select for the best alternative A^*, namely the alternative with the maximum overall best value.

Example 3.7

Compare bus rapid transit, light rail, and metro by the maximax method. Take all the necessary data from the Example 3.4. The normalized values a_{ij} are given in Table 3.4.
The best attribute values for the alternatives equal:

A_1 :   1

A_2 :   0.7

Table 3.4 Maximax method: The normalized a_{ij} values

	X_1	X_2	X_3	X_4	X_5	X_6
A_1	1	1	0.45	0.7	1	1
A_2	0.6	0.22	0.33	0.7	0.43	0.37
A_3	0.21	0.17	1	1	0.09	0.6

$A_3:$ 1

The highest value on the best attribute values equals $\max(1, 0.7, 1) = 1$. This value corresponds to the alternative A_1 and the alternative A_3. We conclude that the best alternative A^*, according to the maximax method, is jointly the alternative A_1 and the alternative A_3.

3.1.7 Conjunctive method

The conjunctive method is based on the principle that an alternative must satisfy minimal acceptable levels for all attributes. The alternative that does not satisfy one or more established attribute levels is rejected. The conjunctive method contains the following steps:

Step 1: Establish acceptable levels for all attributes.
Step 2: Check whether alternatives satisfy the acceptable attribute levels.
Step 3: Retain the alternatives that satisfy the acceptable levels. Reject the alternatives that do not satisfy acceptable levels for at least one attribute.

Example 3.8

The characteristics of the bus rapid transit, light rail, and metro systems are given in Table 3.5.
where:

A_1 – Bus rapid transit
Alternative A_2 – Light rail
Alternative A_3 – Metro
X_1 – Construction time (years)
X_2 – Minimum headway (s)
X_3 – Line capacity (passengers/direction/hour)
X_4 – Maximum speed (kph)
X_5 – Capital costs (monetary units)
X_6 – Operating costs (monetary units)

Table 3.5 Comparison of bus rapid transit, light rail, and metro by the conjunctive method

	X_1	X_2	X_3	X_4	X_5	X_6
A_1	1.5	20	27,000	70	9	3
A_2	2.5	90	20,000	70	21	8
A_3	7	120	60,000	100	104	5

Table 3.6 Ability of alternatives to satisfy criterion thresholds

	X_1	X_2	X_3	X_4	X_5	X_6
A_1	Satisfies	Satisfies	Satisfies	Satisfies	Satisfies	Satisfies
A_2	Satisfies	Satisfies	Satisfies	Satisfies	Satisfies	Does not satisfy
A_3	Does not satisfy	Satisfies	Satisfies	Satisfies	Does not satisfy	Does not satisfy

The following are the acceptable attribute levels set by the decisionmaker:

Construction time:	4 years
Minimum headway:	150 s
Line capacity:	15,000 (passengers/direction/hour)
Maximum speed:	60 (kph)
Capital costs:	80 (monetary units)
Operating costs:	4 (monetary units)

Compare bus rapid transit, light rail, and metro by the conjunctive method.

SOLUTION:

The results related to the satisfaction of the acceptable attribute levels are given in Table 3.6.

We conclude that only alternative A_1 satisfies all established attribute levels.

3.1.8 Disjunctive method

The disjunctive method establishes a desired level for each attribute. These levels are used to select the alternatives that equal or exceed those levels in anyone of the attributes. The disjunctive method contains the following steps:

Step 1: Set up the desired level for each attribute.
Step 2: Check if alternative's attribute values equals or exceeds the desirable level.
Step 3: If any alternative's attribute value equals or exceeds the desirable level, retain the alternatives. Otherwise, reject the alternative.

Example 3.9

The characteristics of the bus rapid transit, light rail, and metro systems are given in Table 3.7.

Table 3.7 Comparison of bus rapid transit, light rail, and metro by the disjunctive method

	X_1	X_2	X_3	X_4	X_5	X_6
A_1	1.5	20	27,000	70	9	3
A_2	2.5	90	20,000	70	21	8
A_3	7	120	60,000	100	104	5

where:

Alternative A_1– Bus rapid transit
Alternative A_2– Light rail
Alternative A_3– Metro
X_1– Construction time (years)
X_2– Minimum headway (s)
X_3– Line capacity (passengers/direction/hour)
X_4– Maximum speed (kph)
X_5– Capital costs (monetary units)
X_6– Operating costs (monetary units)

The following are the desirable attribute levels set up by the decisionmaker:

Construction time:	2 years
Minimum headway:	60 s
Line capacity:	40,000 (passengers/direction/hour)
Maximum speed:	80 (kph)
Capital costs:	15 (monetary units)
Operating costs:	5 (monetary units)

Compare bus rapid transit, light rail, and metro by the disjunctive method.

SOLUTION:

We check if alternatives satisfy the desired attribute levels. The results of the check are given in Table 3.8.

We see that the alternatives A_1 and A_3 are equal to or exceed the desired levels for several attributes. The alternative A_2 has no attribute

Table 3.8 Ability of the alternatives to satisfy the desired attribute thresholds

	X_1	X_2	X_3	X_4	X_5	X_6
A_1	1.5	20	$27,000 < 40,000$	$70 < 80$	9	3
A_2	$2.5 > 2$	$90 > 60$	$20,000 < 40,000$	$70 < 80$	$21 > 15$	$8 > 5$
A_3	$7 > 2$	$120 > 60$	60,000	100	$104 > 15$	5

values that exceed the desired levels. Consequently, we reject the alternative A_2.

3.1.9 Lexicographic method

The lexicographic method establishes, in the first step, the order of importance of the attributes. In the second step, this method compares the considered alternatives by using the defined order of attribute importance. The lexicographic method contains the following steps:

Step 1: Rank the attributes from the most important to the least important attribute.

Step 2: Compare the alternatives according to the most important attribute. Choose the alternative with the highest value of the most important attribute.

Step 3: In the case when there is more than one alternative with the highest value of the most important attribute, compare these alternatives according to the second-most-important attribute. Choose the alternative with the highest value of the second-most-important attribute.

Step 4: Continue in this way until only one alternative is left, or until every one of the attributes has been taken into account.

Example 3.10

Transit network design problem is a well-known transportation engineering problem, where the best possible set of bus routes should be identified. There are various criteria that should be used for evaluation of the solutions generated. The most important criteria are: percentage of unsatisfied demand, average travel time, percentage of passengers that can travel without transfers, percentage of passengers that can travel with one transfer, percentage of passengers that can travel with two transfers. Passengers that arrive at their destination with more than two transfers belong to the group whose demand is unsatisfied. Table 3.9 shows the values of these criteria in the case of the six transit networks generated. Rank the generated networks by the lexicographic method, assuming that the importance of the criteria is the same as their order.

SOLUTION:

Unsatisfied demand is the most important criterion. According to this criterion, the better solution is for it to have a smaller value. By analyzing the values with respect to unsatisfied demand, we note that all networks, except for network 5, have the same value. Consequently, network 5 will be ranked as the worst and put in the 6th place. In order to rank the other networks, we consider the other criteria.

The second criterion in terms of importance is the average travel time. According to this criterion, the better solution is the one with the

Table 3.9 The characteristics of the six transit networks

	Unsatisfied demand [%]	Average travel time [min]	Travel without transfers [%]	Travel with one transfer [%]	Travel with two transfers [%]
Network 1	0	13	70	29.5	0.5
Network 2	0	13	70	27	3
Network 3	0	12	87	12	1
Network 4	0	11	93	7	0
Network 5	6.5	10	80	12	0.5
Network 6	0	11	92	7	1

smaller value. We compare networks 1, 2, 3, 4, and 6. We note that the best value corresponds to networks 4 and 6. The next, according to this criterion, is network 3, and, finally, networks 1 and 2. In the next step, we compare networks 4 and 6 according to the percentage of passengers that can travel without transfers. We conclude that network 4 is better than network 6. Network 4 will be ranked1 and network 6 will be ranked2.

According to the second criterion, next in the rankings is network 3. This network will be given rank 3. According to the second criterion, networks 1 and 2 have the same value. Consequently, we have to compare them according to the third criterion. We note that both networks also have the same value according to the third criterion, so. we continue the procedure and compare these networks according to the fourth criterion. According to the fourth criterion (percentage of passengers that can travel from origin to destination with one transfer), we note that network1 has the better value (29.5% is better than 27%). Network 1 will therefore have rank 4 and network 2 will have rank 5. The final ranking of the transit networks is presented in Table 3.10.

Table 3.10 The end solution obtained by the lexicographic method

Rank	Solution	Unsatisfied demand [%]	Average travel time [min]	Travel without transfers [%]	Travel with one transfer [%]	Travel with two transfers [%]
1	Solution 4	0	11	93	7	0
2	Solution 6	0	11	92	7	1
3	Solution 3	0	12	87	12	1
4	Solution 1	0	13	70	29.5	0.5
5	Solution 2	0	13	70	27	3
6	Solution 5	6.5	10	80	12	0.5

3.1.10 Simple additive weighting method (SAW)

The SAW method is a multi-attribute decision-making method widely used in engineering and management. Various traffic, technical, economic, or environmental criteria are converted to a common scale before applying the SAW method. Within the SAW method, the score of each alternative under consideration is obtained by adding the contributions from each attribute. The final score of the alternative is obtained by multiplying the rating for each attribute by the attribute weight and then summing these products over all the attributes.

The SAW method translates a multi-criterion problem into a single-dimensional system. The weighted score V_i of the alternative A_i equals:

$$V_i = \sum_{j=1}^{n} w_j \cdot a_{ij}$$

where:
   w_j– weight of the criteria X_j
   a_{ij}–normalized value of x_{ij}

The alternative with the highest weighted score is selected by the decisionmaker.

Example 3.11

where:
   Alternative A_1– Bus rapid transit
   Alternative A_2– Light rail
   Alternative A_3– Metro
   X_1– Construction time (years)
   X_2– Minimum headway (s)
   X_3– Line capacity (passengers/direction/hour)
   X_4– Maximum speed (kph)
   X_5– Capital costs (monetary units)
   X_6– Operating costs (monetary units)

The following are the weights set up by the decisionmaker:

Construction time:	$w_1 = 0.1$
Minimum headway:	$w_1 = 0.1$
Line capacity:	$w_1 = 0.2$
Maximum speed:	$w_1 = 0.1$
Capital costs:	$w_5 = 0.35$
Operating costs:	$w_6 = 0.15$

Table 3.11 Comparison of bus rapid transit, light rail, and metro by the SAW

	X_1	X_2	X_3	X_4	X_5	X_6
A_1	1.5	20	27,000	70	9	3
A_2	2.5	90	20,000	70	21	8
A_3	7	120	60,000	100	104	5

Table 3.12 SAW method: The normalized values a_{ij}

	X_1	X_2	X_3	X_4	X_5	X_6
A_1	1	1	0.45	0.7	1	1
A_2	0.6	0.22	0.33	0.7	0.43	0.37
A_3	0.21	0.17	1	1	0.09	0.6

Compare bus rapid transit, light rail, and metro by the SAW method (Table 3.11).

SOLUTION (TABLE 3.12):

The weighted scores V_i of the alternatives A_i are:

$$V_1 = 0.1 \cdot 1 + 0.1 \cdot 1 + 0.2 \cdot 0.45 + 0.1 \cdot 0.7 + 0.35 \cdot 1 + 0.15 \cdot 1 = 0.86$$

$$V_2 = 0.1 \cdot 0.6 + 0.1 \cdot 0.22 + 0.2 \cdot 0.33 + 0.1 \cdot 0.7 + 0.35 \cdot 0.43 + 0.15 \cdot 0.37$$
$$= 0.3640$$

$$V_2 = 0.1 \cdot 0.21 + 0.1 \cdot 0.17 + 0.2 \cdot 1 + 0.1 \cdot 1 + 0.35 \cdot 0.09 + 0.15 \cdot 0.6 = 0.3785$$

$$V_3 = 0.1 \cdot 0.21 + 0.1 \cdot 0.17 + 0.2 \cdot 1 + 0.1 \cdot 1 + 0.35 \cdot 0.09 + 0.15 \cdot 0.6 = 0.3785$$

The alternative A_1 has the highest weighted score and should be selected.

3.1.11 TOPSIS

The technique for Order Preference by Similarity to Ideal Solution (TOPSIS) is based on the idea that the selected alternative should have the shortest distance from the *positive-ideal solution* and the longest distance from the *negative-ideal solution*.

The decision matrix D reads:

$$D = \begin{array}{c} A_1 \\ A_2 \\ \vdots \\ A_m \end{array} \begin{array}{cccc} X_1 & X_2 & \cdots & X_n \\ \begin{vmatrix} x_{11} & x_{12} & \cdots & x_{1n} \\ x_{21} & x_{22} & \cdots & x_{2n} \\ \vdots & \vdots & \vdots & \vdots \\ x_{m1} & x_{m2} & \cdots & x_{mn} \end{vmatrix} \end{array}$$

We denote the total number of alternatives by m, and the total number of criteria, according to which the alternatives under consideration are compared, by n.

Normalized values r_{ij} are calculated as

$$r_{ij} = \frac{x_{ij}}{\sqrt{\sum_{k=1}^{m} x_{kj}^2}} \qquad i = 1,2,\ldots,m \qquad j = 1,2,\ldots,n$$

In the next step, each column's elements in matrix R are multiplied by weight w_j (reflecting the significance of the criterion), corresponding to a particular column. In this manner, matrix V is obtained such that the values of its elements express the weights (significance) of individual criteria as well. Matrix V is found to be

$$V = \begin{bmatrix} v_{11} & \cdots & v_{1j} & \cdots & v_{1n} \\ \vdots & & \vdots & & \vdots \\ v_{i1} & \cdots & v_{ij} & \cdots & v_{in} \\ \vdots & & \vdots & & \vdots \\ v_{m1} & \cdots & v_{mj} & \cdots & v_{mn} \end{bmatrix} = \begin{bmatrix} w_1 \cdot r_{11} & \cdots & w_1 \cdot r_{1j} & \cdots & w_1 \cdot r_{1n} \\ \vdots & & \vdots & & \vdots \\ w_2 \cdot r_{i1} & \cdots & w_2 \cdot r_{ij} & \cdots & w_n \cdot r_{in} \\ \vdots & & \vdots & & \vdots \\ w_m \cdot r_{m1} & \cdots & w_m \cdot r_{mj} & \cdots & w_n \cdot r_{mn} \end{bmatrix}$$

On calculating the elements of matrix V, the positive ideal solution A^*, and the negative ideal solution A^- are determined. These solutions are defined as:

$$A^* = \left\{ \left(\max_i v_{ij} \mid j \in J \right), \left(\min_i v_{ij} \mid j \in J' \right) \mid i = 1,2,\ldots,m \right\} = \left\{ v_1^*, v_2^*, \ldots, v_j^*, \ldots, v_n^* \right\}$$

$$A^- = \left\{ \left(\min_i v_{ij} \mid j \in J \right), \left(\max_i v_{ij} \mid j \in J' \right) \mid i = 1,2,\ldots,m \right\} = \left\{ v_1^-, v_2^-, \ldots, v_j^-, \ldots, v_n^- \right\}$$

where

$$J = \left\{ j = 1,2,\ldots,n \mid j \text{ belongs to the benefit criteria} \right\}$$

$$J' = \left\{ j = 1,2,\ldots,n \mid j \text{ belongs to the cost criteria} \right\}$$

A positive ideal solution A^* represents the ideal alternative, expressing the best values for all the criteria. The ideal solution usually does not exist in real life. The decision makers try to choose the alternative which is as close as possible to an ideal solution. The negative-ideal solution is composed of all the worst attribute ratings (Figure 3.2).

Please note that the benefit criteria are understood to be those for which an alternative is better if it exhibits a greater value. As far as the cost

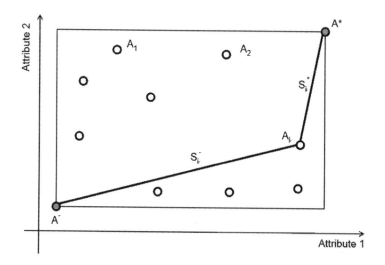

Figure 3.2 Alternatives, positive-ideal and negative-ideal solutions in two-dimensional space

criteria are concerned, an alternative is better if, in terms of these criteria, it exhibits a lower value. The distance S_i^* of each alternative from the ideal alternative is:

$$S_i^* = \sqrt{\sum_{j=1}^{n}\left(v_{ij} - v_j^*\right)^2} \qquad i = 1,2,\ldots,m$$

The distance S_i^- of each alternative from the negative ideal solution is:

$$S_i^- = \sqrt{\sum_{j=1}^{n}\left(v_{ij} - v_j^-\right)^2} \qquad i = 1,2,\ldots,m$$

Relative closeness C_i^* of the alternative A_i to the ideal solution A^* is:

$$C_i^* = \frac{S_i^-}{S_i^* + S_i^-} \qquad 0 \le C_i^* \le 1 \quad i = 1,2,\ldots,m$$

Since $C_i^* = 1$ if $A_i = A^*$, and $C_i^* = 0$ if $A_i = A^-$, the alternative A_i is better if C_i^* is closer to 1. It is clear that, from the set of alternatives $A_1, A_2, \ldots, A_m$, the best alternative is A_i with the largest value of C_i^*.

Example 3.12

The characteristics of the bus rapid transit, light rail, and metro systems are given in Table 3.13.

Table 3.13 Comparison of bus rapid transit, light rail, and metro by TOPSIS

	Construction time (years) X_1	Minimum headway (s) X_2	Line capacity (Passenger/ direction/h) X_3	Maximum speed (kph) X_4	Capital costs (monetary units) X_5	Operating costs (monetary units) X_6
Bus rapid transit A_1	1.5	20	27,000	70	9	3
Light rail A_2	2.5	90	20,000	70	21	8
Metro A_3	7	120	60,000	100	104	5

The decision maker decided that the criterion weights are:

Construction time:	$w_1 = 0.1$
Minimum headway:	$w_2 = 0.1$
Line capacity:	$w_3 = 0.2$
Maximum speed:	$w_4 = 0.1$
Capital costs:	$w_5 = 0.35$
Operating costs:	$w_6 = 0.15$

Compare bus rapid transit, light rail, and metro by the TOPSIS method.

SOLUTION:

We first calculate normalized values r_{ij} as:

$$r_{ij} = \frac{x_{ij}}{\sqrt{\sum_{k=1}^{m} x_{kj}^2}} \qquad i = 1, 2, \ldots, m \qquad j = 1, 2, \ldots, n$$

For example, r_{23} equals:

$$r_{21} = \frac{x_{21}}{\sqrt{\sum_{k=1}^{3} x_{k1}^2}} = \frac{2.5}{\sqrt{1.5^2 + 2.5^2 + 7^2}} = 0.33$$

The normalized values r_{ij} are given in Table 3.14 (matrix R).

We now multiply each column's elements in matrix R by the weight w_j corresponding to a particular column. For example, element v_{23} equals:

$$v_{23} = r_{23} \cdot w_3 = 0.29 \cdot 0.2 = 0.058$$

Table 3.14 TOPSIS: The normalized values r_{ij}

	Construction time (years) X_1	Minimum headway (s) X_2	Line capacity (Passengers/ direction/h) X_3	Maximum speed (kph) X_4	Capital costs (monetary units) X_5	Operating costs (monetary units) X_6
Bus rapid transit A_1	0.20	0.13	0.39	0.50	0.08	0.30
Light rail A_2	0.33	0.59	0.29	0.50	0.20	0.81
Metro A_3	0.92	0.79	0.87	0.71	0.98	0.51

Table 3.15 Matrix V

	Construction time (years) X_1	Minimum headway (s) X_2	Line capacity (Passengers/ direction/h) X_3	Maximum speed (kph) X_4	Capital costs (monetary units) X_5	Operating costs (monetary units) X_6
Bus rapid transit A_1	0.020	0.013	0.078	0.050	0.028	0.045
Light rail A_2	0.033	0.059	0.058	0.050	0.070	0.1215
Metro A_3	0.092	0.079	0.174	0.071	0.343	0.0765

We obtain the matrix V, such that the values of its elements express the weights (significance) of the individual criteria as well. Matrix V is shown in Table 3.15.

The positive ideal solution A^* and the negative ideal solution A^- are:

$$A^* = \{0.020, \ 0.013, \ 0.174, \ 0.071, \ 0.028, \ 0.045\}$$

$$A^- = \{0.092, \ 0.079, \ 0.058, \ 0.050, \ 0.343, \ 0.1215\}$$

The distance S_i^* of each alternative from the ideal alternative is:

$$S_i^* = \sqrt{\sum_{j=1}^{n} \left(v_{ij} - v_j^* \right)^2} \qquad i = 1, 2, \ldots, m$$

Table 3.16 The distances of each alternative from the ideal alternative (S_i^*) and from the negative ideal alternative (S_i^-), and the relative closeness (C_i^*)

Alternative	S_i^*	S_i^-	$C_i^* = \dfrac{S_i^-}{S_i^* + S_i^-}$
Bus rapid transit A_1	0.098	0.339	0.775
light Rail A_2	0.154	0.280	0.645
Metro A_3	0.331	0.126	0.276

For example, the distance S_2^* of the alternative $A_2 = \{0.033,\ 0.059,\ 0.058,\ 0.050,\ 0.070,\ 0.1215\}$ from the ideal alternative $A^* = \{0.020,\ 0.013,\ 0.174,\ 0.071,\ 0.028,\ 0.045\}$ equals:

$$\sum_{j=1}^{n}\left(v_{ij}-v_j^*\right)^2 = \left(0.033-0.020\right)^2 + \left(0.059-0.013\right)^2 + \left(0.058-0.174\right)^2 +$$

$$+\left(0.050-0.071\right)^2 + \left(0.070-0.028\right)^2 + \left(0.1215-0.045\right)^2 = 0.0238$$

We obtain:

$$S_2^* = \sqrt{\sum_{j=1}^{n}\left(v_{ij}-v_j^*\right)^2} = \sqrt{0.0238} = 0.154$$

The distance of each alternative from the ideal alternative and from the negative ideal alternative are calculated and shown in Table 3.16.

The third column in the Table shows the relative closeness C_i^* of the alternative A_i to the ideal solution A^*. We see from the Table that the alternative A_1 has the highest value of relative closeness. Consequently, we conclude that the alternative A_1 is the best alternative, according to the TOPSIS method.

3.2 DATA ENVELOPMENT ANALYSIS (DEA)

Transportation engineers and analysts frequently face the problem of comparing the efficiency of airports, hubs, terminals, ports, airline routes, and bus lines, as well as the problem of measuring their performances. Most frequently, when performing such an analysis, the engineers use *ratios*. Ratios are obtained by dividing some output measure (e.g. number of passengers processed, number of aircraft operations, cargo volumes, etc.) by some input measure (number of runways, number of check-in desks, etc.).

Table 3.17 Representation of the data for the DEA method

	Input 1	Input 2	$\cdots$	Input m	Output 1	Output 2	$\cdots$	Output s
DMU 1	x_{11}	x_{21}	$\cdots$	x_{m1}	y_{11}	y_{21}	$\cdots$	y_{s1}
DMU 2	x_{12}	x_{22}	$\cdots$	x_{m2}	y_{12}	y_{22}	$\cdots$	y_{s2}
$\vdots$	$\vdots$	$\vdots$		$\vdots$	$\vdots$	$\vdots$		$\vdots$
DMU n	x_{1n}	x_{2n}	$\cdots$	x_{mn}	y_{1n}	y_{2n}	$\cdots$	y_{sn}

Various ratios can generate different conclusions about the efficiency of the transportation facilities under comparison.

The *data envelopment analysis* (DEA) is a measurement technique that is used for evaluating the relative efficiency of decision-making units (DMUs). By using the DEA, the analyst can evaluate the efficiency of any number of DMUs. The analyzed DMUs could have any number of inputs and outputs. The DMUs in the area of traffic and transportation could be airports, airline routes, intersections, high-occupancy vehicle (HOV) lanes, ports, networks, park and ride facilities, etc.

The DEA was initially suggested by Charnes, Cooper, and Rhodes in 1978. Each DMU represents the entity that changes inputs into outputs. In a general case, we have n DMUs, m inputs, and s outputs. Also, we denote with x_{ij} the amount of input i $(i = 1,\ldots,m)$ that has DMU j $(j = 1,\ldots,n)$, and with y_{rj} the amount of output $r(r = 1,\ldots,s)$ that has DMU j $(j = 1,\ldots,n)$. Table 3.17 shows an example of representative data.

The DEA defines the relative efficiency in the following way:

$$\text{Efficiency} = \frac{\text{Weighted sum of outputs}}{\text{Weighted sum of inputs}}$$

i.e.:

$$\text{Efficiency} = \frac{u_1 \cdot y_{1j} + u_2 \cdot y_{2j} + \cdots + u_s \cdot y_{sj}}{v_1 \cdot x_{1j} + v_2 \cdot x_{2j} + \cdots + v_m \cdot x_{mj}}$$

where:

v_i - weight of the input i $(i = 1,\ldots,m)$
u_r - weight of the output r $(r = 1,\ldots,s)$

The DEA defines the efficiency for every DMU as a weighted sum of outputs divided by a weighted sum of inputs. The efficiency of any DMU is within the range [0, 1], or [0%, 100%]. Frequently, different DMU's have different goals. For example, some airports could try to maximize the number of served passengers, while some others could try to maximize cargo volumes. When calculating the efficiency of a specific DMU by the DEA technique, the weights of a DMU are chosen to present the DMU under consideration in the best possible light.

The efficiency e_{j_0} of the DMU j_0 could be obtained by solving the following fractional programming problem:

Maximize

$$e_{j_0} = \frac{\sum_{r=1}^{s} u_r \cdot y_{rj_0}}{\sum_{i=1}^{m} v_i \cdot x_{ij_0}}$$

subject to:

$$\frac{\sum_{r=1}^{s} u_r \cdot y_{rj}}{\sum_{i=1}^{m} v_i \cdot x_{ij}} \leq 1 \qquad \forall j = 1,\ldots,n$$

$$v_i, u_r \geq \varepsilon \qquad \forall i = 1,\ldots,m; r = 1,\ldots,s$$

By solving the fractional programming problem, the analyst obtains the input weights v_i, as well as output weights u_r. In the case of n DMUs to be evaluated, the analyst has to perform n optimizations (one for every DMU to be evaluated). The fractional program could be replaced by the following equivalent linear program:

Maximize

$$e_{j_0} = \sum_{r=1}^{s} u_r \cdot y_{rj_0}$$

subject to:

$$\sum_{i=1}^{m} v_i \cdot x_{ij_0} = 1$$

$$\sum_{r=1}^{s} u_r \cdot y_{rj} - \sum_{i=1}^{m} v_i \cdot x_{ij} \leq 0 \qquad \forall j = 1,\cdots,n$$

$$v_i, u_r \geq \varepsilon \qquad \forall i = 1,\cdots,m; r = 1,\cdots,s$$

The inputs and outputs in the DEA have different units. The DEA enables the analyst to directly compare DMUs under consideration with their peers.

This previously mentioned DEA model is known in the literature as the CCR model, because it was proposed by the authors Charnes, Cooper, and Rhodes in 1978. Various other DEA models have been proposed, but we will not consider them in this book.

Table 3.18 Input data for evaluation of efficiencies

Company	Number of trucks	Annual profit (monetary units)
1	10	360,000
2	12	388,800
3	15	720,000
4	5	138,000
5	20	840,000

Let us take into consideration examples four cases with the following numbers of inputs and outputs:

1. one input and one output,
2. two inputs and one output,
3. one input and two outputs, and
4. more inputs and more outputs.

Example 3.13

Determine the efficiencies of the five companies according to the data given in Table 3.18. For each company, input is the number of trucks, and output is the annual profit.

SOLUTION:

In this example the companies are the DMUs, the number of trucks is the input and the annual profit is the output. Examples with one input and one output can be solved in one of three ways. The first one is applying previously mentioned mathematical models and solving one linear program for each DMU. The second is also a numerical method, based on the determination of the ratio $\frac{\text{output}}{\text{input}}$, and the third one is a graphical method.

If we want to determine efficiencies in the first way, we have to solve one linear program for each DMU. The linear programs are given in Table 3.19.

The second way to determine the efficiencies is very easy. Let us determine the ratio of the annual profit and the number of trucks for each company. These values are given in the second column of Table 3.20. The efficiencies are normalized values. Normalization should be made by dividing the calculated values by the highest calculated value. In this example, the highest value equals 48,000. Efficiencies of the companies are:

$$\text{Efficiency of company } 1 = \frac{36,000}{48,000} = 0.75$$

Table 3.19 Linear programs for all companies

Linear program for company 1:
Maximize
$$e_1 = 360,000 \cdot u_1$$
subject to:
$$10 \cdot v_1 = 1$$
$$360,000 \cdot u_1 - 10 \cdot v_1 \leq 0$$
$$388,800 \cdot u_1 - 12 \cdot v_1 \leq 0$$
$$720,000 \cdot u_1 - 15 \cdot v_1 \leq 0$$
$$138,000 \cdot u_1 - 5 \cdot v_1 \leq 0$$
$$840,000 \cdot u_1 - 20 \cdot v_1 \leq 0$$
$$v_1 \geq 0.000001$$
$$u_1 \geq 0.000001$$

Linear program for company 3:
Maximize
$$e_3 = 720,000 \cdot u_1$$
subject to:
$$15 \cdot v_1 = 1$$
$$360,000 \cdot u_1 - 10 \cdot v_1 \leq 0$$
$$388,800 \cdot u_1 - 12 \cdot v_1 \leq 0$$
$$720,000 \cdot u_1 - 15 \cdot v_1 \leq 0$$
$$138,000 \cdot u_1 - 5 \cdot v_1 \leq 0$$
$$840,000 \cdot u_1 - 20 \cdot v_1 \leq 0$$
$$v_1 \geq 0.000001$$
$$u_1 \geq 0.000001$$

Linear program for company 5:
Maximize:
$$e_1 = 840,000 \cdot u_1$$
subject to
$$20 \cdot v_1 = 1$$
$$360,000 \cdot u_1 - 10 \cdot v_1 \leq 0$$
$$388,800 \cdot u_1 - 12 \cdot v_1 \leq 0$$
$$720,000 \cdot u_1 - 15 \cdot v_1 \leq 0$$
$$138,000 \cdot u_1 - 5 \cdot v_1 \leq 0$$
$$840,000 \cdot u_1 - 20 \cdot v_1 \leq 0$$
$$v_1 \geq 0.000001$$
$$u_1 \geq 0.000001$$

Linear program for company 2:
Maximize
$$e_2 = 388,000 \cdot u_1$$
subject to:
$$12 \cdot v_1 = 1$$
$$360,000 \cdot u_1 - 10 \cdot v_1 \leq 0$$
$$388,800 \cdot u_1 - 12 \cdot v_1 \leq 0$$
$$720,000 \cdot u_1 - 15 \cdot v_1 \leq 0$$
$$138,000 \cdot u_1 - 5 \cdot v_1 \leq 0$$
$$840,000 \cdot u_1 - 20 \cdot v_1 \leq 0$$
$$v_1 \geq 0.000001$$
$$u_1 \geq 0.000001$$

Linear program for company 4:
Maximize
$$e_4 = 138,000 \cdot u_1$$
subject to:
$$5 \cdot v_1 = 1$$
$$360,000 \cdot u_1 - 10 \cdot v_1 \leq 0$$
$$388,800 \cdot u_1 - 12 \cdot v_1 \leq 0$$
$$720,000 \cdot u_1 - 15 \cdot v_1 \leq 0$$
$$138,000 \cdot u_1 - 5 \cdot v_1 \leq 0$$
$$840,000 \cdot u_1 - 20 \cdot v_1 \leq 0$$
$$v_1 \geq 0.000001$$
$$u_1 \geq 0.000001$$

$$\text{Efficiency of company } 2 = \frac{32,400}{48,000} = 0.675$$

$$\vdots$$

$$\text{Efficiency of company } 5 = \frac{42,000}{48,000} = 0.875$$

Table 3.20 Efficiencies of the five companies

Company	Annual profit / Number of trucks	Efficiency
1	36,000	0.75
2	32,400	0.675
3	48,000	1
4	27,600	0.575
5	42,000	0.875

The values of efficiencies are given in Table 3.20 in the third column.

The graphical method is the third way in which we can calculate the efficiencies of DMUs. The arrangements of the DMUs are given in Figure 3.3.

Now we have to determine the efficiency frontier. For that purpose, we draw one line through each point of DMUs, as is shown in Figure 3.4. We take a straight line through DMU 3 as the frontier, since it has the greatest angle with the x-axis. The other straight lines should be discarded. Figure 3.5 shows the efficiency frontier for the example under consideration.

From Figure 3.5 we notice that only DMU 3 is efficient (it has an efficiency equal to 1). The other DMUs (companies) are not efficient, with efficiencies between 0 and 1. We can determine their efficiencies by using the graphical method. Let us illustrate this procedure in the case of DMU 1. First, note nodes A, B, and C in Figure 3.6. To calculate the efficiency of DMU 1 we have to divide the length between nodes A and B by the length between nodes A and C, i.e.:

$$\text{efficiency of company } 1 = \frac{\text{length}(A, B)}{\text{length}(A, C)}$$

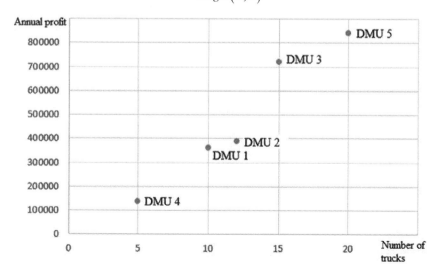

Figure 3.3 Graphical representation of data

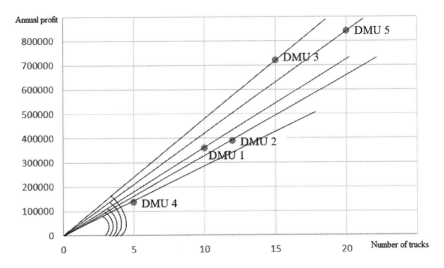

Figure 3.4 Determination of the efficiency frontier

In a similar way, we can calculate the efficiencies of the other companies.

Let us consider the example that has two inputs and one output. In this case, we can solve the problem graphically or by solving the linear programs for each DMU.

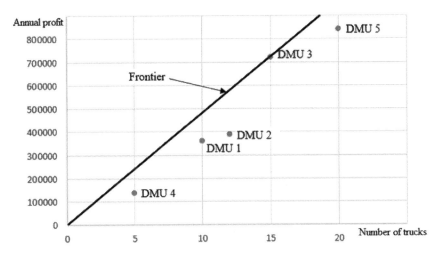

Figure 3.5 Efficiency frontier for the problem under consideration

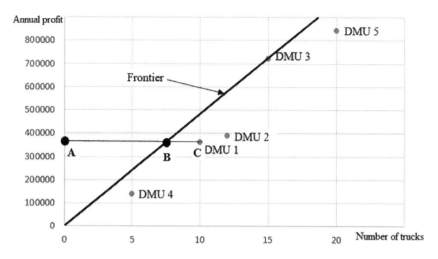

Figure 3.6 Determination of efficiency of DMU 1

Example 3.14

Determine the efficiencies of metro systems of the world regions, according to data given in Table 3.21.

SOLUTION:

In this example, the world regions are the DMUs, kilometers of tracks and numbers of carriages are the inputs, and the annual number of passengers is the output. Efficiencies can be determined by solving one linear program for each region (DMU) or from a graphical presentation of data, i.e. by taking into consideration the distance from the efficiency frontier. The first approach means that we have to solve 6 linear programs. These linear programs are given in Table 3.22.

Solving the linear programs generates the solution given in Table 3.23. From the results we can notice that only Latin America has efficiency 1. It follows Eurasia, Middle East/North Africa (MENA), etc. The region with the lowest value of efficiency is North America.

Table 3.21 Data for metro systems (UITP, World Metro Figures 2018)

Region	Track (km)	Carriages	Annual number of passengers (in millions)
North America	1,544	14,200	3,730
Latin America	943	9,000	5,915
Europe	2,921	25,800	10,750
MENA	464	3,300	1,990
Eurasia	813	8,100	4,700
Asia-Pacific	7,218	53,700	26,690

Table 3.22 Linear programs for calculation of efficiencies of metro systems for each region

Linear program for North America is:
Maximize

$$e_1 = 3,730 \cdot u_1$$

subject to:

$$1544 \cdot v_1 + 14,200 \cdot v_2 = 1$$
$$3,730 \cdot u_1 - 1,544 \cdot v_1 - 14,200 \cdot v_2 \leq 0$$
$$5,915 \cdot u_1 - 943 \cdot v_1 - 9,000 \cdot v_2 \leq 0$$
$$10,750 \cdot u_1 - 2,921 \cdot v_1 - 25,800 \cdot v_2 \leq 0$$
$$1,990 \cdot u_1 - 464 \cdot v_1 - 3,300 \cdot v_2 \leq 0$$
$$4,700 \cdot u_1 - 813 \cdot v_1 - 8,100 \cdot v_2 \leq 0$$
$$26,690 \cdot u_1 - 7,218 \cdot v_1 - 53,700 \cdot v_2 \leq 0$$
$$v_1 \geq 0.00001$$
$$v_2 \geq 0.00001$$
$$u_1 \geq 0.00001$$

Linear program for Latin America is:
Maximize

$$e_2 = 5,915 \cdot u_1$$

subject to

$$943 \cdot v_1 + 9,000 \cdot v_2 = 1$$
$$3,730 \cdot u_1 - 1,544 \cdot v_1 - 14,200 \cdot v_2 \leq 0$$
$$5,915 \cdot u_1 - 943 \cdot v_1 - 9,000 \cdot v_2 \leq 0$$
$$10,750 \cdot u_1 - 2921 \cdot v_1 - 25,800 \cdot v_2 \leq 0$$
$$1,990 \cdot u_1 - 464 \cdot v_1 - 3,300 \cdot v_2 \leq 0$$
$$4,700 \cdot u_1 - 813 \cdot v_1 - 8,100 \cdot v_2 \leq 0$$
$$26,690 \cdot u_1 - 7,218 \cdot v_1 - 53,700 \cdot v_2 \leq 0$$
$$v_1 \geq 0.00001$$
$$v_2 \geq 0.00001$$
$$u_1 \geq 0.00001$$

Linear program for Europe is:
Maximize

$$e_3 = 10,750 \cdot u_1$$

subject to:

$$2,921 \cdot v_1 + 25,800 \cdot v_2 = 1$$
$$3,730 \cdot u_1 - 1,544 \cdot v_1 - 14,200 \cdot v_2 \leq 0$$
$$5,915 \cdot u_1 - 943 \cdot v_1 - 9,000 \cdot v_2 \leq 0$$
$$10,750 \cdot u_1 - 2,921 \cdot v_1 - 25,800 \cdot v_2 \leq 0$$
$$1,990 \cdot u_1 - 464 \cdot v_1 - 3,300 \cdot v_2 \leq 0$$
$$4,700 \cdot u_1 - 813 \cdot v_1 - 8,100 \cdot v_2 \leq 0$$
$$26,690 \cdot u_1 - 7,218 \cdot v_1 - 53,700 \cdot v_2 \leq 0$$
$$v_1 \geq 0.00001$$
$$v_2 \geq 0.00001$$
$$u_1 \geq 0.00001$$

Linear program for the region Middle East and North Africa (MENA) is:
Maximize

$$e_4 = 1,990 \cdot u_1$$

subject to

$$:464 \cdot v_1 + 3,300 \cdot v_2 = 1$$
$$3,730 \cdot u_1 - 1,544 \cdot v_1 - 14,200 \cdot v_2 \leq 0$$
$$5,915 \cdot u_1 - 943 \cdot v_1 - 9,000 \cdot v_2 \leq 0$$
$$10,750 \cdot u_1 - 2,921 \cdot v_1 - 25,800 \cdot v_2 \leq 0$$
$$1,990 \cdot u_1 - 464 \cdot v_1 - 3,300 \cdot v_2 \leq 0$$
$$4,700 \cdot u_1 - 813 \cdot v_1 - 8,100 \cdot v_2 \leq 0$$
$$26,690 \cdot u_1 - 7,218 \cdot v_1 - 53,700 \cdot v_2 \leq 0$$
$$v_1 \geq 0.00001$$
$$v_2 \geq 0.00001$$
$$u_1 \geq 0.00001$$

Linear program for the region Eurasia is:
Maximize

$$e_5 = 4,700 \cdot u_1$$

subject to:

$$813 \cdot v_1 + 8,100 \cdot v_2 = 1$$
$$3,730 \cdot u_1 - 1,544 \cdot v_1 - 14,200 \cdot v_2 \leq 0$$
$$5,915 \cdot u_1 - 943 \cdot v_1 - 9,000 \cdot v_2 \leq 0$$
$$10,750 \cdot u_1 - 2,921 \cdot v_1 - 25,800 \cdot v_2 \leq 0$$
$$1,990 \cdot u_1 - 464 \cdot v_1 - 3,300 \cdot v_2 \leq 0$$
$$4,700 \cdot u_1 - 813 \cdot v_1 - 8,100 \cdot v_2 \leq 0$$
$$26,690 \cdot u_1 - 7,218 \cdot v_1 - 53,700 \cdot v_2 \leq 0$$
$$v_1 \geq 0.00001$$
$$v_2 \geq 0.00001$$
$$u_1 \geq 0.00001$$

Linear program for the region Asia-Pacific is:
Maximize

$$e_6 = 26,690 \cdot u_1$$

subject to

$$:7,218 \cdot v_1 + 53,700 \cdot v_2 = 1$$
$$3,730 \cdot u_1 - 1,544 \cdot v_1 - 14,200 \cdot v_2 \leq 0$$
$$5,915 \cdot u_1 - 943 \cdot v_1 - 9,000 \cdot v_2 \leq 0$$
$$10,750 \cdot u_1 - 2,921 \cdot v_1 - 25,800 \cdot v_2 \leq 0$$
$$1,990 \cdot u_1 - 464 \cdot v_1 - 3,300 \cdot v_2 \leq 0$$
$$4,700 \cdot u_1 - 813 \cdot v_1 - 8,100 \cdot v_2 \leq 0$$
$$26,690 \cdot u_1 - 7,218 \cdot v_1 - 53,700 \cdot v_2 \leq 0$$
$$v_1 \geq 0.00001$$
$$v_2 \geq 0.00001$$
$$u_1 \geq 0.00001$$

Table 3.23 Efficiencies of regional metro systems

Region	Efficiency
North America	0.399
Latin America	1
Europe	0.633
MENA	0.916
Eurasia	0.919
Asia-Pacific	0.744

According to the data given in Table 3.21, regions can be graphically represented in the way shown in Figure 3.7. At the x-axis are values: $\dfrac{\text{input 1}}{\text{output}} = \dfrac{\text{km of track}}{\text{annually number of passengers}}$, and at the y-axis: $\dfrac{\text{input 2}}{\text{output}} = \dfrac{\text{carriages}}{\text{annually number of passengers}}$.

From Figure 3.7, we note that only Latin America is efficient (has an efficiency value of 1). Efficiencies of the other regions can be calculated in the way shown in Figure 3.8 for Europe. We note points A, B, and C. Efficiency is the quotient of the distance between A and B, and the distance between A and C, i.e.:

$$\text{efficiency of Europe} = \frac{\text{distance}\,(A,B)}{\text{distance}\,(A,C)}$$

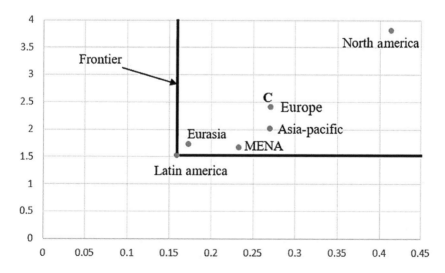

Figure 3.7 Efficiency frontier in a case with two inputs and one output

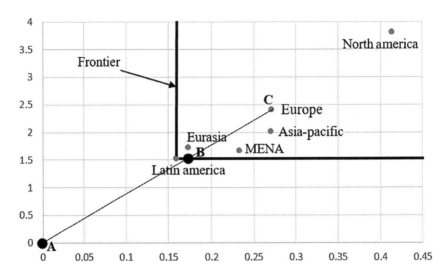

Figure 3.8 Calculation of efficiency for regions in Europe

Example 3.15

Evaluate the traffic safety of the countries according to the data given in Table 3.24 by CCR DEA.

SOLUTION:

From the data given in Table 3.24:

- Input is passenger-kilometers, and
- Outputs are:

Table 3.24 Data for evaluation of traffic safety (Wu and Goh, 2010.)

Country	Passenger-kilometers (pkm) x 10⁶	Number of the road injuries (nᵢ)	Number of the road fatalities (nᵣ)
Bulgaria	8,588	8,466	611
Czech Republic	88,921.4	27,680	656
Finland	74,800	5,277	234
France	815,582	69,887	3,248
North Macedonia	11,698	5,860	133
Hungary	82,606.61	21,999	633
Italy	826,284	242,621	3,325
Slovak Republic	34,699	6,915	260
Sweden	126,480	18,501	324
United Kingdom	707,980.3788	167,261	1,837

- Number of road injures, and
- Number of road fatalities.

We note that it is better if we have more passenger-kilometers for the same outputs. This is the opposite of the basic idea of the DEA method, where it is better if we have less input for the same amount of output. We take the reciprocal value of passenger-kilometers, as well as the reciprocal values of numbers of injuries and number of the fatalities. Table 3.25 shows the reciprocal values.

Figure 3.9 shows spatial arrangement of the countries according to the data given in Table 3.25. In the x-axis, the values obtained are:

$$\frac{\text{Output 1}}{\text{Input}} = \frac{\dfrac{1}{\text{number of the road injuries}}}{\dfrac{1}{\text{passenger-kilometers}}} = \frac{\text{passenger-kilometers}}{\text{number of the road injuries}}$$

In the y-axis, the values obtained are:

$$\frac{\text{Output 2}}{\text{Input}} = \frac{\dfrac{1}{\text{number of the road fatalities}}}{\dfrac{1}{\text{passenger-kilometers}}} = \frac{\text{passenger-kilometers}}{\text{number of the road fatalities}}$$

Figure 3.10 shows the efficiency frontier for the case with one input and two outputs. We note from Figure 3.10 that Sweden and Finland are efficient (their efficiencies equal 1). For the other countries, we can calculate the efficiencies, taking into account distances from the frontier. For example, if we want to determine efficiency of the Slovak Republic, we should note points A, B, and C (see Figure 3.11) and

Table 3.25 Data prepared for evaluation of efficiencies

DMU	Country	Input $\dfrac{1}{pkm}$	Output 1 $\dfrac{1}{n_i} \cdot 10^6$	Output 2 $\dfrac{1}{n_f} \cdot 10^6$
1	Bulgaria	116.442	118.120	1,636.661
2	Czech Republic	11.246	36.127	1,524.390
3	Finland	13.369	189.502	4,273.504
4	France	1.226	14.309	307.882
5	North Macedonia	85.485	170.648	7,518.797
6	Hungary	12.106	45.457	1,579.779
7	Italy	1.210	4.122	300.752
8	Slovak Republic	28.819	144.613	3,846.154
9	Sweden	7.906	54.051	3,086.420
10	United Kingdom	1.412	5.979	544.366

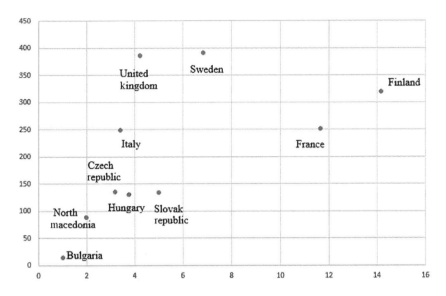

Figure 3.9 Spatial arrangement of the countries

the distances between A and B, and between A and C. Efficiency is obtained by dividing these two distances, i.e.:

$$\text{efficiency of Slovak Republic} = \frac{\text{distance}\,(A,B)}{\text{distance}\,(A,C)}$$

In a similar way, we can calculate the efficiencies of the other countries.

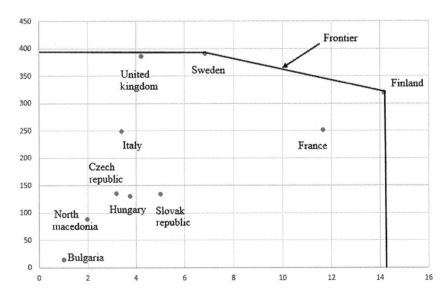

Figure 3.10 Efficiency frontier for the example in question

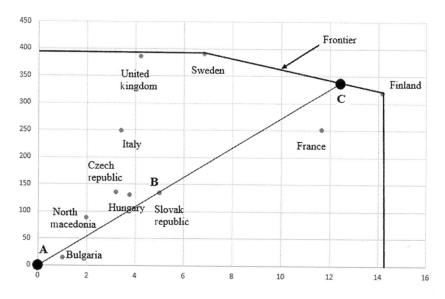

Figure 3.11 Calculation efficiency for the Slovak Republic

Let us demonstrate how efficiencies can be determined by solving linear programs. The linear program for the first country can be written in the following way:

Maximize

$$e_1 = 118.120 \cdot u_1 + 1{,}636.661 \cdot u_2$$

subject to:

$$116.442 \cdot v_1 = 1$$

$$118.120 \cdot u_1 + 1{,}636.661 \cdot u_2 - 116.442 \cdot v_1 \leq 0$$

$$36.127 \cdot u_1 + 1{,}524.390 \cdot u_2 - 11.246 \cdot v_1 \leq 0$$

$$189.502 \cdot u_1 + 4{,}273.504 \cdot u_2 - 13.369 \cdot v_1 \leq 0$$

$$14.309 \cdot u_1 + 307.882 \cdot u_2 - 1.226 \cdot v_1 \leq 0$$

$$170.648 \cdot u_1 + 7{,}518.797 \cdot u_2 - 85.485 \cdot v_1 \leq 0$$

$$45.457 \cdot u_1 + 1{,}579.779 \cdot u_2 - 12.106 \cdot v_1 \leq 0$$

$$4.122 \cdot u_1 + 300.752 \cdot u_2 - 1.210 \cdot v_1 \leq 0$$

$$144.613 \cdot u_1 + 3{,}846.154 \cdot u_2 - 28.819 \cdot v_1 \leq 0$$

$$54.051 \cdot u_1 + 3,086.420 \cdot u_2 - 7.906 \cdot v_1 \leq 0$$

$$5.979 \cdot u_1 + 544.366 \cdot u_2 - 1.412 \cdot v_1 \leq 0$$

$$v_1 \geq 0.00001$$

$$u_1 \geq 0.00001$$

$$u_2 \geq 0.00001$$

The linear program for the second country can be written in the following way:
Maximize

$$e_2 = 36.127 \cdot u_1 + 1,524.390 \cdot u_2$$

subject to:

$$11.246 \cdot v_1 = 1$$

$$118.120 \cdot u_1 + 1,636.661 \cdot u_2 - 116.442 \cdot v_1 \leq 0$$

$$36.127 \cdot u_1 + 1,524.390 \cdot u_2 - 11.246 \cdot v_1 \leq 0$$

$$189.502 \cdot u_1 + 4,273.504 \cdot u_2 - 13.369 \cdot v_1 \leq 0$$

$$14.309 \cdot u_1 + 307.882 \cdot u_2 - 1.226 \cdot v_1 \leq 0$$

$$170.648 \cdot u_1 + 7,518.797 \cdot u_2 - 85.485 \cdot v_1 \leq 0$$

$$45.457 \cdot u_1 + 1,579.779 \cdot u_2 - 12.106 \cdot v_1 \leq 0$$

$$4.122 \cdot u_1 + 300.752 \cdot u_2 - 1.210 \cdot v_1 \leq 0$$

$$144.613 \cdot u_1 + 3,846.154 \cdot u_2 - 28.819 \cdot v_1 \leq 0$$

$$54.051 \cdot u_1 + 3,086.420 \cdot u_2 - 7.906 \cdot v_1 \leq 0$$

$$5.979 \cdot u_1 + 544.366 \cdot u_2 - 1.412 \cdot v_1 \leq 0$$

$$v_1 \geq 0.00001$$

$$u_1 \geq 0.00001$$

$$u_2 \geq 0.00001$$

The linear program for the last country can be written in the following way:
Maximize

$$e_{10} = 5.979 \cdot u_1 + 544.366 \cdot u_2$$

subject to:

$$1.412 \cdot v_1 = 1$$

$$118.120 \cdot u_1 + 1,636.661 \cdot u_2 - 116.442 \cdot v_1 \leq 0$$

$$36.127 \cdot u_1 + 1,524.390 \cdot u_2 - 11.246 \cdot v_1 \leq 0$$

$$189.502 \cdot u_1 + 4,273.504 \cdot u_2 - 13.369 \cdot v_1 \leq 0$$

$$14.309 \cdot u_1 + 307.882 \cdot u_2 - 1.226 \cdot v_1 \leq 0$$

$$170.648 \cdot u_1 + 7,518.797 \cdot u_2 - 85.485 \cdot v_1 \leq 0$$

$$45.457 \cdot u_1 + 1,579.779 \cdot u_2 - 12.106 \cdot v_1 \leq 0$$

$$4.122 \cdot u_1 + 300.752 \cdot u_2 - 1.210 \cdot v_1 \leq 0$$

$$144.613 \cdot u_1 + 3,846.154 \cdot u_2 - 28.819 \cdot v_1 \leq 0$$

$$54.051 \cdot u_1 + 3,086.420 \cdot u_2 - 7.906 \cdot v_1 \leq 0$$

$$5.979 \cdot u_1 + 544.366 \cdot u_2 - 1.412 \cdot v_1 \leq 0$$

$$v_1 \geq 0.00001$$

$$u_1 \geq 0.00001$$

$$u_2 \geq 0.00001$$

The final values of the efficiencies are given in Table 3.26. From the results, we note that Finland and Sweden have efficiencies equal to 1, the United Kingdom has an efficiency equal to 0.9872, France has

Table 3.26 The final values of the efficiencies

Country	Efficiency
Bulgaria	0.0613
Czech Republic	0.3649
Finland	1
France	0.8231
North Macedonia	0.2349
Hungary	0.3653
Italy	0.6366
Slovak Republic	0.3985
Sweden	1
United Kingdom	0.9872

Table 3.27 Characteristics of the ports (Wu and Goh, 2010.)

Ports of the country	Terminal area (ha)	Total quay length (m)	No. of pieces of equipment	No. of containers
USA	647.7	9278	143	7,484,624
France	205	6075	90	2,118,509
Italy	130	3155	138	3,160,981
Canada	158.3	4019	60	1,767,379
Germany	531.8	9248	326	8,087,545
Japan	102.1	4016	95	3,593,071
UK	177.6	4026	125	2,750,000
China	617	7542	343	18,084,000
Russia	59.9	2203	59	1,119,346
Brazil	69.1	2160	33	2,267,921
India	68.8	1280	71	2,666,703
Bangladesh	1.5	450	25	783,353
Egypt	60	1050	29	1,621,066
Indonesia	165.6	3192	133	3,281,580
Iran	10.7	6190	43	1,292,962
Korea	392.2	12,610	252	11,843,151
Mexico	31.8	2205	28	873,976
Pakistan	36.9	1200	26	850,000
Philippines	184.5	8102	77	2,665,015
Turkey	78.3	5620	68	1,185,768
Vietnam	142.2	3731	64	1,911,016

an efficiency equal to 0.8231, etc. The lowest efficiency value is for Bulgaria, with an efficiency value t equal to 0.0613.

Example 3.16

Determine the efficiencies of container ports of the countries given in Table 3.27 by applying the CCR DEA model.

SOLUTION:

In this example, the ports of the countries are the DMUs. Bearing that in mind, we label USA as DMU 1, France as DMU 2, ..., and Vietnam as DMU 21. From Table 3.27, we observe that terminal area (second column), total quay length (third column), and the number of pieces of equipment (fourth column) are inputs, and the number of containers is the output. For each country, we should solve one linear program. The linear program for the first country can be written in the following way:

Maximize

$$e_1 = 7,484,624 \cdot u_1$$

subject to:

$$647.7 \cdot v_1 + 9,278 \cdot v_2 + 143 \cdot v_3 = 1$$

$$7,484,624 \cdot u_1 - 647.7 \cdot v_1 - 9,278 \cdot v_2 - 143 \cdot v_3 \le 0$$

$$2,118,509 \cdot u_1 - 205 \cdot v_1 - 6,075 \cdot v_2 - 90 \cdot v_3 \le 0$$

$$3,160,981 \cdot u_1 - 130 \cdot v_1 - 3,155 \cdot v_2 - 138 \cdot v_3 \le 0$$

$$1,767,379 \cdot u_1 - 158.3 \cdot v_1 - 4,019 \cdot v_2 - 60 \cdot v_3 \le 0$$

$$8,087,545 \cdot u_1 - 531.8 \cdot v_1 - 9,248 \cdot v_2 - 326 \cdot v_3 \le 0$$

$$3,593,071 \cdot u_1 - 102.1 \cdot v_1 - 4,016 \cdot v_2 - 95 \cdot v_3 \le 0$$

$$2,750,000 \cdot u_1 - 177.6 \cdot v_1 - 4,026 \cdot v_2 - 125 \cdot v_3 \le 0$$

$$18,084,000 \cdot u_1 - 617 \cdot v_1 - 7,542 \cdot v_2 - 343 \cdot v_3 \le 0$$

$$1,119,346 \cdot u_1 - 59.9 \cdot v_1 - 2,203 \cdot v_2 - 59 \cdot v_3 \le 0$$

$$2,267,921 \cdot u_1 - 69.1 \cdot v_1 - 2,160 \cdot v_2 - 33 \cdot v_3 \le 0$$

$$2,666,703 \cdot u_1 - 68.8 \cdot v_1 - 1,280 \cdot v_2 - 71 \cdot v_3 \le 0$$

$$783,353 \cdot u_1 - 1.5 \cdot v_1 - 450 \cdot v_2 - 25 \cdot v_3 \le 0$$

$$1,621,066 \cdot u_1 - 60 \cdot v_1 - 1,050 \cdot v_2 - 29 \cdot v_3 \le 0$$

$$3,281,580 \cdot u_1 - 165.6 \cdot v_1 - 3,192 \cdot v_2 - 133 \cdot v_3 \le 0$$

$$1,292,962 \cdot u_1 - 10.7 \cdot v_1 - 6,190 \cdot v_2 - 43 \cdot v_3 \le 0$$

$$11,843,151 \cdot u_1 - 392.2 \cdot v_1 - 12,610 \cdot v_2 - 252 \cdot v_3 \le 0$$

$$873,976 \cdot u_1 - 31.8 \cdot v_1 - 2,205 \cdot v_2 - 28 \cdot v_3 \le 0$$

$$850,000 \cdot u_1 - 36.9 \cdot v_1 - 1,200 \cdot v_2 - 26 \cdot v_3 \le 0$$

$$2,665,015 \cdot u_1 - 184.5 \cdot v_1 - 8,102 \cdot v_2 - 77 \cdot v_3 \le 0$$

$$1,185,768 \cdot u_1 - 78.3 \cdot v_1 - 5,620 \cdot v_2 - 68 \cdot v_3 \le 0$$

$$1,911,016 \cdot u_1 - 142.2 \cdot v_1 - 3,731 \cdot v_2 - 64 \cdot v_3 \le 0$$

$$v_1 \ge 0.000001$$

$$v_2 \geq 0.000001$$

$$v_3 \geq 0.000001$$

$$u_1 \geq 0.000001$$

The linear program for the second country can be written in the following way:
  Maximize

$$e_2 = 2,118,509 \cdot u_1$$

subject to:

$$205 \cdot v_1 + 6,075 \cdot v_2 + 90 \cdot v_3 = 1$$

$$7,484,624 \cdot u_1 - 647.7 \cdot v_1 - 9,278 \cdot v_2 - 143 \cdot v_3 \leq 0$$

$$2,118,509 \cdot u_1 - 205 \cdot v_1 - 6,075 \cdot v_2 - 90 \cdot v_3 \leq 0$$

$$3,160,981 \cdot u_1 - 130 \cdot v_1 - 3,155 \cdot v_2 - 138 \cdot v_3 \leq 0$$

$$1,767,379 \cdot u_1 - 158.3 \cdot v_1 - 4,019 \cdot v_2 - 60 \cdot v_3 \leq 0$$

$$8,087,545 \cdot u_1 - 531.8 \cdot v_1 - 9,248 \cdot v_2 - 326 \cdot v_3 \leq 0$$

$$3,593,071 \cdot u_1 - 102.1 \cdot v_1 - 4,016 \cdot v_2 - 95 \cdot v_3 \leq 0$$

$$2,750,000 \cdot u_1 - 177.6 \cdot v_1 - 4,026 \cdot v_2 - 125 \cdot v_3 \leq 0$$

$$18,084,000 \cdot u_1 - 617 \cdot v_1 - 7,542 \cdot v_2 - 343 \cdot v_3 \leq 0$$

$$1,119,346 \cdot u_1 - 59.9 \cdot v_1 - 2,203 \cdot v_2 - 59 \cdot v_3 \leq 0$$

$$2,267,921 \cdot u_1 - 69.1 \cdot v_1 - 2,160 \cdot v_2 - 33 \cdot v_3 \leq 0$$

$$2,666,703 \cdot u_1 - 68.8 \cdot v_1 - 1,280 \cdot v_2 - 71 \cdot v_3 \leq 0$$

$$783,353 \cdot u_1 - 1.5 \cdot v_1 - 450 \cdot v_2 - 25 \cdot v_3 \leq 0$$

$$1,621,066 \cdot u_1 - 60 \cdot v_1 - 1,050 \cdot v_2 - 29 \cdot v_3 \leq 0$$

$$3,281,580 \cdot u_1 - 165.6 \cdot v_1 - 3,192 \cdot v_2 - 133 \cdot v_3 \leq 0$$

$$1,292,962 \cdot u_1 - 10.7 \cdot v_1 - 6,190 \cdot v_2 - 43 \cdot v_3 \leq 0$$

$$11,843,151 \cdot u_1 - 392.2 \cdot v_1 - 12,610 \cdot v_2 - 252 \cdot v_3 \leq 0$$

$$873,976 \cdot u_1 - 31.8 \cdot v_1 - 2,205 \cdot v_2 - 28 \cdot v_3 \leq 0$$

$$850,000 \cdot u_1 - 36.9 \cdot v_1 - 1,200 \cdot v_2 - 26 \cdot v_3 \leq 0$$

$$2,665,015 \cdot u_1 - 184.5 \cdot v_1 - 8,102 \cdot v_2 - 77 \cdot v_3 \leq 0$$

$$1,185,768 \cdot u_1 - 78.3 \cdot v_1 - 5,620 \cdot v_2 - 68 \cdot v_3 \leq 0$$

$$1,911,016 \cdot u_1 - 142.2 \cdot v_1 - 3,731 \cdot v_2 - 64 \cdot v_3 \leq 0$$

$$v_1 \geq 0.000001$$

$$v_2 \geq 0.000001$$

$$v_3 \geq 0.000001$$

$$u_1 \geq 0.000001$$

In a similar way, we can write linear programs for the other countries. The linear program for the last country can be written in the following way:

Maximize

$$e_{21} = 1,911,016 \cdot u_1$$

subject to:

$$142.2 \cdot v_1 + 3,731 \cdot v_2 + 64 \cdot v_3 = 1$$

$$7,484,624 \cdot u_1 - 647.7 \cdot v_1 - 9,278 \cdot v_2 - 143 \cdot v_3 \leq 0$$

$$2,118,509 \cdot u_1 - 205 \cdot v_1 - 6,075 \cdot v_2 - 90 \cdot v_3 \leq 0$$

$$3,160,981 \cdot u_1 - 130 \cdot v_1 - 3,155 \cdot v_2 - 138 \cdot v_3 \leq 0$$

$$1,767,379 \cdot u_1 - 158.3 \cdot v_1 - 4,019 \cdot v_2 - 60 \cdot v_3 \leq 0$$

$$8,087,545 \cdot u_1 - 531.8 \cdot v_1 - 9,248 \cdot v_2 - 326 \cdot v_3 \leq 0$$

$$3,593,071 \cdot u_1 - 102.1 \cdot v_1 - 4,016 \cdot v_2 - 95 \cdot v_3 \leq 0$$

$$2,750,000 \cdot u_1 - 177.6 \cdot v_1 - 4,026 \cdot v_2 - 125 \cdot v_3 \leq 0$$

$$18,084,000 \cdot u_1 - 617 \cdot v_1 - 7,542 \cdot v_2 - 343 \cdot v_3 \leq 0$$

$$1,119,346 \cdot u_1 - 59.9 \cdot v_1 - 2,203 \cdot v_2 - 59 \cdot v_3 \leq 0$$

$$2,267,921 \cdot u_1 - 69.1 \cdot v_1 - 2,160 \cdot v_2 - 33 \cdot v_3 \leq 0$$

$$2,666,703 \cdot u_1 - 68.8 \cdot v_1 - 1,280 \cdot v_2 - 71 \cdot v_3 \leq 0$$

$$783,353 \cdot u_1 - 1.5 \cdot v_1 - 450 \cdot v_2 - 25 \cdot v_3 \leq 0$$

$$1,621,066 \cdot u_1 - 60 \cdot v_1 - 1,050 \cdot v_2 - 29 \cdot v_3 \leq 0$$

$$3,281,580 \cdot u_1 - 165.6 \cdot v_1 - 3,192 \cdot v_2 - 133 \cdot v_3 \leq 0$$

$$1,292,962 \cdot u_1 - 10.7 \cdot v_1 - 6,190 \cdot v_2 - 43 \cdot v_3 \leq 0$$

$$11,843,151 \cdot u_1 - 392.2 \cdot v_1 - 12,610 \cdot v_2 - 252 \cdot v_3 \leq 0$$

$$873,976 \cdot u_1 - 31.8 \cdot v_1 - 2,205 \cdot v_2 - 28 \cdot v_3 \leq 0$$

$$850,000 \cdot u_1 - 36.9 \cdot v_1 - 1,200 \cdot v_2 - 26 \cdot v_3 \leq 0$$

$$2,665,015 \cdot u_1 - 184.5 \cdot v_1 - 8,102 \cdot v_2 - 77 \cdot v_3 \leq 0$$

$$1,185,768 \cdot u_1 - 78.3 \cdot v_1 - 5,620 \cdot v_2 - 68 \cdot v_3 \leq 0$$

$$1,911,016 \cdot u_1 - 142.2 \cdot v_1 - 3,731 \cdot v_2 - 64 \cdot v_3 \leq 0$$

$$v_1 \geq 0.000001$$

$$v_2 \geq 0.000001$$

$$v_3 \geq 0.000001$$

$$u_1 \geq 0.000001$$

Table 3.28 shows the efficiencies of the countries obtained by solving the linear programs for each country. From the results, we see that China, Brazil, and Bangladesh have efficiencies of 1, followed by Egypt, India, etc., with the lowest efficiency value being from Turkey.

Example 3.17

Evaluate the efficiencies of 10 airports in Spain, using the DEA method, according to the data given in Table 3.29. The airports are the DMUs, inputs are the total runway area, the number of check-in counters and the number of gates, and outputs are the annual number of passengers and the number of operations.

Table 3.28 Efficiencies of the ports in different countries

DMU	Country	Efficiency
1	USA	0.764
2	France	0.342
3	Italy	0.533
4	Canada	0.429
5	Germany	0.471
6	Japan	0.757
7	UK	0.429
8	China	1.000
9	Russia	0.395
10	Brazil	1.000
11	India	0.963
12	Bangladesh	1.000
13	Egypt	0.964
14	Indonesia	0.527
15	Iran	0.859
16	Korea	0.810
17	Mexico	0.610
18	Pakistan	0.591
19	Philippines	0.502
20	Turkey	0.338
21	Vietnam	0.452

Table 3.29 Data for evaluation of efficiencies of airports in Spain (Lozano and Gutiérrez, 2009)

Airport	Total runway area $[m^2]$	Number of check-in counters	Number of gates	Annual number of passengers (in thousands)	Number of operations
Madrid Barajas	927,000	484	230	45,530.01	435.018
Barcelona	475,020	143	65	30,008.152	327.636
Palma de Mallorca	295,650	204	68	22,408.302	190.28
Malaga	144,000	85	30	13,076.252	127.769
Gran Canaria	139,500	86	38	10,286.635	114.938
Logrono	90,000	5	2	55.427	3.333
La Gomera	45,000	5	2	38.846	3.384
Salamanca	150,000	4	2	29.308	8.656
Cordoba	62,100	1	1	19.568	9.212
Albacete	162,000	4	2	17.52	1.347

SOLUTION:

We solve this example by solving 10 linear programs, one for each airport. Linear program for the first airport reads:
Maximize

$$e_1 = 45,530.01 \cdot u_1 + 435.018 \cdot u_2$$

subject to

$$927,000 \cdot v_1 + 484 \cdot v_2 + 230 \cdot v_3 = 1$$

$$45,530.01 \cdot u_1 + 435.018 \cdot u_2 - 927,000 \cdot v_1 - 484 \cdot v_2 - 230 \cdot v_3 \leq 0$$

$$30,008.152 \cdot u_1 + 327.636 \cdot u_2 - 475.02 \cdot v_1 - 143 \cdot v_2 - 65 \cdot v_3 \leq 0$$

$$22,408.302 \cdot u_1 + 190.28 \cdot u_2 - 295,650 \cdot v_1 - 204 \cdot v_2 - 68 \cdot v_3 \leq 0$$

$$13,076.252 \cdot u_1 + 127.769 \cdot u_2 - 144,000 \cdot v_1 - 85 \cdot v_2 - 30 \cdot v_3 \leq 0$$

$$10,286.635 \cdot u_1 + 114.938 \cdot u_2 - 139,500 \cdot v_1 - 86 \cdot v_2 - 38 \cdot v_3 \leq 0$$

$$55.427 \cdot u_1 + 3.333 \cdot u_2 - 90,000 \cdot v_1 - 5 \cdot v_2 - 2 \cdot v_3 \leq 0$$

$$38.846 \cdot u_1 + 3.384 \cdot u_2 - 45,000 \cdot v_1 - 5 \cdot v_2 - 2 \cdot v_3 \leq 0$$

$$29.308 \cdot u_1 + 8.656 \cdot u_2 - 150,000 \cdot v_1 - 4 \cdot v_2 - 2 \cdot v_3 \leq 0$$

$$19.568 \cdot u_1 + 9.212 \cdot u_2 - 62,100 \cdot v_1 - 1 \cdot v_2 - 1 \cdot v_3 \leq 0$$

$$17.52 \cdot u_1 + 1.347 \cdot u_2 - 162,000 \cdot v_1 - 4 \cdot v_2 - 2 \cdot v_3 \leq 0$$

$$v_1 \geq 0.000001$$

$$v_2 \geq 0.000001$$

$$v_3 \geq 0.000001$$

$$u_1 \geq 0.000001$$

$$u_2 \geq 0.000001$$

The linear program for the second airport reads:
Maximize

$$e_2 = 30,008.152 \cdot u_1 + 327.636 \cdot u_2$$

subject to

$$475.02 \cdot v_1 + 143 \cdot v_2 + 65 \cdot v_3 = 1$$

$$45{,}530.01 \cdot u_1 + 435.018 \cdot u_2 - 927{,}000 \cdot v_1 - 484 \cdot v_2 - 230 \cdot v_3 \le 0$$

$$30{,}008.152 \cdot u_1 + 327.636 \cdot u_2 - 475.02 \cdot v_1 - 143 \cdot v_2 - 65 \cdot v_3 \le 0$$

$$22{,}408.302 \cdot u_1 + 190.28 \cdot u_2 - 295{,}650 \cdot v_1 - 204 \cdot v_2 - 68 \cdot v_3 \le 0$$

$$13{,}076.252 \cdot u_1 + 127.769 \cdot u_2 - 144{,}000 \cdot v_1 - 85 \cdot v_2 - 30 \cdot v_3 \le 0$$

$$10{,}286.635 \cdot u_1 + 114.938 \cdot u_2 - 139{,}500 \cdot v_1 - 86 \cdot v_2 - 38 \cdot v_3 \le 0$$

$$55.427 \cdot u_1 + 3.333 \cdot u_2 - 90{,}000 \cdot v_1 - 5 \cdot v_2 - 2 \cdot v_3 \le 0$$

$$38.846 \cdot u_1 + 3.384 \cdot u_2 - 45{,}000 \cdot v_1 - 5 \cdot v_2 - 2 \cdot v_3 \le 0$$

$$29.308 \cdot u_1 + 8.656 \cdot u_2 - 150{,}000 \cdot v_1 - 4 \cdot v_2 - 2 \cdot v_3 \le 0$$

$$19.568 \cdot u_1 + 9.212 \cdot u_2 - 62{,}100 \cdot v_1 - 1 \cdot v_2 - 1 \cdot v_3 \le 0$$

$$17.52 \cdot u_1 + 1.347 \cdot u_2 - 162{,}000 \cdot v_1 - 4 \cdot v_2 - 2 \cdot v_3 \le 0$$

$$v_1 \ge 0.000001$$

$$v_2 \ge 0.000001$$

$$v_3 \ge 0.000001$$

$$u_1 \ge 0.000001$$

$$u_2 \ge 0.000001$$

In a similar way, we can write linear programs for the other airports. The linear program for the last airport is written in the following way:
  Maximize

$$e_{10} = 17.52 \cdot u_1 + 1.347 \cdot u_2$$

subject to

$$162{,}000 \cdot v_1 + 4 \cdot v_2 + 2 \cdot v_3 = 1$$

$$45{,}530.01 \cdot u_1 + 435.018 \cdot u_2 - 927{,}000 \cdot v_1 - 484 \cdot v_2 - 230 \cdot v_3 \le 0$$

$$30{,}008.152 \cdot u_1 + 327.636 \cdot u_2 - 475.02 \cdot v_1 - 143 \cdot v_2 - 65 \cdot v_3 \le 0$$

$$22,408.302 \cdot u_1 + 190.28 \cdot u_2 - 295,650 \cdot v_1 - 204 \cdot v_2 - 68 \cdot v_3 \leq 0$$

$$13,076.252 \cdot u_1 + 127.769 \cdot u_2 - 144,000 \cdot v_1 - 85 \cdot v_2 - 30 \cdot v_3 \leq 0$$

$$10,286.635 \cdot u_1 + 114.938 \cdot u_2 - 139,500 \cdot v_1 - 86 \cdot v_2 - 38 \cdot v_3 \leq 0$$

$$55.427 \cdot u_1 + 3.333 \cdot u_2 - 90,000 \cdot v_1 - 5 \cdot v_2 - 2 \cdot v_3 \leq 0$$

$$38.846 \cdot u_1 + 3.384 \cdot u_2 - 45,000 \cdot v_1 - 5 \cdot v_2 - 2 \cdot v_3 \leq 0$$

$$29.308 \cdot u_1 + 8.656 \cdot u_2 - 150,000 \cdot v_1 - 4 \cdot v_2 - 2 \cdot v_3 \leq 0$$

$$19.568 \cdot u_1 + 9.212 \cdot u_2 - 62,100 \cdot v_1 - 1 \cdot v_2 - 1 \cdot v_3 \leq 0$$

$$17.52 \cdot u_1 + 1.347 \cdot u_2 - 162,000 \cdot v_1 - 4 \cdot v_2 - 2 \cdot v_3 \leq 0$$

$$v_1 \geq 0.000001$$

$$v_2 \geq 0.000001$$

$$v_3 \geq 0.000001$$

$$u_1 \geq 0.000001$$

$$u_2 \geq 0.000001$$

By solving 10 linear programs, we obtained the efficiencies that are presented in Table 3.30. We see, from the obtained results, that airports at Barcelona, Malaga, and Cordoba have efficiency values of 1, followed by Gran Canaria (with efficiency of 0.921), Palma de Mallorca (with efficiency of 0.835), etc. The airport with the lowest efficiency is Albacete (0.076).

Table 3.30 Efficiencies of the airports in Spain

Airport	Efficiency
Madrid Barajas	0.546
Barcelona	1
Palma de Mallorca	0.835
Malaga	1
Gran Canaria	0.927
Logrono	0.211
La Gomera	0.273
Salamanca	0.462
Cordoba	1
Albacete	0.076

PROBLEMS

3.1. A customer wants to select one of five operators for future container transportation between two cities. The following are the criteria for operator selection: price, delivery time, and the number of container handlings. Table 3.31 shows the values of the alternatives for each criterion. Determine the best company according to:
a. Simple additive weighting (SAW) and
b. TOPSIS method.
The criteria weights are: $w_1 = 0.5$, $w_2 = 0.3$ and $w_3 = 0.2$.

3.2. A transportation company wants to employ an engineer. The selection criteria are age, level of education, and work experience. The desirable attribute values set up by the decision maker are:
• Age: 45 or less,
• Education: master's degree (or PhD)
• Working experience: 3 years or more.
Table 3.32 shows a list of eight candidates who applied. Make a list of candidates for further consideration, according to the conjunctive method.

3.3. Evaluate the efficiencies of 5 warehouses, according to the data given in Table 3.33.

3.4. Determine the efficiencies of five bus operators, according to the data given in Table 3.34.

Table 3.31 Values for each selection criterion for operator alternatives

Alternative	Price ($)	Delivery time (hrs)	Number of container handlings (TEU)
Company 1	1200	15	2
Company 2	1100	18	3
Company 3	1000	19	3
Company 4	1050	17	4
Company 5	950	20	4

Table 3.32 Characteristics of the job candidates

Candidate	Age	Education	Working experience
1	25	Bachelor's degree	2
2	28	Master's degree	3
3	32	Master's degree	3
4	27	Master's degree	2
5	35	PhD	8
6	47	PhD	19
7	29	Master's degree	4
8	26	Bachelor's degree	3

Table 3.33 Data for evaluation of warehouse efficiencies

Warehouse	Number of workers	Capacity [pallets]	Profit [thousand $]
1	20	500	150
2	25	1000	250
3	30	1500	450
4	22	1600	300
5	35	2000	500

Table 3.34 Data for evaluation of efficiencies of bus operators

Company	Number of buses	Number of passengers
1	10	4500
2	15	6000
3	5	3200
4	8	3000
5	16	9000

Table 3.35 Data for evaluation of efficiencies of distribution operators

Company	Number of workers	Number of trucks	Warehouse capacities [pallets]	Profit [monetary units]
1	25	20	700	350
2	40	25	1000	450
3	60	30	1500	750
4	50	22	1600	550
5	65	35	2000	800

3.5. Determine the efficiencies of five distribution operators, according to the data given in Table 3.35.

BIBLIOGRAPHY

Charnes, A. and Cooper, W. W., Preface to Topics in Data Envelopment Analysis, *Annals of Operations Research*, 2, 59–94, 1985.
Charnes, A., Cooper, W.W., and Rhodes, E., Measuring the Efficiency of Decision Making Units, *European Journal of Operational Research*, 2, 429–444, 1978.
Charnes, A., Cooper, W.W., and Rhodes, E., Short Communication; Measuring the Efficiency of Decision Making Units, *European Journal of Operational Research*, 3, 339, 1979.

Charnes, A., Cooper, W.W., and Rhodes, E., Evaluating Program and Managerial Efficiency: An Application of Data Envelopment Analysis to Program Follow Through, *Management Science*,27, 668–697, 1981.

Chen, C.T., Extension of the TOPSIS for Group Decision Making under Fuzzy Environment, *Fuzzy Sets and Systems*, 114, 1–9, 2000.

Chen, S.J. and Hwang, C.L., *Fuzzy Multiple Attribute Decision Making-methods and Applications.* Lecture Notes in Economics and Mathematical Systems, Springer, New York, 1992.

Hillier, F. and Lieberman, G., *Introduction to Operations Research*, McGraw–Hill Publishing Company, New York, 1990.

Hwang, C. L. and Yoon, K., *Multiple Attribute Decision Making*, Springer-Verlag, Berlin, 1981.

Köksalan, M., Wallenius, J., and Zionts, S., *Multiple Criteria Decision Making: From Early History to the 21st Century*, World Scientific, Singapore, 2011.

Lootsma, F.A., *Multi-criteria Decision Analysis via Ratio and Difference Judgement*, Kluwer Academic Publishers, Dordrecht, 1999.

Lozano, S. and Gutiérrez, E., Efficiency Analysis and Target Setting of Spanish Airports, *Networks and Spatial Economics*, 11, 139–157, 2009.

Pjevčević, D., Vladisavljević, I., Vukadinović, K., and Teodorović, D., Application of DEA to the Analysis of AGV Fleet Operations in a Port Container Terminal, *Procedia-Social and Behavioral Sciences*, 20, 816–825, 2011.

Rakas, J., Teodorović, D. and Kim, T., Multi-objective Modeling for Determining Location of Undesirable Facilities, *Transportation Research D*,9, 125–138, 2004.

Roy, B. and Vincke, P., Multicriteria Analysis: Survey and New Directions, *European Journal of Operational Research*,8, 207–218, 1981.

Saaty, T.L., *The Analytic Hierarchy Process*, McGraw Hill, New York, 1980.

Sheth, C., Triantis, K., and Teodorović, D., Performance Evaluation of Bus Routes: A Provider and Passenger Perspective, *Transportation Research E*, 43, 453–478, 2006.

Taha, H., *Operations Research*, MacMillan Publishing Co., Inc., New York, 1982.

Teodorović, D., Multicriteria Ranking of Air Shuttle Alternatives, *Transportation Research*, 19 B, 63–72, 1985.

Teodorović, D. and Janić, M., *Transportation Engineering: Theory, Practice and Modeling*, Elsevier, New York, 2016.

UITP, World metro figures 2018, www.uitp.org/sites/default/files/cck-focus-papers-files/Statistics Brief - World metro figures 2018V4_WEB.pdf.

Winston, W., *Operations Research*, Duxbury Press, Belmont, CA, 1994.

Wu, Y.C.J. and Goh, M., Container Port Efficiency in Emerging and More Advanced Markets, *Transportation Research Part E*, 46 (6): 1030–1042, 2010.

Yoon, K.P. and Hwang, C.L., *Multiple Attribute Decision Making*. SAGE Publications, Beverly Hills, CA, 1995.

Yu, P.L., A Class of Solutions for Group Decision Problems, *Management Science*,19, 936–946, 1973.

Zanakis, S., Solomon, A., Wishart, N., and Dublish, S., Multi-attribute Decision Making: A Comparison of Select Methods, *European Journal of Operational Research*, 107, 507–529, 1998.

Zeleny, M., *The Theory of the Displayed Ideal in Multiple Criteria Decision Making*, Kyoto 1975, Springer-Verlag, Berlin, 1976.

Zeleny, M., *Multiple Criteria Decision Making*, McGraw Hill, New York, 1982.

Chapter 4

Probability theory

Drivers and passengers, that travel from one place to another, create a variety of traffic flows. Flows of aircraft on airport's taxiways, flows of cars on streets and highways, flows of bicycles on streets, flows of peoples in metro stations, and flows of pedestrians on pedestrian crossings fluctuate over the time of day, day of the week, and month of the year. Every trip-maker makes his/her own decision independently of all other trip-makers concerning the day, time, route, and transportation mode he/she wishes to travel. The greater part of traffic and transportation activities are characterized by uncertainty related to time of occurrence. It is practically impossible to exactly predict the time of occurrence of many traffic activities. Various traffic transportation activities or acts could be treated as *experiments*. Every experiment has at least two possible outcomes, and the result of an experiment cannot be predicted with complete certainty.

Numerous independent random factors affect various traffic phenomena (travel time, the total number of cars on a specific urban transportation network link, the total number of passengers on a specific flight, demand at nodes of a distribution system, demand (time and location) for emergency help from urban emergency services, etc.).

Let us consider the following example. An air carrier plans to perform n flights every Tuesday over a period of time. Canceled or delayed flights can be caused by meteorological reasons, technical reasons, late or absent crew members, etc. One or more aircraft from an airline fleet might be taken out of operation due to technical reasons, and the airline has to operate with a reduced number of planes. Every Tuesday, we record the number of canceled flights. The number of canceled flights on Tuesdays (the *outcome* of our experiment) is unknown. In other words, the outcome of our experiment is subject to *chance*. There could be 0, 1, 2, 3, ..., or n canceled flights. In this example, the number of possible outcomes is *finite*. The number of possible outcomes could also be *infinite*. For example, we can measure everyday travel time between our home and the university. The outcomes in this case may take any nonnegative real value. Obviously, the number of potential outcomes in the case of travel time measurement is infinite.

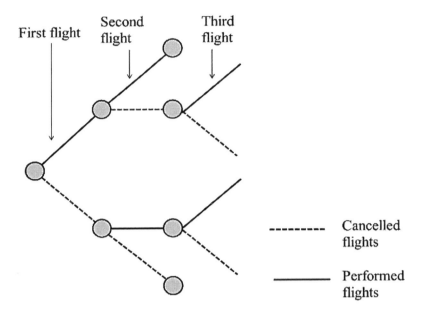

Figure 4.1 Tree diagram that represents all of the outcomes related to the number of canceled flights

Traffic phenomena are stochastic processes (described by random variables). Travel times between specific nodes, waiting times at the intersections, number of cars on a link, number of passengers on a plane, etc., are *random variables* whose values are *unknown* in advance. Obviously, there is a strong need for a probabilistic analysis of traffic phenomena. In the subsequent sections, we describe the basics of probabilistic modeling.

Example 4.1

Let us consider again the air carrier who plans to perform *n* flights every Tuesday during a particular period of time. Canceled or delayed flights can be caused by meteorological reasons, technical reasons, late or absent crew members, etc. The number of canceled flights on Tuesdays (*outcome* of our experiment) is unknown. There could be 0, 1, 2, 3, ..., or *n* canceled flights. All of the outcomes from this experiment can be shown in a *tree diagram* (Figure 4.1).

4.1 SAMPLE SPACE, EVENTS, VENN DIAGRAMS AND THE ALGEBRA OF EVENTS

Practically, every traffic phenomenon that has probabilistic characteristics could be viewed as an experiment. A *sample space* is composed of all possible experiment outcomes. For example, in our experiment of study-ing the number of canceled flights, the sample space is $\{1, 2, 3, ..., n\}$. In the case of travel time measurement, the sample space is composed of all values from the interval $\{0, \infty\}$. An *event* represents a combination of out-comes from the sample space. For example, the event could be "3 canceled flights". The event could also be "5 canceled flights", "8 canceled flights", etc. Usually, we use capital letters A, B, C, ... to denote events.

The impossible events are events that cannot occur. We denote an impos-sible event by V. For example, an event where the speed of a car is 2000 mph is an impossible event.

A sure event is an event which always occurs. We denote sure events by U. For example, it is a sure event that the number of canceled flights is less than or equal to n.

Let us consider events A and B. If all possible outcomes of event A are also outcomes in B, then event A is contained in event B, i.e. (Figure 4.2):

$A \subset B$

The following is satisfied for any event A:

$V \subset A$

$A \subset U$

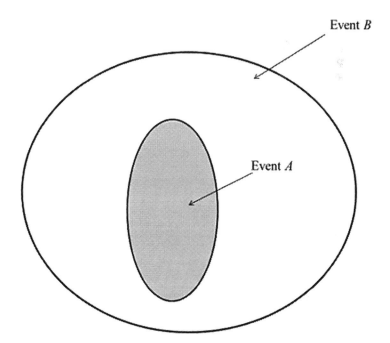

Figure 4.2 Event A that is contained in event B

An event that corresponds to a single possible outcome of the experiment is called a simple event. An event that corresponds to more than a single possible outcome of the experiment is called a compound event. For example, a traveler can go from home to work by bike, by car, by tram, or by bus. The sample space S is:

$$S = \{\text{Bicycle, Car, Tram, Bus}\}$$

The event B, when the traveler chooses to use a bicycle is given by $B = \{\text{Bicycle}\}$. The event B is a simple event. The event P, when the traveler chooses public transit, is given by $P = \{\text{Tram, Bus}\}$. The event P is a compound event.

The complementary event of the event A, $\bar{A}$ (non-A), contains all the outcomes that are not in A. The complementary event $\bar{A}$ shows practically the opposite side of event A (Figure 4.3). Let us consider the situation where a vehicle approaches an intersection. The sample space S, relating to the traffic light at the intersection, is $S = \{\text{green, yellow, red}\}$. Let the event A represent the situation when the vehicle passes the intersection without being stopped, i.e. $A = \{\text{green}\}$. The event $\bar{A} = \{\text{yellow, red}\}$ is the complementary event $\bar{A}$ of the event A and represents the situation where the vehicle that wants to pass the intersection is stopped.

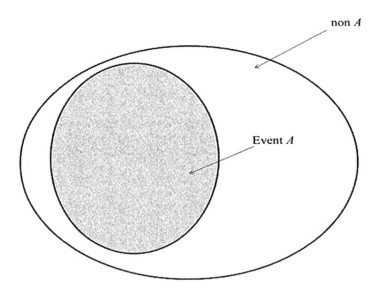

Figure 4.3 The event A and the complementary event $\bar{A}$ (non-A)

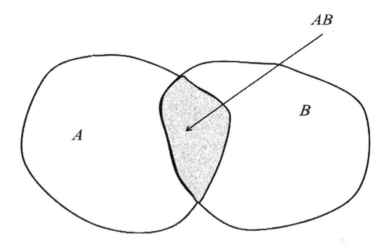

Figure 4.4 Intersection of events

The *intersection* $A \cap B$ of event A and event B means that both A and B are realized (Figure 4.4). The *union* $A \cup B$ of events A and B means that A or B, or both of them, happen (Figure 4.5). The intersection is also denoted as "AB", while "$A+B$" is also used to denote "union".

The union, intersection, and the complement are related to the following axioms:

$$A \cup B = B \cup A$$

$$A \cup (B \cup C) = (A \cup B) \cup C$$

$$A \cap (B \cup C) = (A \cap B) \cup (A \cap C)$$

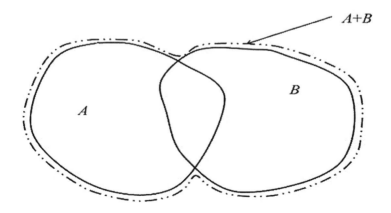

Figure 4.5 Union of events

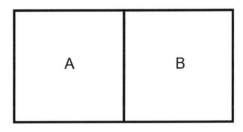

Figure 4.6 Mutually exclusive events

$non(nonA) = A$

$non\,(A \cap B) = non\,A \cup non\,B$

$A \cap non\,A = V$

$A \cap U = A$

Commutative, associative, and distributive laws are described by the points 1, 2, and 3, respectively. Point 4 describes the situation where the complement of the complement of the event is the initial event. Point 5 states that the complement of the intersection of two events equals the union of their complements. Point 6 describes that the intersection between an event and its complement equals an impossible event. Point 7 states that the intersection between an event and a sure event is the original event.

Two events, *A* and *B*, are *mutually exclusive*, if the occurrence of one event means nonoccurrence of the other. In other words, mutually exclusive events *A* and *B* cannot happen simultaneously (Figure 4.6).

Example 4.2

A container ship arrives at the port to have loading and unloading operations carried out. If all berths at the port are occupied, the ship has to wait until any one of the berths is going to become idle. The total time that the ship spends at the port consist of waiting time and service time. Let us denote with *A* an event that the total time which the ship spends at the port is less or equal to 10 hours. A mutually exclusive event of event *A* is that the total time the ship spends at the port is more than 10 hours.

4.2 PROBABILITY OF A RANDOM EVENT

What is a *probability*? Probability is a non-negative real number not greater than one. Could probability be equal to zero? Yes. Could probability be equal to one? Yes. Could probability be greater than one? No. How could we calculate the probability of a specific event? Usually, we repeat the experiment many times, and we count the number of trials m that describe our event. Let us denote by n the total number of trials. By performing the experiment, we observe that m trials out of n trials describe our event. We denote by $P(A)$ the probability of event A. The probability of the event A equals:

$$P(A) = \frac{m}{n}$$

Probability theory has its roots in the work of *Pierre de Fermat* and *Blaise Pascal* in the seventeenth century. *Andrey Kolmogorov*, "father" of the modern probability theory, presented the axiom system for the probability theory in 1933.

When we write that $P(A) = \dfrac{m}{n}$, we practically say that the probability of an event is the ratio of the number of cases beneficial to the event, to the number of all cases possible. It should be noted that we expect that all possible cases are equally possible. The probability of any event is always between zero and one:

$$0 \leq P(A) \leq 1$$

When event A is impossible, i.e. $= V$, then $P(A) = 0$. Impossible events never happen. For example, even in cases of extremely high traffic demand, the number of cars in the left-turn bay will never be equal to 47, since the capacity of the left-turn bay equals 8. We denote by B the following event: "the number of cars in the left-turn bay equals 47". We can write that $P(B) = 0$. Sure events always happen. We denote by C the following event: "travel time by car between home and the university is greater than zero". We write that $P(C) = 1$, since event C is sure.

The addition law (Figure 4.5) reads:

$$P(A + B) = P(A) + P(B) - P(AB)$$

In the case of mutually exclusive events A and B, we have:

$$P(AB) = 0$$

and:

$$P(A + B) = P(A) + P(B)$$

The following is satisfied in the case of event A and the complementary event $\bar{A}$:

$$P(A\bar{A}) = 0$$

and

$$P(A + \bar{A}) = P(A) + P(\bar{A})$$

Since:

$$P(A + \bar{A}) = P(U) = 1$$

we get:

$$P(\bar{A}) = 1 - P(A)$$

Example 4.3

Event A is that the aircraft arrives at the airport with a delay. Event $\bar{A}$ is that the aircraft arrives at the airport without a delay (i.e. it arrives on time). These two events are mutually exclusive because it is not possible that both of them are going to happen together. Because of that, if the probability $P(A) = 0.4$, then:

$$P(\bar{A}) = 1 - P(A) = 1 - 0.4 = 0.6$$

$$P(A\bar{A}) = 0$$

$$P(A + \bar{A}) = P(A) + P(\bar{A}) = 0.4 + 0.6 = 1$$

4.3 CONDITIONAL PROBABILITY

Let us denote by T the event that a commuter arrives at her office on time. Let us assume that the probability of this event is $(T) = 0.95$. We denote by A the event that a traffic accident happened on the commuter's route. In this case, the probability that the commuter arrives at her office on time equals, for example, 0.5. We denote this *conditional probability* as $P(T|A) = 0.5$.

The additional information, that a traffic accident happened, decreased the probability from 0.95 to 0.5.

The probability that event A will happen, knowing that event B already happened, is usually denoted as $P(A|B)$. The conditional probability law helps us to compute the probability $P(A|B)$ of event A, given the event B, i.e.:

$$P(A|B) = \frac{P(AB)}{P(B)}$$

Since:

$$P(B|A) = \frac{P(AB)}{P(A)}$$

we get:

$$P(AB) = P(A) \cdot P(B|A) = P(B) \cdot P(A|B)$$

Let us clarify this last relationship in words. The probability that both A and B happen is equal to the probability that event B happens multiplied by the conditional probability of A, given that event B happened.

The two events, A and B, are independent when:

$$P(A|B) = P(A)$$

In the case of independent events A and B, we have:

$$P(AB) = P(A) \cdot P(B)$$

Example 4.4

Event A is that there is a traffic jam. Event B is that the vehicle suffers a delay. Probabilities $P(A)$ and $P(B|A)$ are equal to 0.5 and 0.6, respectively. Calculate the probability that there is a traffic jam and that the vehicle suffers a delay $(P(AB))$.

SOLUTION:

In this example, we have conditional probability where the probability $P(AB)$ is equal:

$$P(AB) = P(A) \cdot P(B|A) = 0.5 \cdot 0.6 = 0.3$$

Example 4.5

There are two cities. Denote by A an event that there is a traffic jam in the first city. Denote by B an event that a vehicle in city 2 suffers a delay. The probabilities of events A and B are $P(A) = 0.4$ and $P(B) = 0.35$, respectively. Calculate the probability $P(AB)$.

SOLUTION:

Taking into consideration that a traffic jam in city 1 does not have an influence on vehicle travel time in city 2, we can say that events A and B are independent. Because of that, $P(AB)$ equals:

$$P(AB) = P(A) \cdot P(B) = 0.4 \cdot 0.35 = 0.14$$

4.4 BAYES' FORMULA

Let us consider the events H_1, H_2, ..., H_n related to the specific experiment. Let these events be pairwise disjoint events (Figure 4.7).

Their union is the entire sample space of an experiment, i.e.:

$$H_1 + H_2 + \cdots + H_n = U$$

The following is satisfied for any event A of the same probability space:

$$A = U \cdot A = (H_1 + H_2 \cdots + H_n) \cdot A = H_1 \cdot A + H_2 \cdot A + \cdots + H_n \cdot A$$

Consequently:

$$P(A) = P(H_1 \cdot A) + P(H_2 \cdot A) + \cdots + P(H_n \cdot A)$$

$$P(A) = P(H_1) \cdot P(A|H_1) + P(H_2) \cdot P(A|H_2) + \cdots + P(H_n) \cdot P(A|H_n)$$

The last relation is known in probability theory as the law (or formula) of total probability. The probability is also called "average probability" or "overall probability".

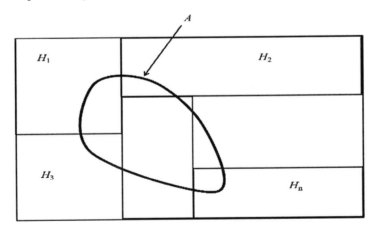

Figure 4.7 Event A and the pairwise disjoint events $H_1, H_2, \ldots H_n$

The conditional probability $P(H_i \mid A)$ of the event H_i, given the event A, equals:

$$P\left(H_i \middle| A\right) = \frac{P\left(H_i \cdot A\right)}{P\left(A\right)}, i = 1, 2, \ldots, n$$

$$P\left(H_i \middle| A\right) = \frac{P\left(H_i \cdot A\right)}{P\left(H_1 \cdot A\right) + P\left(H_2 \cdot A\right) + \cdots + P\left(H_n \cdot A\right)}, i = 1, 2, \ldots, n$$

$$P\left(H_i \middle| A\right) = \frac{P\left(H_i \cdot A\right)}{P\left(H_1\right) \cdot P\left(A \middle| H_1\right) + P\left(H_2\right) \cdot P\left(A \middle| H_2\right) + \cdots + P\left(H_n\right) \cdot P\left(A \middle| H_n\right)}, i = 1, 2, \ldots, n$$

The last relation is known as the Bayes' theorem (or Bayes' law, or Bayes' rule).

Example 4.6

An event where the customer received a shipment with some damage is denoted by A. Damage could have occurred in the warehouse, during the loading/unloading processes, or during the transportation process. In this place, damages could occur if it happened due to some unexpected event. Denote such unexpected events with B, C, and D. The probability that an unexpected event happened in all of these places are respectively $P(B) = 0.1$, $P(C) = 0.3$ and $P(D) = 0.2$. The probabilities that the shipment would be damaged if unexpected events happen are $P\left(A \middle| B\right) = 0.2$, $P\left(A \middle| C\right) = 0.5$ and $P\left(A \middle| D\right) = 0.6$. If event A has happened, determine the probability $P(C \mid A)$.

SOLUTION:

Probability $P(C \mid A)$ is calculated as:

$$P\left(C \middle| A\right) = \frac{P\left(AC\right)}{P\left(A\right)}$$

where:

$$P\left(AC\right) = P\left(C\right) \cdot P\left(A \middle| C\right) = 0.3 \cdot 0.5 = 0.15$$

$$P\left(A\right) = P\left(B\right) \cdot P\left(A \middle| B\right) + P\left(C\right) \cdot P\left(A \middle| C\right) + P\left(D\right) \cdot P\left(A \middle| D\right)$$

$$= 0.1 \cdot 0.2 + 0.3 \cdot 0.5 + 0.2 \cdot 0.6 = 0.01 + 0.15 + 0.12$$

$$= 0.28$$

Now, we have:

$$P(C|A) = \frac{P(AC)}{P(A)} = \frac{0.15}{0.28} = 0.536$$

4.5 RANDOM VARIABLES AND PROBABILITY DISTRIBUTIONS

The numerical outcomes of the observed experiment are represented by a random variable. For example, let us assume that passengers, who travel a 10-kmdistance in a city, could choose for their trip a private car (C), or public transport (P). By assigning 0 to C and 1 to P, the potential passengers' choices (outcomes of the experiment) could be presented as a random variable. A random variable could be *discrete*, or *continuous*. A discrete random variable takes on specific values at discrete points on the real line. In the case of left-turn bays at the intersection whose capacity equals 8, the number of vehicles *in* the bay could be 0, 1, 2, 3, 4, 5, 6, 7, or 8. The number of vehicles in the bay cannot, for example, be 4.32, or 6.17. In the case of continuous variables, the variable can take any value between two specified values.

A continuous random variable takes on an infinitely large number of values on the real line. There is a function $f(x)$ that assigns probability measures to the random variable values x. The function $f(x)$ is called a *probability density function* (*pdf*). Let us explain the concept of the probability density function by using the following example.

The number of cars waiting for a right-of-way through the intersection has been recorded at a specific time of a day for 365 days. Table 4.1 shows the distribution of number of cars waiting for a right-of-way through the intersection.

The total number of days under observation equals 365. We transform all the data in Table 4.2 to probabilities by dividing by this total.

From these data, we can obtain the bar graph (Figure 4.8). We call this bar graph a *probability distribution histogram*.

The probability distribution histogram enables us to calculate different probabilities. The probability that the number of cars waiting to pass through the intersection is between 6 and 12 is shown in Figure 4.9. In this way, we can calculate various probabilities by summing up corresponding areas.

Table 4.1 The distribution of numbers of cars waiting for a right-of-way through an intersection

Number of cars	1–3	4–6	7–9	10–12	13–15	16–18	19–21	22–24
Number of days	38	52	70	55	45	40	40	25

Table 4.2 The distribution of numbers of cars waiting for a right-of-way through an intersection

Number of cars	1–3	4–6	7–9	10–12	13–15	16–18	19–21	22–24
Number of days	0.11	0.14	0.19	0.15	0.12	0.11	0.11	0.07

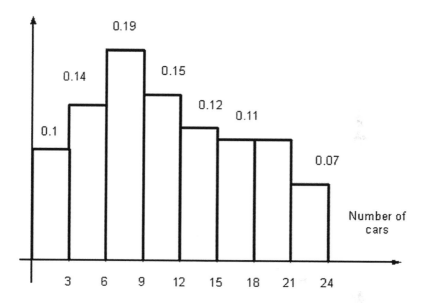

Figure 4.8 Probability distribution histogram of the number of cars

The probability distribution histogram shown can be replaced by a continuous curve shown in Figure 4.10. The curve shown in Figure 4.10 is known as a *probability density function*.

A probability density function (in the case of a continuous random variable) has the following properties:

$$f(x) \geq 0 \qquad \forall x$$

$$\int_{-\infty}^{\infty} f(x)\,dx = 1$$

In the case of discrete random variables, we denote the probability density function by $P(x)$. The $P(x)$ defines the probability that x takes a given value. The $P(x)$ must satisfy the following:

$$0 \leq P(x) \leq 1 \qquad \text{for all } x$$

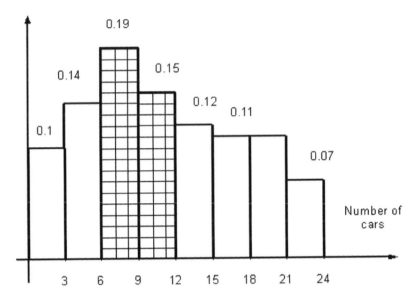

Figure 4.9 Probability calculation by summing up corresponding areas

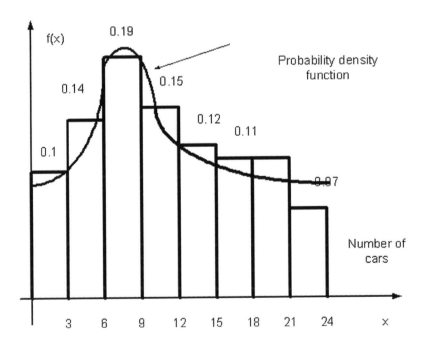

Figure 4.10 Probability density function

$$\sum_{\text{all } x} P(x) = 1$$

We see that the probability $P(x)$ must be between 0 and 1. Additionally, the sum of all probabilities in the probability distribution of the discrete random variable X must be equal to 1.

The probability $P(a \le x \le b)$ that the continuous random variable X will take a value from the interval $[a, b]$ is given by (Figure 4.11):

$$P(a \le x \le b) = \int_a^b f(x)dx \backslash\backslash$$

As we see, in the case of continuous random variables, the probability $P(a \le x \le b)$ that the random variable X will take a value from the interval $[a,b]$ is represented by the area under the probability density curve, between a and b. Figure 4.12 shows three shaded areas that represent various probabilities.

The *cumulative density function F(x)* is defined as the probability that the observed value of the random variable X will be less than or equal to x, i.e.:

$$F(x) = P(X \le x) = \int_{-\infty}^{x} f(x)dx$$

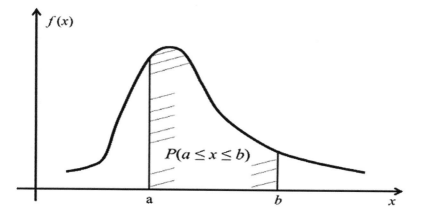

Figure 4.11 The probability $P(a \le x \le b)$ that the random variable X will take a value from the interval $[a, b]$

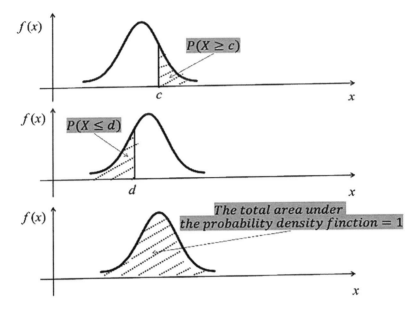

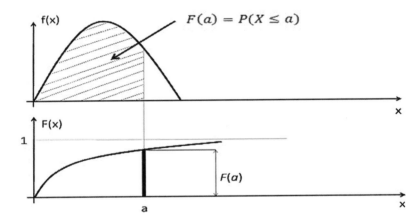

Figure 4.12 The shaded areas that represent various probabilities

Figure 4.13 Probability density function $f(x)$ and the corresponding cumulative density function $F(x)$

Probability density function and the corresponding cumulative density function are shown in Figure 4.13.

We see that, as a shifts from the left to the right, we "accumulate" more and more probability. The probability $P(a \leq x \leq b)$, that the random variable X will take a value from the interval $[a, b]$ could be also calculated, using cumulative density function:

$$P(a \le x \le b) = \int_{a}^{b} f(x) dx = \int_{-\infty}^{b} f(x) dx - \int_{-\infty}^{a} f(x) dx = F(b) - F(a)$$

4.5.1 Distribution shape

The first distribution shown in Figure 4.14 is a symmetrical distribution. Skewed distributions are distributions that are not symmetric. The distribution in the middle of Figure 4.14 is a distribution skewed to the right. The third distribution shown in Figure 4.14 is a distribution skewed to the left.

The tails of the distribution could be light or heavy (Figure 4.15).

4.5.2 Measures of central tendency

Mean

We denote by $E(X)$ the *expected value* (mean value or mean) of the random variable X. The mean measures the central tendency of the distribution. In the case of discrete random variables, the expected value $E(X)$ equals:

$$E(x) = \sum_{i=1}^{n} x_i p_i$$

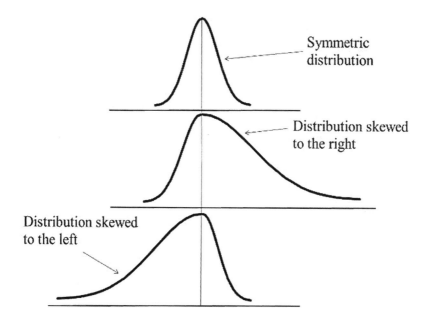

Figure 4.14 Symmetric distribution, distribution skewed to the right, and distribution skewed to the left

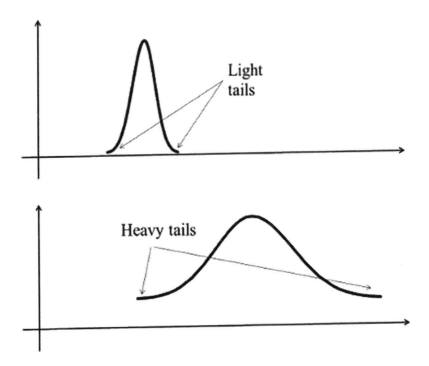

Figure 4.15 The distribution with light tails and the distribution with heavy tails

x_i – i-th possible value of the random variable X
p_i – probability that the random variable X will take the value x_i
n – the total number of the possible values of the random variable X.

In the case of continuous random variables, the expected value $E(x)$ is calculated as follows:

$$E(x) = \int_{-\infty}^{\infty} x \cdot f(x) \cdot dx$$

where:
  $f(x)$ – the probability density function of the random variable X.
The mean is expressed in the same units as the random variable.

Example 4.7

Prove the following:

(a) $E(C) = C \cdot 1 = C$

(b) $E(C \cdot X) = C \cdot E(X)$

SOLUTION:

(a) The constant C should be viewed as a discrete random variable that always takes a value equal to C, with the probability equal to 1. Consequently:

$$E(C) = C \cdot 1 = C$$

(b) $E(C \cdot X) = C \cdot x_1 \cdot p_1 + C \cdot x_2 \cdot p_2 C + \cdots + C \cdot x_n \cdot p_n = C \cdot E(X)$

Example 4.8

We denote by X the random variable that represents passenger waiting time at the terminal's check-in point at the airport. The probability density function of this random variable (Figure 4.16) is:

$$f(x) = \begin{cases} 0 & \text{for } x < a \\ \dfrac{1}{b-a} & \text{for } a < x < b \\ 0 & \text{for } x > b \end{cases}$$

Calculate the expected value of the passenger waiting time.

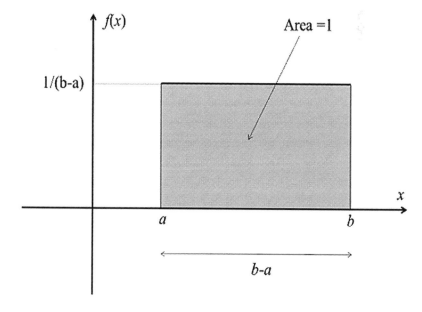

Figure 4.16 Uniform distribution

SOLUTION:

The expected value equals:

$$E(X) = \int_a^b x \cdot \frac{1}{b-a} dx = \frac{1}{b-a} \cdot \frac{b^2 - a^2}{2} = \frac{a+b}{2}$$

4.5.3 Measures of variability

Variance

The *variance* is a measure of dispersion of the distribution around its expected value. The variance is defined as:

$$\mathrm{Var}(X) = E[X - E(X)]^2$$

$$\mathrm{Var}(X) = E([X^2 - 2X \cdot E(X) + E^2(X)])$$

$$\mathrm{Var}(X) = E(X^2) - 2 \cdot E^2(X) + E^2(X)$$

Finally, variance equals:

$$\mathrm{Var}(X) = E(X^2) - E^2(X)$$

The standard deviation σ_x is also a measure of dispersion of the distribution around its expected value. The standard deviation σ_x is defined as:

$$\sigma_x = \sqrt{\mathrm{Var}(X)}$$

The standard deviation σ_x is expressed in the same units as the random variable.

The coefficient of variation c_v represents the ratio of the standard deviation to the mean, i.e.:

$$c_v = \frac{\sigma_x}{E(X)} \cdot 100\%$$

The c_v is a relative measure of dispersion of the distribution around its expected value.

Example 4.9

The random variable X denotes the number of heavy vehicles that will appear during a time interval of 1 minute at a specific point on the highway. The probability mass function of this random variable reads:

$$X = \begin{Bmatrix} 0 & 1 \\ 0.9 & 0.1 \end{Bmatrix}$$

Calculate the dispersion of the random variable X.

SOLUTION:

The expected value $E(X)$ of the random variable X equals:

$$E(X) = 0 \cdot 0.9 + 1 \cdot 0.1 = 0.1$$

The variance equals:

$$\text{Var}(X) = (0-0.1)^2 \cdot 0.9 + (1-0.1)^2 \cdot 0.1 = 0.009 + 0.081 = 0.09$$

Example 4.10

Let C be the constant. Calculate: (a) $\text{Var}(C)$; (b) $\text{Var}(C \cdot X)$

SOLUTION:

$$\text{Var}(C) = E\left\{ (C - E(C))^2 \right\} = E(C - C)^2 = 0$$

$$\text{Var}(C \cdot X) = E\left\{ (C \cdot X - E(C \cdot X))^2 \right\}$$

$$\text{Var}(C \cdot X) = E\left[C \cdot X - C \cdot E(X) \right]^2$$

$$\text{Var}(C \cdot X) = C^2 \cdot E\left[X - E(X) \right]^2$$

$$\text{Var}(C \cdot X) = C^2 \cdot \text{Var}(X)$$

4.5.4 Bernoulli distribution

Swiss mathematician Jacob Bernoulli (1655–1705) introduced a simple random variable X by considering the following experiment. In the case of experimental success, the random variable takes the value equal to 1. In the case of experimental failure, the random variable is equal to zero.

The probability of experimental success is equal to p. The probability of failure is equal to q $(p+q=1)$. The Bernoulli random variable can take on two values, 1 and 0. The probability mass function reads:

$$X = \begin{Bmatrix} 0 & 1 \\ q & p \end{Bmatrix}$$

This can also be expressed as:

$$P(X = x) = p^x \cdot q^{1-x} \quad x = 0,1$$

The mean and variance of the Bernoulli random variable are equal, respectively, to:

$$E(X) = 0 \cdot q + 1 \cdot p = p$$

$$\text{var}(X) = E(X^2) - (E(X))^2 = 1^2 \cdot p - p^2 = p - p^2 = p \cdot (1-p) = p \cdot q$$

The Bernoulli process represents a sequence of independent experiments, where the probability of success in each trial is equal to p. Binomial distributions and geometric distributions are associated with the Bernoulli process.

4.5.5 Binomial distribution

Let us perform the following experiment under the following conditions:

(a) We perform a fixed number of trials, this number being denoted by n.
(b) There are two possible outcomes in each trial – success (A) or failure (B).
(c) The probability of success in each success is equal to p.
(d) The trials are independent; this indicates that the outcome of one trial does not impact on the outcome of any other trial).

The random variable that represents the number of successes in n trials is denoted by X. The random variable X can take the following values: $0,1,2,\ldots,n$. Let us consider the event where the random variable X takes the value x. This means that success happens x times in n trials. A combination represents a selection of items from a collection, such that the order of selection is not relevant. The sequence:

$$AAAAAAAAAABBBBBBBBBB$$

$$x \text{ times} \qquad (n-x) \text{ times}$$

represents one possible combination that success happens x times in n trials. The total number of combinations is equal to:

$$\binom{n}{x} = \frac{n!}{x!(n-x)!}$$

Each of these combinations can occur with a probability equal to $p^x \cdot q^{n-x}$. Consequently, the probability of an event, where success happens x times in n trials equals:

$$P(X = x) = \frac{n!}{x!(n-x)!} \cdot p^x \cdot q^{n-x}$$

The mean and variance of the binomial random variable are, respectively, equal to:

$$E(X) = n \cdot p$$

$$\text{var}(X) = n \cdot p(1-p)$$

Example 4.11

Let us denote with X a binomial random variable, which represents the number of delayed aircraft. The probability that the aircraft will arrive delayed at the airport is 0.35. The total number of aircraft is 10. Calculate the probability that:

a) 0 aircraft will have a delay
b) 3 aircrafts will have a delay and
c) 6 aircrafts will have a delay

SOLUTION:

$$n = 10$$

$$p = 0.35$$

$$q = 1 - p = 1 - 0.35 = 0.65$$

$$P(X = x) = \frac{n!}{x!(n-x)!} \cdot p^x \cdot q^{n-x}$$

$$P(X = 0) = \frac{10!}{0!(10-0)!} \cdot 0.35^0 \cdot 0.65^{10} = 1 \cdot 1 \cdot 0.0135 = 0.0135$$

This value can be obtained using Excel function: "$BINOMDIST(0,10,0.35,FALSE)$".

$$P(X = 3) = \frac{10!}{3!(10-3)!} \cdot 0.35^3 \cdot 0.65^7 = \frac{3628800}{6 \cdot 5040} \cdot 0.0429 \cdot 0.04902 = 0.252$$

This value can be obtained using Excel function: "$BINOMDIST(3,10,0.35,FALSE)$".

$$P(X=6) = \frac{10!}{6!\,(10-6)!} \cdot 0.35^6 \cdot 0.65^4 = \frac{3628800}{720 \cdot 24} \cdot 0.00184 \cdot 0.1785 = 0.06891$$

This value can be obtained using Excel function: "$BINOMDIST(6,10,0.35,FALSE)$".

4.5.6 Geometric distribution

Let us consider the experimental scheme where there are two possible outcomes in each trial – success (A) or failure (B).The probability of success in each trial is equal to p and the trials are independent. In the case of a binomial distribution, we analyzed the random variable that represented the number of successes in n trials.

Let us now consider the random variable X that represents the number of trials until the first success. The random variable X can take the following values: 1, 2, 3,... The probability $P(X = x)$ that the random variable X will take the value x equals:

$$P(X = x) = q^{x-1} \cdot p \qquad x = 1,2,3,\ldots$$

It is important to emphasize that geometric probability mass function has a no-memory property. This means that the number of trials carried out until the first success occurs in no way affects how many more trials will need to be carried out until the next success.

The mean and variance are, respectively, equal:

$$E(X) = \frac{1}{p}$$

$$\mathrm{var}(X) = \frac{q}{p^2} = \frac{1-p}{p^2}$$

4.5.7 Negative binomial distribution

Let us again consider the experimental scheme when there are two possible outcomes in each trial: success (A) or failure (B).The probability of success in each trial is equal to p and the trials are independent. We denote by X the random variable that represents the number of trials until the success (event A) happens k times. We also denote by Y the following random variable:

$$Y = X - k$$

The random variable Y represents the number of additional trials (in addition to the k necessary trials) performed until the success (event A) happens k times. The probability $P(Y = y)$ that the random variable Y will take the value y equals:

$$P(Y = y) = P(X - k = y) = P(X = y + k) = \binom{y+k-1}{k-1} \cdot p^k \cdot q^y$$

Since:

$$\binom{y+k-1}{k-1} = \frac{(y+k-1)!}{(k-1)!(y+k-1-k+1)!} = \frac{(y+k-1)!}{(k-1)!(y)!} = \binom{y+k-1}{y}$$

We get:

$$P(Y = y) = \binom{y+k-1}{y} \cdot p^k \cdot q^y \quad y = 0,1,2,\dots$$

Since:

$$\binom{y+k-1}{y} = (-1)^y \cdot \binom{-k}{y}$$

where:

$$\binom{-k}{y} = \frac{(-k) \cdot (-k-1) \cdots \cdots (-k-y+1)}{y!}$$

We get the probability mass function:

$$P(Y = y) = \binom{-k}{y} \cdot p^k \cdot (-q)^y \qquad y = 0,1,2,\dots$$

This distribution is called negative binomial distribution. The mean and the variance are, respectively, equal to:

$$E(Y) = k \cdot \frac{q}{p}$$

$$\text{var}(Y) = k \cdot \frac{q}{p^2}$$

Example 4.12

Suppose that a disruption at the airport arises when 5 or more aircraft are delayed. The probability that an aircraft is delayed is 0.3. Determine the probabilities that 5, 10, and 15 aircraft are going to arrive at the airport before the disruption happens.

SOLUTION:

$$k = 5, \ p = 0.3, \ y_1 = 5, \ y_2 = 10, \ y_3 = 15$$

$$P(Y = y_1) = P(Y = 5) = \binom{y_1 + k - 1}{k - 1} \cdot p^k \cdot q^{y_1}$$

$$= \binom{9}{4} \cdot 0.3^5 \cdot 0.7^5 = \frac{9 \cdot 8 \cdot 7 \cdot 6}{4 \cdot 3 \cdot 2 \cdot 1} \cdot 0.00243 \cdot 0.16807$$

$$= 126 \cdot 0.000408 = 0.05146$$

$$P(Y = y_2) = P(Y = 10) = \binom{y_2 + k - 1}{k - 1} \cdot p^k \cdot q^{y_2} = \binom{14}{4} \cdot 0.3^5 \cdot 0.7^{10}$$

$$= \frac{14 \cdot 13 \cdot 12 \cdot 11}{4 \cdot 3 \cdot 2 \cdot 1} \cdot 0.00243 \cdot 0.0282 = 1001 \cdot 6.86 \cdot 10^{-5} = 0.06871$$

$$P(Y = y_3) = P(Y = 15) = \binom{y_3 + k - 1}{k - 1} \cdot p^k \cdot q^{y_3}$$

$$= \binom{19}{4} \cdot 0.3^5 \cdot 0.7^{15} = \frac{19 \cdot 18 \cdot 17 \cdot 16}{4 \cdot 3 \cdot 2 \cdot 1} \cdot 0.00243 \cdot 0.004748$$

$$= 3876 \cdot 1.15 \cdot 10^{-5} = 0.0447$$

These values of probabilities can be obtained by Excel. The Excel function for the probability mass function is:

"$= NEGBINOMDIST(y, k, p)$".

In this example, probability $P(Y = 5)$ can be obtained by Excel function:

"$= NEGBINOMDIST(5, 5, 0.3)$".

Probability $P(Y = 10)$ can be obtained by Excel function:

"$= NEGBINOMDIST(10, 5, 0.3)$",

and the probability $P(Y = 15)$ can be obtained by Excel function:

"$= NEGBINOMDIST(15, 5, 0.3)$".

4.5.8 Poisson distribution

Measurements in many transportation systems show that the client arrivals pattern could be described by the *Poisson distribution*. It has been shown that Poisson distribution describes many real-life situations. The Poisson random variable represents counts of the number of times that an observed event happens within a given time interval. This distribution was introduced by French mathematician Siméon Denis Poisson (1781–1840).

If the discrete random variable X has a Poisson distribution with a parameter λ, then the probability $P(X = k)$ is:

$$P(X = k) = \frac{\lambda^k}{k!} \cdot e^{-\lambda}$$

The expected value $E(x)$ and the variance var(x) are equal in the case of a Poisson distribution, i.e.:

$$E(X) = \lambda$$

$$\mathrm{var}(X) = \lambda$$

Let us consider specific points at the highway (Figure 4.17)

Vehicles randomly show up and pass. We count every vehicle. The number of vehicle arrivals X during a specific time interval represents the random variable. In other words, it can happen that, during the specific time interval, no vehicles arrive, one vehicle arrives, two vehicles arrive, etc. Now, suppose that the vehicle arrival rate is λ [vehicles/hour], and the observed time interval is t [hour]. The random variable X is distributed according to the *Poisson* distribution, with the parameter $\lambda \cdot t$ [vehicles]. Probability $P(X = k)$ is:

$$P(X = k) = \frac{(\lambda t)^k}{k!} \cdot e^{-\lambda t}$$

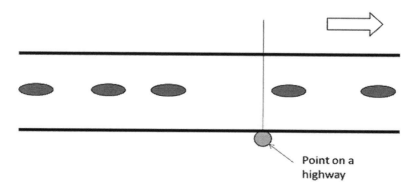

Point on a highway

Figure 4.17 Point on a highway

The latter relation describes the probability $P(k)$ that the total number of vehicle arrivals X happening in a time interval of length t is equal to k. The expected value $E(x)$ and the variance var(x) are equal in the case of a Poisson distribution, i.e.:

$$E(X) = \lambda t$$

$$\mathrm{var}(X) = \lambda t$$

By observing the collected statistical data, and by calculating mean and variance, one can easily get an impression of the observed traffic phenomena. In the case that the calculated mean and variance are approximately equal, it is likely that a Poisson distribution describes the traffic phenomenon being studied.

Example 4.13

Consider a set of non-signalized intersections in a city. The number of accidents at a given intersection during an observed period of time is denoted by X. Most frequently, the occurrence of this random variable is Poisson distributed, i.e.:

$$P(X = k) = \frac{(\lambda t)^k}{k!} \cdot e^{-\lambda t}$$

where:
λ–accident rate (mean number of accidents per intersection during the observed period)

The accident rate is usually obtained as a mean of the number of accidents at the intersections in the city during the observed time period. Calculate the probability that an observed intersection will have more than m accidents.

SOLUTION:

The probability that an observed intersection will have more than m accidents equals:

$$P(X > m) = \sum_{m+1}^{\infty} \frac{(\lambda t)^k}{k!} \cdot e^{-\lambda t}$$

Consequently,

$$1 - P(X > m) = \sum_{0}^{m} \frac{(\lambda t)^k}{k!} \cdot e^{-\lambda t}$$

$$P(X > m) = 1 - \sum_{0}^{m} \frac{(\lambda t)^k}{k!} \cdot e^{-\lambda t}$$

The traffic engineer could treat all intersections that have a number of traffic accidents higher than m during the observed time period as hazardous. The value of m depends on the subjective judgment of the traffic engineer.

Example 4.14

A total of 720 vehicles wanting to merge onto the highway appeared during one hour (Figure 4.18). It has been shown that the vehicle arrival pattern could be described by the Poisson process. The assumption is that the vehicle arrival pattern will be unchanged during the following days.
  Calculate:

(a) mean time between arrivals;
(b) Probabilities of having 3, 4, 5 vehicles during 30-second intervals;
(c) Percentage of the 10-second intervals with no vehicles.

SOLUTION:

We express the average vehicle arrival rate in [vehicles/s], and the mean time between vehicle arrivals in [s]. The average vehicle arrival rate equals:

$$q = \frac{720 \text{ vehicles}}{1 \text{ hour}} = \frac{720 \text{ vehicles}}{3,600 \text{ seconds}} = 0.2 [\text{veh/sec}]$$

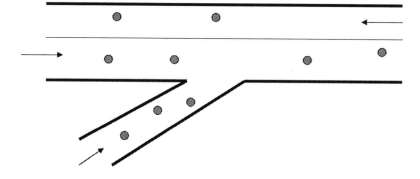

Figure 4.18 Vehicles merging onto highway

The mean time between arrivals equal:

$$\frac{1}{q} = \frac{1}{0.2[\text{veh/sec}]} = 5 \text{ seconds}$$

Probability of having n vehicles arriving during a 30-s interval equals:

$$P(X = n) = \frac{\left(0.2[\text{veh/sec}] \cdot 30[\text{sec}]\right)^n \cdot e^{-0.2[\text{veh/sec}] \cdot 30[\text{sec}]}}{n!}$$

$$P(X = n) = \frac{(6)^n \cdot e^{-6}}{n!}$$

$$P(X = 3) = \frac{(6)^3 \cdot e^{-6}}{3!} 0.089244$$

$$P(X = 4) = \frac{(6)^4 \cdot e^{-6}}{4!} = 0.133866$$

$$P(X = 5) = \frac{(6)^5 \cdot e^{-6}}{5!} = 0.1606392$$

The percentage of the 10-s intervals with no vehicles represents the probability of the event that no one vehicle will show up during a 10-s period. This probability equals:

$$P(X = 0) = \frac{\left(0.2[\text{veh/sec}] \cdot 30[\text{sec}]\right)^0 \cdot e^{-0.2[\text{veh/sec}] \cdot 10[\text{sec}]}}{0!}$$

$$P(X = 0) = e^{-0.2[\text{veh/sec}] \cdot 10[\text{sec}]} = e^{-2} = 0.13534$$

We conclude that, in 13.534% cases, no vehicle will appear in a 10-s period.

Example 4.15

East-westbound moving vehicles (vehicles moving along the major street) have right-of-way (Figure 4.19). Vehicles coming from the minor street must wait for an acceptable gap in traffic in order to cross the major street. The intersection between the major and minor street is an unsignalized intersection. The vehicle arrival pattern along the major street could be described by the Poisson distribution. Measurement

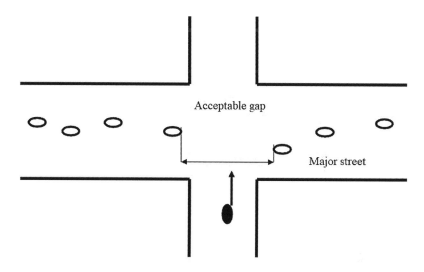

Figure 4.19 Acceptable gap

shows that 900 vehicles pass through the intersection along the major street in a one-hour period.

Assume that the minimum acceptable gap equals 4 seconds and calculate the expected number of acceptable gaps during a one-hour period.

(a) The acceptable gap equals 8 seconds in the case of senior citizens. In this case, calculate the expected number of acceptable gaps during a one-hour period.
(b) The time between vehicle arrivals in a Poisson distribution is random variable T, that has an exponential distribution, with parameter q:

$$f(t) = qe^{-qt}$$

In our case, parameter q equals:

$$q = 900 \left[\frac{veh}{hr} \right] = 0.25 \left[\frac{veh}{sec} \right]$$

The probability that the next vehicle arrival will happen after t equals e^{-qt}. The situation when it is safe to cross the major street happens when the gap in the major vehicle flow is greater than 4 seconds. The probability that the random variable T takes a value greater than 4 seconds equals:

$$P(T > 4) = e^{-0.25 \cdot (4)} = e^{-1} = 0.367$$

We know that 900 vehicles pass through the intersection along the major street in one hour. This means that there are 899 gaps in the vehicle flow along the main street in one hour, so that the expected number of acceptable gaps during one hour equals:

$$899 \cdot P(T > 4) = 899 \cdot (0.367) = 330$$

We conclude that there is an acceptable gap in the main vehicle flow in $\dfrac{330}{899}(100)\% = 36.7\%$ cases.

(b) The probability that the random variable T takes the value greater than 8 seconds equals:

$$P(T > 8) = e^{-0.25(8)} = e^{-2} = 0.134$$

The expected number of acceptable gaps in this case equals:

$$899 \cdot P(T > 8) = 899 \cdot (0.134) = 120$$

The acceptable gap in the main flow in $\dfrac{120}{899}(100\ \%) = 13.3\ \%$ cases.

If a random variable X has a Poisson distribution with parameter λ, probability $P(X = k)$ can be obtained in Excel, using the following function: "$= POISSONDIST(k, \lambda, FALSE)$".

Excel function "$POISSONDIST(k, \lambda, TRUE)$" can be used to obtain a cumulative probability, $P(X \le k)$.

Example 4.16

Let us suppose that random variable X has a Poisson distribution and represents the number of vehicles which arrive in the terminal in a period $t = 0.5$ hours. If we have that vehicle arrival rate at the terminal is $= 10 \left[\dfrac{\text{vehicles}}{\text{hour}} \right]$, let us calculate the following probabilities: $P(X = 0)$, $P(X = 1)$, $P(X = 5)$, $P(X = 7)$, $P(X \le 0)$, $P(X \le 1)$, $P(X \le 5)$ and $P(X \le 7)$.

SOLUTION:

$$P(X = 0) = \frac{(\lambda \cdot t)^k}{k!} \cdot e^{-\lambda \cdot t} = \frac{(10 \cdot 0.5)^0}{0!} \cdot e^{-10 \cdot 0.5} = \frac{5^0}{1} \cdot e^{-5} = e^{-5} = 0.006738$$

This value can also be obtained using the following Excel function: "$= POISSON(0, 5, FALSE)$".

$$P(X = 1) = \frac{(\lambda \cdot t)^k}{k!} \cdot e^{-\lambda \cdot t} = \frac{(10 \cdot 0.5)^1}{1!} \cdot e^{-10 \cdot 0.5} = \frac{5^1}{1} \cdot e^{-5} = 5 \cdot e^{-5} = 0.03369$$

This value can be obtained by the Excel function: "$= POISSON(1, 5, FALSE)$".

$$P(X = 5) = \frac{(\lambda \cdot t)^k}{k!} \cdot e^{-\lambda \cdot t} = \frac{(10 \cdot 0.5)^5}{5!} \cdot e^{-10 \cdot 0.5} = \frac{5^5}{120} \cdot e^{-5} = 26.042 \cdot e^{-5} = 0.1755$$

This value can be obtained by the Excel function "$= POISSON(5, 5, FALSE)$".

$$P(X = 7) = \frac{(\lambda \cdot t)^k}{k!} \cdot e^{-\lambda \cdot t} = \frac{(10 \cdot 0.5)^7}{7!} \cdot e^{-10 \cdot 0.5} = 15.5 \cdot e^{-5} = 0.1044$$

This value can be obtained by the Excel function: "$= POISSON(7, 5, FALSE)$".

Probability $P(X \leq 0)$ equals:

$$P(X \leq 0) = \sum_{k=0}^{0} P(X = k) = P(X = 0) = 0.006738$$

Probability $P(X \leq 1)$ equals:

$$P(X \leq 1) = \sum_{k=0}^{1} P(X = k) = P(X = 0) + P(X = 1) = 0.006738 + 0.3369 = 0.0404$$

This value can be also obtained by the Excel function "$POISSON(1, 5, TRUE)$".

$$P(X \leq 5) = \sum_{k=0}^{5} P(X = k)$$

Using the Excel function "$POISSON(5, 5, TRUE)$", it follows that:

$$P(X \leq 5) = \sum_{k=0}^{5} P(X = k) = 0.61596$$

Probability $P(X \leq 7)$ equals:

$$P(X \leq 7) = \sum_{k=0}^{7} P(X = k)$$

Using Excel function "$POISSON(7,5,TRUE)$", it obtains that:

$$P(X \le 7) = \sum_{k=0}^{7} P(X = k) = 0.8666$$

4.5.9 Uniform distribution

The probability density function of the uniform distribution is:

$$f(x) = \begin{cases} 0 & \text{for } x < a \\ \dfrac{1}{b-a} & \text{for } a < x < b \\ 0 & \text{for } x > b \end{cases}$$

This probability density function is shown in Figure 4.20.

The mean $E(x)$ and the variance var(X) of the uniform distribution are, respectively, equal to:

$$E(x) = \frac{a+b}{2}$$

$$\text{var}(x) = \frac{(b-a)^2}{12}$$

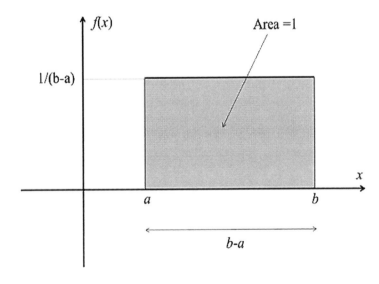

Figure 4.20 Uniform random variable on the interval [a, b]

Example 4.17

Consider a bus route with a turnaround time $T = 45$ minutes. The number of buses which operate on the route is $N = 5$. If passenger waiting time has a uniform distribution $T_w \sim U(0, T/N)$, determine the average passenger waiting time, variance, and the probability $P(T_w < 5 \text{ min})$.

SOLUTION:

Average waiting time and variance can be determined in the following way:

$$E(T_w) = \frac{0 + \dfrac{45}{5}}{2} = 4.5 \text{ min}$$

$$\text{var}(T_w) = \frac{\left(\dfrac{45}{5} - 0\right)^2}{12} = \frac{9^2}{12} = \frac{81}{12} = 6.75$$

Probability $P(T_w < 5 \text{ min})$ is:

$$P(T_w < 5 \text{ min}) = F(5) = \frac{5 - 0}{\dfrac{45}{5} - 0} = \frac{5}{9} = 0.556$$

4.5.10 Exponential distribution

Let us again consider a point on the highway (Figure 4.17). The time interval between the appearances of successive vehicles (headway) could be, for example, 5 s, 10 s, 11 s, 14 s, etc. In other words, the time interval between vehicle arrivals (headway) is a random variable, that frequently has an *exponential* distribution. This continuous random variable T has an *exponential* distribution with the parameter λ:

$$f(t) = \lambda \cdot e^{-\lambda \cdot t}$$

The exponential distribution of the time between vehicle arrivals is shown in Figure 4.21.

The probability that the random variable T is greater than or equal to the specific value h equals:

$$P(T \geq h) = 1 - P(T < h)$$

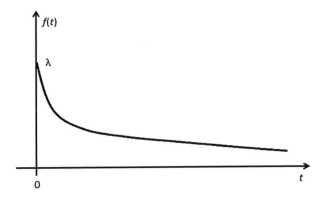

Figure 4.21 Exponential distribution of the time between vehicle arrivals

$$P(T \geq h) = 1 - \int_{0}^{h} \lambda \cdot e^{-\lambda \cdot t} \cdot dt$$

$$P(T \geq h) = 1 - \lambda \cdot \left[\frac{e^{-\lambda \cdot t}}{-\lambda} \right]_{0}^{h}$$

$$P(T \geq h) = 1 + e^{-\lambda \cdot t} \Big|_{0}^{h}$$

$$P(T \geq h) = 1 + e^{-\lambda \cdot h} - 1$$

$$P(T \geq h) = e^{-\lambda \cdot h}$$

On the other hand, the probability density function of the Poisson distribution equals:

$$P(X = k) = \frac{(\lambda t)^{k}}{k!} \cdot e^{-\lambda t}$$

The probability that no one vehicle arrives during the time interval h is equal to the probability that the headway is greater than or equal to h. For that reason,

$$P(X = 0) = \frac{(\lambda t)^{0}}{0!} \cdot e^{-\lambda h} = e^{-\lambda h} = P(T \geq h)$$

In a case of low-volume traffic flow, where there is no significant interaction between vehicles, and when minimum headway is directed primarily by safety reasons, the headways have exponential distribution. As long as the vehicle interarrival time is exponential, the number of vehicle arrivals during time interval t is Poisson.

Example 4.18

The average interarrival time between two vehicles on a highway equals 20 s. Assume that the vehicle interarrival times are distributed according to an exponential distribution. Calculate the percentage of cases when the interarrival time is less than 10 s.

SOLUTION:

The probability density function of vehicle interarrival times equals:

$$f(t) = \lambda \cdot e^{-\lambda t}$$

The expected interarrival time value equals:

$$E(t) = \int_{-\infty}^{\infty} t \cdot \lambda \cdot e^{-\lambda t} dt = \int_{0}^{\infty} t \cdot \lambda \cdot e^{-\lambda t} dt$$

After solving the integral, we obtain the following:

$$E(t) = \frac{1}{\lambda}$$

Since the average interarrival time between two vehicles on a highway equals 20 s, we have:

$$\frac{1}{\lambda} = 20$$

$$\lambda = 0.05 \left[\frac{\text{vehicles}}{\text{sec}} \right]$$

The probability of events, where the interarrival time is less than 10 s, equals:

$$P(0 \le T \le 10) = \int_{0}^{10} f(t) \cdot dt = \int_{0}^{10} 0.05 \cdot e^{-0.05 \cdot t} \cdot dt$$

$$P(0 \le T \le 10) = 0.393$$

The probability of event that the interarrival time is less than 10 seconds represents the percentage of cases when interarrival time is less than 10 seconds. We conclude that, in 39.3% of cases, interarrival time will be less than 10 seconds.

Example 4.19

If vehicle interarrival time (T) has an exponential distribution, and an average interarrival time is 5 min, determine the probability $P(T < 3$ min) and $P(T < 6$ min).

SOLUTION:

Taking into account that $\bar{t} = 5$ min and $E(T) = \dfrac{1}{\lambda} = \bar{t}$, we can calculate λ as:

$$\lambda = \frac{1}{\bar{t}} = \frac{1}{5} = 0.4 \left[\min^{-1} \right]$$

Probability $P(T < 3$ min) is:

$$P(T < 3) = F(t = 3) = 1 - e^{-\lambda t} = 1 - e^{-0.2 \cdot 3} = 1 - 0.5488 = 0.4512$$

This value of probability can also be obtained by the Excel function $= EXPONDIST(3, 0.2, TRUE)$".
    Probability $P(T < 6)$ is:

$$P(T < 6) = F(t = 6) = 1 - e^{-\lambda t} = 1 - e^{-0.2 \cdot 6} = 1 - 0.3012 = 0.6988$$

This value can be obtained by the Excel function: "$= EXPONDIST(6, 0.2, TRUE)$".

4.5.11 Erlang distribution

The Erlang distribution was introduced by Agner Krarup Erlang (1878–1929), Danish mathematician and engineer, while working for the Copenhagen Telephone Company. A random variable has an Erlang distribution, if its probability density function is given by:

$$f_k = \frac{\lambda^k \cdot x^{k-1}}{(k-1)!} \cdot e^{-\lambda \cdot x} \qquad x > 0, \quad \lambda > 0, \quad k \text{ positive integer}$$

where:
    −k the shape parameter
    λ–the rate parameter.

When the shape parameter k is equal to 1, the Erlang distribution is the same as the exponential distribution, i.e.:

$$f_0 = \frac{\lambda \cdot x^0}{0!} \cdot e^{-\lambda \cdot x} = \lambda \cdot e^{-\lambda \cdot x}$$

The mean $E(x)$ and the variance var(X) of the Erlang distribution are, respectively:

$$E(x) = \frac{k}{\lambda}$$

$$var(x) = \frac{k}{\lambda^2}$$

Example 4.20

There are two trucks at the terminal. The second truck should be served after the first truck is served. The service times of each truck (T_1 and T_2) has an exponential distribution, with the parameter $\lambda = 1\,h^{-1}$. Determine the probability that the total service time ($T = T_1 + T_2$) for both trucks will be less than 3 hours.

SOLUTION:

T_1, T_2 – service time of truck 1 and truck 2.
   T_1 and T_2 have exponential distribution with the parameter λ.

$$T = T_1 + T_2$$

Because T_1 and T_2 have exponential distribution with parameter λ, then T has an Erlang distribution with shape parameter 2 and rate parameter λ. Now, probability $P(T < 3)$ equals:

$$P(T < 3) = \int_0^3 \frac{\lambda^k \cdot x^{k-1}}{(k-1)!} \cdot e^{-\lambda \cdot x}\, dx = \int_0^3 \frac{1^2 \cdot x^{2-1}}{(2-1)!} \cdot e^{-1 \cdot x}\, dx = \int_0^3 x \cdot e^{-x}\, dx$$

An integration by parts where $u = x$ and $dv = e^{-x}\, dx$ yields $du = dx$ and $v = -e^{-x}$. Now, we have:

$$P(T < 3) = -x \cdot e^{-x} \Big|_0^3 + \int_0^3 e^{-x}\, dx = \left(-3 \cdot e^{-3} + 0 \cdot e^0\right) - e^{-x} \Big|_0^3 = -3 \cdot e^{-3} - \left(e^{-3} - e^0\right)$$

$$= -3 \cdot e^{-3} - e^{-3} + 1 = -4 \cdot e^{-3} + 1 = -4 \cdot 0.04979 + 1 = 0.801$$

This can be obtained by Excel using the function for Gamma distribution. This is possible to do as, if parameter k has a positive integer value, the random variable follows both Gamma and Erlang distribution. Because of that, in our example, we can use the following function: "$= GAMMA.DIST(3,2,1,TRUE)$", and in this way obtain the value of probability 0.801.

4.5.12 Normal distribution

A normal distribution was introduced and studied by the French mathematician Abraham de Moivre (1667–1754) and the German mathematician Johann Carl Friedrich Gauss (1777–1855). This distribution has a bell-shaped frequency distribution curve (Figure 4.22). It has been shown that many phenomena in nature and society follow a normal distribution. In a normal distribution, data show an inclination to gather around the mean. The greater the distance of the specific data point from the mean value, the lower the probability that it will occur.

The probability density function of the normal distribution equals:

$$f(x) = \frac{1}{\sigma\sqrt{2 \cdot \pi}} \exp\left[-\frac{(x-\mu)^2}{2 \cdot \sigma^2} \right] \qquad -\infty < x < \infty$$

where parameter μ denotes the mean, and σ^2 denotes the variance of the distribution.

The line $x = \mu$ is the line of symmetry that divides normal distribution into two mirror-image halves. Consequently, the median is also equal to μ.

Since the point of maximum has coordinates $\left(\mu, \dfrac{1}{\sigma \cdot \sqrt{2\pi}} \right)$, the mode is also equal to μ.

The cumulative density function of the normal distribution equals:

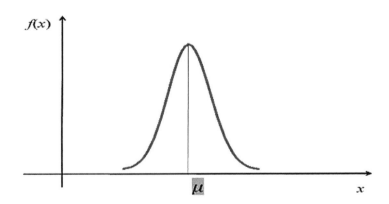

Figure 4.22 The probability density function of the normal distribution

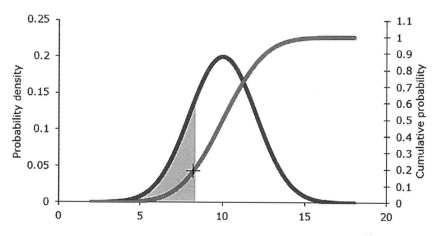

Figure 4.23 Probability density function and the cumulative density function of the normal distribution

$$F(x) = \int_{-\infty}^{x} \frac{1}{\sigma\sqrt{2\cdot\pi}} \exp\left[-\frac{(y-\mu)^2}{2\cdot\sigma^2}\right] dy$$

The probability density function and the cumulative density function of the normal distribution are shown in Figure 4.23.

The standard normal distribution, or Z-distribution, is frequently used in probability and statistics studies. The random variable Z is formulated as:

$$Z = \frac{X-\mu}{\sigma}$$

Figure 4.24 shows transformation of the values on the normal distribution to the values on the Z-distribution. The random variable Z is normally distributed with parameters 0 and 1, i.e. $Z \sim N(0,1)$ (Figure 4.24). A specific Z-distribution value denotes the number of standard deviations the data is greater than or lower than the mean. The Z-distribution values are called z-scores or z-values. For example, $z = 2$ on the Z-distribution (Figure 4.24) denotes a value that is 2 standard deviations greater than the mean. In the same way, $z = -1.5$ denotes a value that is 1.5 standard deviations below the mean.

The probability $P(a < X < b)$ that the random variable $X \sim N(\mu,\sigma)$ has the value in the interval (a,b) is given by:

$$P(a < X < b) = P\left(\frac{a-\mu}{\sigma} < \frac{X-\mu}{\sigma} < \frac{b-\mu}{\sigma}\right) = P(z_1 < Z < z_2) = \frac{1}{\sqrt{2\pi}}\int_{z_1}^{z_2} e^{-\frac{z^2}{2}} dz$$

where:

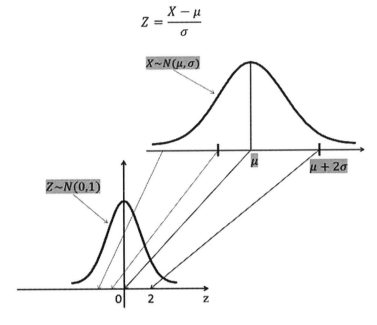

$$Z = \frac{X - \mu}{\sigma}$$

Figure 4.24 Transformation of the values on the normal distribution to the values on the Z-distribution

$$z_1 = \frac{a - \mu}{\sigma}$$

$$z_2 = \frac{b - \mu}{\sigma}$$

The probability $P(Z < z) = \frac{1}{\sqrt{2\pi}} \int_{-\infty}^{z} e^{-\frac{t^2}{2}} dt$ represents the cumulative density function of the standard normal distribution This function is denoted as $\Phi(z)$, i.e. $\Phi(z) = \frac{1}{\sqrt{2\pi}} \int_{-\infty}^{z} e^{-\frac{t^2}{2}} dt$. The values of $\Phi(z)$ are calculated for a broad range of non-negative values of z. These values are given in Table A1 in the Appendix. Table A1 is usually called the Z-table.

Since there is symmetry in the case of the $\Phi(z)$, we can obtain the negative values from the Z-table in the following way (Figure 4.25):

$$\Phi(-x) = P(Z < -x) = P(Z > x) = 1 - \Phi(x)$$

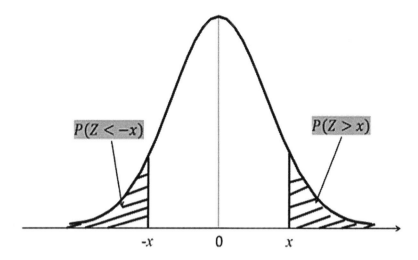

Figure 4.25 The standard normal distribution

Example 4.21

The random variable X is normally distributed with parameters μ and σ, i.e. $X \sim N(\mu, \sigma)$. Calculate the probability $P(\mu - k \cdot \sigma < X < \mu - k \cdot \sigma)$ for the $k = 1, 2, 3$.

SOLUTION:

The probability equals:

$$P(\mu - k \cdot \sigma < X < \mu - k \cdot \sigma) = P\left(-k < \frac{X - \mu}{\sigma} < k\right)$$

$$= P(-k < Z < k) = P(Z < k) - P(Z < -k)$$

$$= \Phi(k) - \left[1 - \Phi(k)\right] = \Phi(k) - 1 + \Phi(k)$$

$$= 2 \cdot \Phi(k) - 1$$

By using data given in Table A1, we obtain the following:
For $k = 1$, probability equals:

$$P(\mu - \sigma < X < \mu - \sigma) = 2 \cdot \Phi(1) - 1 = 2 \cdot 0.8413 - 1 = 1.6826 - 1 = 0.6826$$

For $k = 2$, probability equals:

$$P(\mu - 2 \cdot \sigma < X < \mu - 2 \cdot \sigma) = 2 \cdot \Phi(2) - 1 = 2 \cdot 0.9773 - 1 = 1.9546 - 1 = 0.9546$$

For $k = 3$, probability equals:

$$P(\mu - 3 \cdot \sigma < X < \mu - 3 \cdot \sigma) = 2 \cdot \Phi(3) - 1 = 2 \cdot 0.99865 - 1 = 1.99730 - 1$$
$$= 0.99730$$

These probabilities are graphically shown in Figure 4.26 (a), (b), and (c).

We see that 68.26%, 95.46%, and 99.73% of the values of the random variable that are normally distributed are positioned inside ($\pm$) one, two, and three standard deviations of the mean. This is known as the three-sigma rule.

Example 4.22

The expected demand (number of passengers) for the flight that departs every Tuesday at 08:30 a.m. between city A and city B is a random variable that is normally distributed, with parameters 170 and 15, i.e. $X \sim N(170,15)$. The probability of the event, that the random variable value belongs to the symmetric interval, around the mean, equals 0.95. Determine the borders of this interval.

SOLUTION:

The symmetric interval around the mean is shown in Figure 4.27.

The probability of the event, that the random variable value belongs to the symmetric interval around the mean, equals:

$$P(\mu - k < X < \mu + k) = 0.95$$

$$P(-\frac{k}{\sigma} < \frac{X - \mu}{\sigma} < \frac{k}{\sigma}) = 0.95$$

$$2 \cdot \Phi\left(\frac{k}{\sigma}\right) - 1 = 0.95$$

$$2 \cdot \Phi\left(\frac{k}{\sigma}\right) = 1.95$$

$$\Phi\left(\frac{k}{\sigma}\right) = 0.975$$

We get from the Table A.1 in the Appendix:

$$\frac{k}{\sigma} = 1.96$$

$$k = 1.96 \cdot \sigma = 1.96 \cdot 15 = 29.4$$

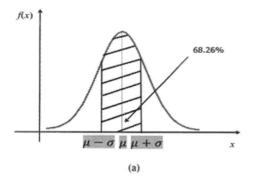

(a)

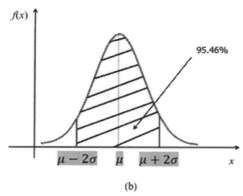

(b)

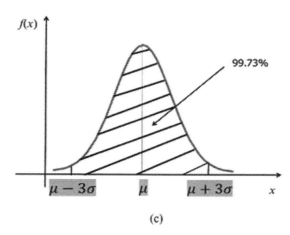

(c)

Figure 4.26 Three-sigma rule

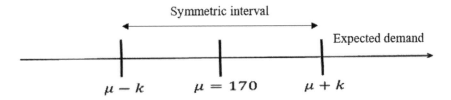

Figure 4.27 The symmetric interval around the mean

Consequently:

$$P(\mu - k < X < \mu + k) = P(170 - 29.4 < X < 170 + 29.4)$$

$$P(140.6 < X < 199.4) \sim P(141 < X < 199) = 0.95$$

We can claim that the expected demand for the flight will be between 141 and 199, with the probability equal to 0.95.

Example 4.23

In a case of high-volume traffic flow on the highway, when the traffic flow is close to capacity, and there is very high interaction between the vehicles, headways are frequently normally distributed.

We obtain the probability $P(X < x)$, where $X \sim N(\mu, \sigma)$, in Excel by using the following function: "$= NORMDIST(x, \mu, \sigma, 1)$".

In the case of standard normal distribution, the probability $P(Z < z)$, where $Z \sim N(0,1)$, in Excel can be obtained by using the following function: "$= NORMSDIST(z)$". The same results can be obtained by using the previous function in the following way: "$NORMDIST(z, 0, 1, 1)$".

For example, if $Z \sim N(0,1)$, determine the probability $P(Z < 2)$.

The value of the probability $P(Z < 2)$ can be obtained by using the following Excel function: "$= NORMSDIST(2)$" or "$= NORMDIST(2, 0, 1, 1)$". By using any of the previous Excel functions, the following value is obtained: $P(Z < 2) = 0.97725$.

Let us consider one more example: if we have a random variable, $X \sim N(130, 8)$ determine the probability: $P(120 < X < 150)$.

The probability equals:

$$P(120 < X < 150) = P(X < 150) - P(X < 120)$$

To determine the probability $P(X < 150)$ we use the following Excel function: "$NORMDIS(150, 130, 8, 1)$". This probability equals 0.9938.

To determine the probability $P(X < 120)$ we use the following Excel function: "$NORMDIST(120,130,8,1)$". The value for this probability is 0.1056.

Now, we have:

$$P(140 < X < 199) = 0.9938 - 0.1056 = 0.8882$$

In expression $P(X < x) = \alpha$, $X \sim N(\mu, \sigma)$, if we have a value for α, than we can obtain x by using the following Excel function: "$= NORMINV(\alpha, \mu, \sigma)$". In a standard normal distribution, where $P(Z < k) = \alpha$, $Z \sim N(0,1)$, the following Excel function is used: "$NORMSINV(\alpha)$". For example, let us assume that we have

$\Phi(3 \cdot k) = 0.975$ and we want to determine k. It can be done in the following way:

$$\Phi(3 \cdot k) = P(Z < 3 \cdot k) = 0.975$$

By using the Excel function "$= NORMSINV(0.975)$" or "$= NORMINV(0.975, 0, 1)$" we obtain the following:

$$3 \cdot k = 1.96$$

$$k = 0.6533$$

The *Poisson*, *Exponential* and *Normal* distributions (Table 4.3) are the distributions that frequently appear in various traffic and transportation engineering problems. The Poisson distribution is discrete, while exponential and normal are continuous distributions (Table 4.3).

Table 4.4 contains descriptions of the other important discrete and continuous distributions.

PROBLEMS

4.1. Two colleagues from the office choose modes of transportation to travel to work. Let us consider the following events:

A – first person chooses private car

B – first person chooses public transit

C – second person chooses private car

D – second person chooses public transit

E – at least one person chooses private car

F – at least one person chooses public transit

G – one person chooses private car and one person chooses public transit

H – two persons choose public transit

K – two persons choose private car

Table 4.3 Probability density functions that frequently appear in traffic and transportation

Name of the distribution	Probability density function, mean, and the variance	Some examples of the random variable distributed according to the probability density function
Poisson	$P(X = k) = \dfrac{\lambda^k}{k!} \cdot e^{-\lambda}$ $E(x) = \lambda$ $var(x) = \lambda$	The number of vehicle arrivals at the intersection during specific time interval The number of calls for emergency help from firefighters during specific time interval
Exponential	$f(x) = \lambda \cdot e^{-\lambda x}$ $E(x) = \dfrac{1}{\lambda}$ $var(x) = \dfrac{1}{\lambda^2}$	The vehicle interarrival times at the toll plaza Passenger interarrival times at the travel agent office
Normal	$f(x) = \dfrac{1}{\sigma \sqrt{2 \cdot \pi}} \exp\left[-\dfrac{(x-\mu)^2}{2 \cdot \sigma^2} \right]$ $E(x) = \mu$ $var(x) = \sigma^2$	The number of passengers in a bus The number of passengers in a plane

Determine events that are equivalent to the following events:

$A + C$, AC, EF, $G + E$, GE, BD, $E + K$

4.2. We consider a specific point on the highway, and we observe the first four vehicles that appear over time. Private cars or heavy vehicles could appear. Let us consider the events:
A – only one private car appeared
B – at least one private car appeared
C – not less than two private cars appeared
D – exactly two private cars appeared
E – exactly three private cars appeared
F – all vehicles that appeared were private cars
What is the meaning of the following events?

$A + B$, AB, $B + C$, BC, $D + E + F$, BF

4.3. If $A = B$, prove that the following relation is satisfied:

$\bar{A} = \bar{B}$

4.4. Simplify the following expression:

$A = (B + C) \cdot (B + \bar{C}) \cdot (\bar{B} + C)$

Table 4.4 Other discrete and continuous distributions

Name of the distribution	Probability density function, mean and variance
Bernoulli	$X = \begin{Bmatrix} 0 & 1 \\ q & p \end{Bmatrix}$ $P(X = x) = p^x \cdot q^{1-x}\, x = 0,1$ $E(X) = p$ $\text{var}(X) = p \cdot q$
Binomial	$P(X = x) = \dfrac{n!}{x!(n-x)!} \cdot p^x \cdot q^{n-x}$ $E(X) = n \cdot p$ $\text{var}(X) = n \cdot p(1-p)$
Geometric	$P(X = x) = q^{x-1} \cdot p$ $x = 1,2,3,\ldots$ $E(X) = \dfrac{1}{p}$ $\text{var}(X) = \dfrac{q}{p^2} = \dfrac{1-p}{p^2}$
Negative binomial	$P(Y = y) = \begin{pmatrix} -k \\ y \end{pmatrix} \cdot p^k \cdot (-q)^y$ $y = 0,1,2,\ldots$ $E(Y) = k \cdot \dfrac{q}{p}$ $\text{var}(Y) = k \cdot \dfrac{q}{p^2}$
Uniform	$f(x) = \dfrac{1}{b-a}$ $a < x < b$ $E(x) = \dfrac{a+b}{2}$ $\text{var}(x) = \dfrac{(b-a)^2}{12}$
Erlang	$f_k = \dfrac{\lambda^k \cdot x^{k-1}}{(k-1)!} \cdot e^{-\lambda \cdot x}$ $x > 0,\ \lambda > 0,\ k$ positive integer $E(x) = \dfrac{k}{\lambda}$ $\text{var}(x) = \dfrac{k}{\lambda^2}$
Gamma	$f(x) = \dfrac{\lambda(\lambda x)^{\alpha-1} \cdot e^{-\lambda x}}{\Gamma(\alpha)}$ $x > 0$ and $\alpha > 0, \lambda > 0$ $E(x) = \dfrac{\alpha}{\lambda}$ $\text{var}(x) = \dfrac{\alpha}{\lambda^2}$

4.5. Simplify the following expression:

$$(A + B) \cdot (A + \bar{B})$$

4.6. Simplify the following expression:

$$(A + B) \cdot (B + C) \cdot (C + A)$$

4.7. There are flight activities that do not involve commercial air transportation of passengers, and cargo (agriculture, photography, surveying, patrol, search and rescue, company own-use flight operations, private travel, etc.). The International Civil Aviation Organization (ICAO) calls such activities General Aviation (GA). A total of n aircraft landed at the airport. We denote by A_i $(i = 1,2,...,n)$ the event that the i-th landed aircraft belongs to class of general aviation. By using the events A_i and the events $\overline{A_i}$ describe the following events:

(a)   no one aircraft belongs to the general aviation
(b)   at least one aircraft belongs to the general aviation
(c)   only one aircraft belongs to the general aviation
(d)   not more than one aircraft that belongs to the general aviation

4.8. Let us consider the following events:

A– first bus from depot A departed on time
B– first bus from depot B departed on time
C– first bus from depot C departed on time

Describe the following events:

(a)   only the bus from depot A departed on time
(b)   buses from depots A and B departed on time, but a bus from depot C did not
(c)   all three buses departed on time
(d)   at least one of the buses departed on time
(e)   at least two buses departed on time
(f)   only one bus departed on time
(g)   two buses departed on time
(h)   no one bus departed on time
(i)   no more than two buses departed on time

4.9. Events A and B are not mutually exclusive events. The addition law reads:

$$P(A+B) = P(A) + P(B) - P(AB)$$

Prove the addition law.

4.10. A dice was rolled three times. Determine the probability of rolling at least one 3.

4.11. Any person could be the owner, or non-owner of a private car. Two persons are randomly chosen for the interview related to travel behavior. Determine the probability that two owners are chosen.

4.12. Six very experienced drivers (more than 10 years of driving experience), 4 moderately experienced drivers (between 3 and 10 years of driving experience), and 5 inexperienced drivers (less than 3 years of driving experience) were driving on the road section. A traffic accident happened. Determine the following probabilities:

(a)   the traffic accident was caused by a very experienced driver
(b)   the traffic accident was caused by a moderately experienced driver
(c)   the traffic accident was caused by an inexperienced driver

(d)   the traffic accident was not caused by a very experienced driver

(e)   the traffic accident was caused by a very experienced or moderately experienced driver

4.13. By sharing *rides*, drivers offer vacant seats in their vehicles to travelers who would like to travel in similar directions. In the observed city zone, there are a man and c women who would like to participate in ride-sharing. In the nearby city zone, there are c man and d women who would like to participate in ride-sharing. Two persons are randomly considered for possible ride-sharing. The first person is from the first zone, and the second person is from the second zone. Determine the following probabilities:

(a)   Both persons are men

(b)   The two persons are of different sexes

4.14. The randomly chosen license plate has a city designation and 6 digits. In every digit position on the license plate could be anyone of the digits: 0,1,2,3,4,5,6,7,8,9. Determine the probability that all digits on the license plate are different.

4.15. Delivery of goods is scheduled between 4 p.m. and 5 p.m. in the parking lot on the highway. The driver that delivers and the driver that receives the goods prefer to wait for one another for not more than 20 minutes. Both drivers can arrive at the parking lot at any moment between 4 p.m. and 5 p.m. Determine the probability that no one driver will wait more than 20 minutes.

4.16. The events A, B, and C are independent. The probabilities of occurrence of these events are:

$$P(A) = \frac{1}{2}$$

$$P(B) = \frac{1}{3}$$

$$P(C) = \frac{2}{3}$$

Determine the probabilities of occurrence of the following events:

$$AB, \quad A+B, \quad C-(A+B), \quad A-BC$$

4.17. The maintenance department reported that, at the regular service, once per year, the following vehicle spare parts needed to be replaced: spare part S_1 in 36% of cases; spare part S_2 in 42% of cases; simultaneously, spare part S_1 and spare part S_2 in 30% of cases. Based on these data, test whether replacing one part causes the replacement of another part.

4.18. We assume that there are 50% men and 50% women in the population. Some evidence suggests that 5% of the men are color blind and
that 0.25% of the women are color blind.
   (a) Determine the probability that a person selected in a random
        manner is color blind.
   (b) Determine the probability that a selected color-blind person is
        male.
4.19. There are three urns. The first contains a white and b black balls. The
second one contains c white and d black balls. The third urn contains only white balls. One ball is drawn randomly from a randomly
selected urn. Determine the probability that the drawn ball is white.
4.20. There are n passengers on the bus. The vehicle is approaching the bus
stop. The typical boarding/alighting dynamics is as follows: Every
passenger leaves the vehicle at the next stop with the probability equal
to p. At the next stop, a maximum of one passenger enters the vehicle. The probability that no one will enter the vehicle is equal to p_0.
Consequently, the probability that one passenger will enter the vehicle
is equal to $1 - p_0$. Determine the probability that the total number on
board passengers equals n, after departing the bus stop.
4.21. Does the function $f(n) = \dfrac{1}{2^n}$ $n = 1, 2, 3, \ldots$ define the probability
density function?
4.22. When driving along the corridor, the driver has to pass through three
intersections. The probability that the driver will be stopped at the
intersection is equal to 0.5. Let us denote by X the number of times
the driver was stopped at the intersections. Determine the associated
probabilities of the discrete random variable X, and the cumulative
density function.
4.23. The probability density function reads:

$$f(x) = \begin{cases} 0 & \text{for } x < a \text{ and } x > b \\ \dfrac{2x}{ab} & \text{for } 0 \le x < a \\ \dfrac{2x}{ab - b^2} + \dfrac{2}{b - a} & \text{for } a \le x \le b \end{cases}$$

Determine the cumulative density function.
4.24. The probability density function has a triangular shape (Figure 4.28).
   (a) Determine the probability density function
   (b) Determine the cumulative density function

Determine the probability $P(\dfrac{a}{2} < x < a)$.

4.25. Approximately 2/3 of a city's population uses public transport.
Private cars are used on a daily basis by 1/3 of a city's population.

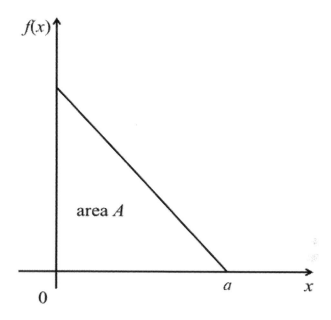

Figure 4.28 Triangular shape of a probability density function

Five citizens are randomly drawn from the population for the survey. The random variable that represents the number of public transport users in the sample of five randomly selected citizens is denoted by X. Determine the mean E(X) and the variance V(X) of the random variable X.

4.26. Vehicles that have various problems arrive at a car repair shop. The vehicle service time is a random variable T. The cumulative density function of this variable equals:

$$F(t) = \begin{cases} 0 & \text{for } t < 0 \\ 1 - e^{-kt} & \text{for } t > 0, \quad (k > 0) \end{cases}$$

Determine the mean E(X) and the variance V(X) of the random variable X.

4.27. There is a traffic light at the entrance to a tunnel in a mountain. A car is approaching the entrance to the tunnel. The duration of the green light equals 1 minute. The green light is followed by a red light that has a duration of ½ minute (30 s). The red light is followed by a green light, etc.

(a) Determine the probability that a car will not be stopped at the entrance to the tunnel.

(b) Determine the mean E(X) and the variance V(X) of the random variable that represents car waiting time at the entrance to the tunnel.

4.28. There is a traffic light at the entrance to the tunnel in a mountain. Ten cars enter the tunnel during the observed period of time. The duration of a green light equals 1 minute. The green light is followed by a red light, that also has a duration of 1 minute. The red light is followed by a green light, etc.

  (a) Determine the probability that eight cars were stopped at the entrance of the tunnel.

4.29. The vehicle interarrival time (T) has an exponential distribution. The average interarrival time is 4 minutes. Determine the probability: $P(T < 2 \text{ min})$.

4.30. It has been shown that the number of drivers under the influence of alcohol in one city area could be described by the Poisson process. It is also known that the number of drivers under the influence on Friday night is approximately 5%. Sixty drivers were stopped by the police officers. Determine the probability that none of the stopped drivers is driving under the influence.

BIBLIOGRAPHY

Ash, C., *The Probability Tutoring Book*, IEEE Press, New York, 1993.

Ayyub, B., *Uncertainty Modeling and Analysis in Civil Engineering*, CRC Press, Boca Raton, FL, 1998.

Brockwell, P.J. and Davis, R.A., *Introduction to Time Series and Forecasting*, Springer Verlag, New York, 1996.

Feller, W., *An introduction to probability theory and its applications*, Vol. II. John Wiley & Sons Inc., New York, 1971.

Haldar, A. and Mahadevan, S., *Probability, Reliability, and Statistical Methods in Engineering Design*, John Wiley & Sons, New York, 2000.

Hillier, F. and Lieberman, G., *Introduction to Operations Research*, McGraw–Hill Publishing Company, New York, 1990.

Kennedy, J. B. and Neville, A.N., *Basic Statistical Methods for Engineers and Scientists*, 3rd ed., Harper & Row, New York, 1986.

Kottegoda, N. and Rosso, R., *Probability, Statistics, and Reliability for Civil and Environmental Engineers*, McGraw-Hill, New York, 1997.

Larson, R. and Odoni, A., *Urban Operations Research*, Prentice Hall, Englewood Cliffs, NJ, 1981.

Newell, G. F., *Applications of Queueing Theory*, 2nd ed., Chapman and Hall, 1982.

Ross, S., *Introduction to Probability and Statistics for Engineers and Scientists*, 2nd ed., Academic Press, 2000.

Ross, S., *A First Course in Probability*, 8th ed., Pearson Prentice Hall, 2010.

Teodorović, D. and Janić, M., *Transportation Engineering: Theory, Practice and Modeling*, Elsevier, New York, 2016.

Taha, H., *Operations Research*, MacMillan Publishing Co., Inc., New York, 1982.

Vukadinović, S., *Introduction to Probability Theory*, Privredni pregled, Belgrade, 1980 (in Serbian).

Walpole, R. E. and Myers, R.H., *Probability and Statistics for Engineers and Scientists*, 7th ed., Prentice Hall, Upper Saddle River, NJ, 2002.

Wesley, B.J., *Statistical Analysis for Engineers and Scientists, A Computer Based Approach*, McGraw-Hill, New York, 1994.

Winston, W., *Operations Research*, Duxbury Press, Belmont, CA, 1994.

Ziemer, R. E., *Elements of Engineering Probability & Statistics*, Prentice Hall, Upper Saddle River, NJ, 1997.

Chapter 5

Statistics

Traffic flows fluctuate over space and over time. Measurements of various parameters help us to better understand traffic flow movement in space and time. By using traffic flow measurements, we can predict queue lengths, estimate level-of-service, and execute actions that could mitigate traffic congestion. Traffic flow measurements are also performed in order to estimate the capacities of transportation facilities needed.

In many cities, it is necessary to perform, from time to time, a mixture of measurements and counting in order to estimate current signal plans at signalized intersections (average queue length, average waiting time of the vehicles in the queue, etc.).

Planners and traffic engineers often organize surveys (by mail, by phone, on web sites, etc.). Surveys represent questionnaires for individual drivers, passengers, etc. These individual drivers or passengers have been chosen from a studied population. For example, surveys are frequently organized in order to determine destinations, transportation mode, chosen paths, and the daily number of trips carried out by members of the households in a specific city zone.

In traffic engineering, we use, in a majority of cases, numerical data. Numerical data represent measurements or counts. Frequently, we are interested in finding the typical value or the central measure of the collected data, as well as to discover how the data are grouped around this center.

We perform statistical analysis in transportation to describe and summarize the collected data. We use statistics to explain the collected data with numbers, pie charts, bar graphs, and histograms. In some cases, data are available. In some other cases, analysts generate data by designing and performing statistical experiments. Statistics help us to analyze data, to make conclusions, to find answers to engineering questions, and to make appropriate decisions.

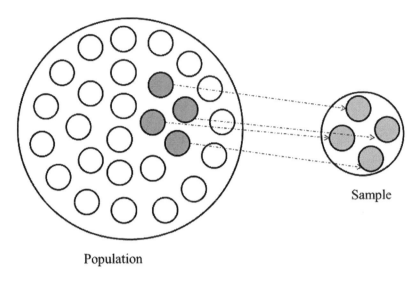

Population

Figure 5.1 Population and sample

5.1 POPULATION AND SAMPLE

Suppose that we stop the drivers at the city exits and ask them for their final trip destination. It would be extremely expensive and timely if we were to try to interview all drivers. In this case, as in many other cases, the studied "population" is too large for analysts to interview all of its members. The usual procedure in the statistics is to create a sample that could represent the entire population (Figure 5.1). In our case, instead of interviewing all drivers, we will interview only a certain number of drivers. These selected drivers form the "sample".

Based on the responses of the chosen drivers, we will derive conclusions about the destinations of all drivers. Usually, based on sample examination, we derive various conclusions about the population. Consequently, the sample must properly represent the population. Most frequently, analysts choose sample members in a completely random way.

5.2 DESCRIPTIVE STATISTICS: WAYS TO DESCRIBE A DATA SET

Descriptive statistics is the statistical field that summarizes and describes collected data. There are different techniques to describe an analyzed data set. We describe a data set by pictures (pie charts, bar graphs, histograms), tables, or by numbers. The numbers that we use to describe a data set are called descriptive statistics. These numbers represent some of the basic characteristics of the analyzed set of data. The most important descriptive statistics measure the center and variability in a data set.

5.2.1 Describing a data set by pictures

Pie charts

The majority of data used in transportation engineering are numerical. On the other hand, we also use various categorical data in transportation analysis. Individuals are put into groups of drivers, or groups of non-drivers. They could be put into the category of private car drivers, or the category of public transport riders. Passengers in intercity transportation could be placed into the category of air passengers, the category of road drivers, the category of intercity train riders, etc.

We use pie charts to show the percentage of individuals that are placed into each of the considered categories. The whole pie represents 100%. Every slice of the pie represents a percentage of the individuals within a specific category.

There are various reasons for urban traffic congestion. A pie chart (Figure 5.2) shows reasons for urban traffic congestion (Source: FHWA, Office of Operations).

From the pie chart, we see that bottlenecks are the primary reason for traffic congestion in 40% of cases, traffic incidents in 25% of cases, bad weather in 15% of cases, etc.

Bar graphs

Bar graphs show analyzed categories side-by-side. Bar graphs contain information about numbers of individuals in each category (or the percentage

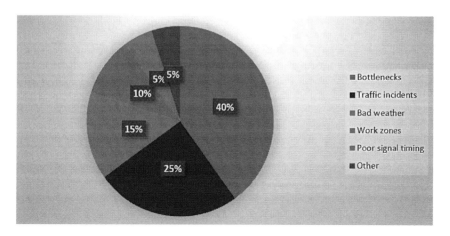

Figure 5.2 Pie chart that shows reasons for urban traffic congestion (FHWA, www.ops .fhwa.dot.gov/congestion_report/executive_summary.htm)

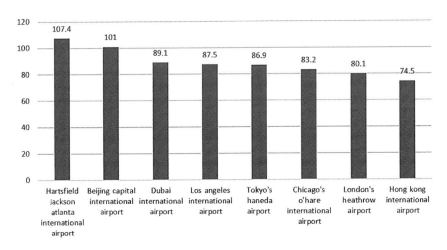

Figure 5.3 Bar graphs that represent international passenger traffic in 2011, 2013, and 2014

that belongs to each category). The bar chart in Figure 5.3 shows the world's busiest airports (in millions of passengers) in 2018.

Time charts

We use time charts in transportation to study and show trends over time. The horizontal axis represents time. Units of time could be minutes, hours, days, weeks, months, or years. On the vertical axis, we measure some quantity (number of cars, number of aircraft, number of passengers, etc.). The value of the observed quantity is represented by a dot at each time period. Tied dots create the time chart.

Figure 5.4 represents a time chart that shows trends in the number of passengers on U.S. airlines from 2003 to 2019 (Source: Bureau of Transportation Statistics).

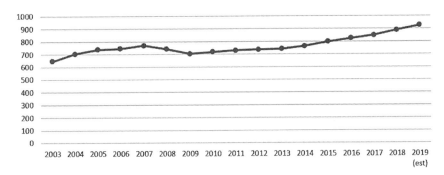

Figure 5.4 Time chart that shows trends in the number of passengers on U.S. airlines from 2003 to 2019. (Source: Bureau of Transportation Statistics)

Scatter diagrams

Frequently, in transportation, we measure a number of variables. For example, when studying traffic flow, we are interested in data related to speed (km/h) and density (veh/km). Scatter diagrams help us to graphically present the possible relationship between the observed variables. We create scatter diagrams in a very simple way. We plot each pair of data points with the first measurement in the pair on the y-axis and the second measurement in the pair on the x-axis. The figure represents a scatter plot of speed versus density. Scatter diagrams are very useful graphical techniques that help an analyst to get the first impressions about the possible relationship between the observed variables and to derive some conclusions. For example, the higher the traffic density (Figure 5.5), the lower the average speed. A scatter plot of speed vs. density also shows that very high traffic density causes traffic jams, where vehicles practically do not move and the speed is close to zero.

Frequency polygons

The speeds of 45 vehicles were measured. The following are values of the measured speeds (in mph):

54, 54, 55, 55, 55, 56, 56, 57, 57, 57, 57, 58, 58, 58, 58, 58, 59, 59, 59, 59, 59, 59, 59, 59, 59, 59, 59, 60, 60, 60, 60, 60, 60, 61, 61, 61, 61, 62, 62, 62, 63, 63, 64, 64

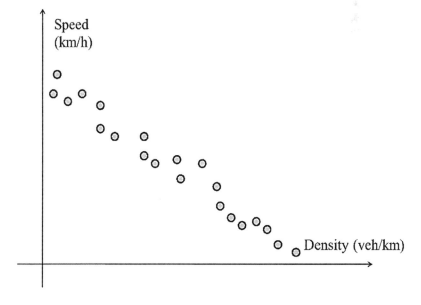

Figure 5.5 Scatterplot of Speed vs. Density

Table 5.1 Frequency table for the data set
that contains 45 measured vehicle
speeds (mph)

Vehicle speed (mph)	Frequency
54	2
55	3
56	2
57	4
58	5
59	12
60	6
61	4
62	3
63	2
64	2

We see that speed equal to 57 mph appeared 4 times, speed of 59 mph appeared 12 times, etc. We can present these speeds and their corresponding number of appearances (frequencies) in a frequency table. Table 5.1 is a frequency table for the data set that contains 45 measured vehicle speeds (in mph).

A frequency polygon is frequently used to graphically present a frequency table. In our case, we have 11 distinct vehicle speeds. Each of these vehicle speed values has a corresponding frequency. There are 11 points with the following coordinates:

(54, 2), (55, 3), (56, 2), (57, 4), (58, 5), (59, 12), (60, 6), (61, 4), (62, 3), (63, 2), (64, 2)

We create a frequency polygon by connecting our 11 points with straight lines (Figure 5.6)

A frequency polygon is mainly used in cases when the total number of observations (measurements) is relatively low. In cases of larger numbers of data ("frequency" in this case), as well as larger numbers of distinct values ("speeds" in this case), analysts use histograms to graphically describe the data set.

Histograms

The one hundred recorded traffic flow values (veh/h) are given in Table 5.2.

It is relatively difficult (especially in cases where the number of distinct values in the data set is large) to get an impression about this data set from the table. Histograms help us to have some basic ideas about the numerical

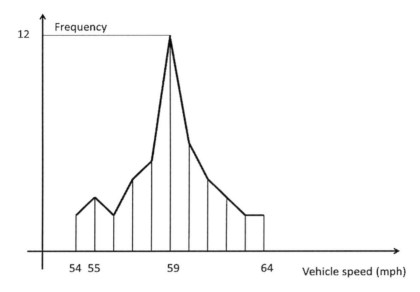

Figure 5.6 Frequency Polygon

Table 5.2 One hundred recorded traffic flow values (veh/h)

189	374	120	8	171
470	165	124	133	227
425	332	308	100	250
122	327	79	50	115
129	374	21	288	200
123	85	331	242	229
90	385	178	158	475
467	425	151	57	373
396	309	204	277	307
430	256	434	462	388
179	206	196	357	26
43	452	448	324	212
489	90	266	131	255
135	172	169	219	114
224	38	53	38	298
286	229	224	249	423
276	275	280	74	181
291	37	247	306	196
432	164	361	209	86
232	207	237	21	231

Table 5.3 Numbers of observations and corresponding numbers of the categories

Number of observations	Number of class intervals
40 to 60	6 to 8
60 to 100	7 to 10
100 to 200	9 to 12
200 to 500	12 to 17
Greater than 500	21

data sets. When creating a histogram, we put individual measurements into numerically ordered groups (class intervals). Practically, a histogram represents a bar graph for numerical data. By inspection of Table 5.2, we see that the highest flow value equals 498 veh/h. The lowest value equals 8 veh/h. The range interval is the interval between the smallest and largest observations. In our case, the range interval is (8, 498). The range of this interval equals $498-8=490$. We put each of 100 traffic flow values into a specific class. There are no strict rules about what the appropriate numbers of the class intervals are. Some empirical rules about the number of class intervals are given in Table 5.3.

Based on the empirical rules, shown in Table 5.3, we decide to have 10 class intervals. The following are our class intervals:

[0, 50), [50, 100), [100, 150), [150, 200), [200, 250), [250, 300), [300, 350), [350, 400), [400, 450), [450, 500)

The class boundaries are the endpoints of a class interval. In this example, we use the left-end inclusion convention. This means that a class interval includes its left-end, but not its right-end boundary point. The data given in Table 5.2 are rearranged and shown in the frequency table (Table 5.4).

Table 5.4 A class frequency table

Class Interval	Number of traffic flow values in the interval (frequency)
0–50	9
50–100	8
100–150	11
150–200	12
200–250	19
250–300	12
300–350	8
350–400	8
400–450	7
450–500	6

For example, the class interval 50–100 includes all flow values that are greater than or equal to 50 and less than 100. The class interval 300–350 contains all traffic flow values that are greater than or equal to 300 and less than 350, etc.

Bar graphs contain information about the number of individuals in each category. Histograms do the same. A histogram is basically a bar graph applied to numerical data. In histograms, class intervals (categories) are arranged in increasing order. Bars in a histogram are indicated on the x-axis by the corresponding values that represent the bar's beginning and the bar's endpoint. The frequency of each class interval is represented by the bar's height. The bar's height could also represent the percentage of individuals in each class interval. In such a case, the histogram shows the relative frequency of each class interval. Histograms help us to get a clearer impression of the data variability, as well as about the center point of the data. Figure 5.7 shows a histogram of traffic flow values, based on data given in Table 5.4.

Relative frequencies

Let us again consider vehicle speed data obtained as a result of field measurements. Table 5.5 is a frequency table for the data set that contains 45 measured vehicle speeds (mph).

The relative frequency of a specific value is defined as the proportion of the data that has that value. In our case, we obtain the relative frequency values by dividing frequency values by 45 (the total number of vehicle speed

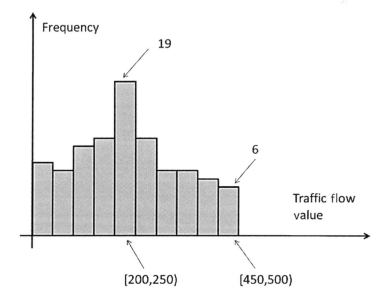

Figure 5.7 Histogram of traffic flow values

Table 5.5 A frequency table for the
data set that contains 45
measured vehicle speeds

Vehicle speed (mph)	Frequency
54	2
55	3
56	2
57	4
58	5
59	12
60	6
61	4
62	3
63	2
64	2

Table 5.6 A relative frequency table for the
data set that contains 45
measured vehicle speeds

Vehicle speed (mph)	Relative frequency
54	2/45 = 0.0444
55	3/45 = 0.0667
56	2/45 = 0.0444
57	4/45 = 0.0889
58	5/45 = 0.1111
59	12/45 = 0.2667
60	6/45 = 0.1333
61	4/45 = 0.0889
62	3/45 = 0.0667
63	2/45 = 0.0444
64	2/45 = 0.0444

measurements). The relative frequencies of the measured vehicle speeds are
shown in Table 5.6.

Cumulative frequency graph

In transportation engineering, we often use cumulative frequency graphs,
or cumulative relative frequency graphs. Possible data values are given
along the (horizontal) x-axis. Each data value has a corresponding vertical
plot. This vertical plot of the specific data value shows the number of the
data, the values of which are less than or equal to the specific data.

Table 5.7 A frequency and cumulative frequency table for the data set that contains 45 measured vehicle speeds

Vehicle speed (mph)	Frequency	Cumulative Frequency
54	2	2
55	3	2+3=5
56	2	2+3+2=7
57	4	2+3+2+4=11
58	5	16
59	12	28
60	6	34
61	4	38
62	3	41
63	2	43
64	2	45

Let us again consider vehicle speeds obtained as a result of field measurements. Table 5.7 is a frequency and cumulative frequency table for the data set that contains 45 measured vehicle speeds (mph).

The cumulative frequency graph for the data set that contains 45 measured vehicle speeds (mph) (Table 5.7) is given in Figure 5.8.

Table 5.8 is a relative frequency and a cumulative relative frequency table for the data set that contains 45 measured vehicle speeds (mph).

The cumulative relative frequency graph for the data set that contains 45 measured vehicle speeds (mph) (Table 5.8) is given in Figure 5.9.

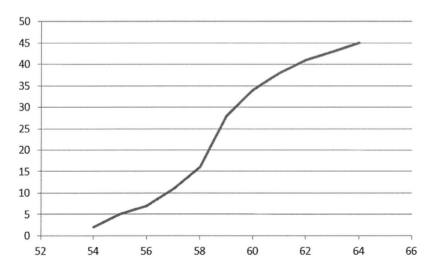

Figure 5.8 Cumulative frequency graph for the data set that contains 45 measured vehicle speeds (mph) (Table 5.7)

Table 5.8 A relative frequency and cumulative relative frequency table for the data set that contains 45 measured vehicle speeds

Vehicle speed (mph)	Relative frequency	Cumulative relative frequency
54	2/45	$2/45 = 0.0444$
55	3/45	$2/45 + 3/45 = 5/45 = 0.1111$
56	2/45	$2/45 + 3/45 + 2/45 = 7/45 = 0.1555$
57	4/45	$7/45 + 4/45 = 11/45 = 0.2444$
58	5/45	$11/45 + 5/45 = 16/45 = 0.3555$
59	12/45	$16/45 + 12/45 = 28/45 = 0.6222$
60	6/45	$28/45 + 6/45 = 34/45 = 0.7555$
61	4/45	$34/45 + 4/45 = 38/45 = 0.8444$
62	3/45	$38/45 + 3/45 = 41/45 = 0.9111$
63	2/45	$41/45 + 2/45 = 43/45 = 0.9555$
64	2/45	$43/45 + 2/45 = 45/45 = 1$

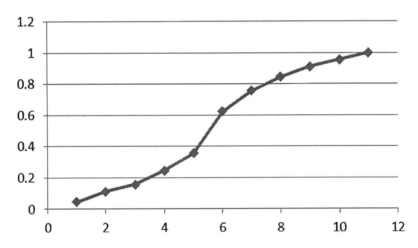

Figure 5.9 The cumulative relative frequency graph for the data set that contains 45 measured vehicle speeds (mph) (Table 5.8)

Cumulative relative frequencies help us to derive various conclusions. For example, from Table 5.8, we can conclude that approximately 15% of vehicles had a speed less than or equal to 56 mph, and that approximately 75% of vehicles had a speed less than or equal to 60 mph, etc.

5.2.2 Describing a data set by numbers

Mean

Frequently, analysts need to know where the data center is. Mean value, or middle of the data are synonyms for the average of the data. In statistics we use mean, median, and mode to determine the data center.

The sample mean $\bar{x}$ of a data set $x_1, x_2, x_3, \ldots, x_n$ is the arithmetic average of these numbers, i.e.:

$$\bar{x} = \frac{x_1 + x_2 + \cdots + x_n}{n} = \frac{1}{n} \sum_{i=1}^{n} x_i$$

The sample mean is easily obtained by adding up all numbers and dividing the obtained sum by the total number of numbers.

Example 5.1

The sample is composed of 10 observations of the number of passengers that wait for the bus at the bus station. These 10 observations are: 10, 12, 9, 8, 11, 9, 9, 10, 7, 11. The sample mean is:

$$\bar{x} = \frac{10 + 12 + 9 + 8 + 11 + 9 + 9 + 10 + 7 + 11}{10} = 9.6$$

We calculate the mean for the data set that is given in the frequency table in the following way. Let us assume, for example, that we measured traffic density n times. Let us suppose that there were k distinct values $x_1, x_2, x_3, \ldots, x_k$ of the traffic density and their corresponding frequencies $f_1, f_2, f_3, \ldots, f_k$. Since value $x_i \left(i = 1, 2, \ldots, n \right)$ appears f_i times, the sample mean of the n measurements equals:

$$\bar{x} = \frac{1}{n} \sum_{i=1}^{k} x_i f_i$$

where:

$$\sum_{i=1}^{k} f_i = n$$

Example 5.2

Let us again consider obtained vehicle speeds as a result of field measurements. Table 5.9 is a frequency table for the data set that contains 45 measured vehicle speeds (mph).

Our task is to find the mean of the 45 measured vehicle speeds. The average speed equals:

Average speed

$$= \frac{54 \cdot 2 + 55 \cdot 3 + 56 \cdot 2 + 57 \cdot 4 + 58 \cdot 5 + 59 \cdot 12 + 60 \cdot 6 + 61 \cdot 4 + 62 \cdot 3 + 63 \cdot 2 + 64 \cdot 2}{45}$$

$$= 56.6 \text{ mph}$$

Table 5.9 A frequency table for the
data set that contains 45
measured vehicle speeds

Vehicle speed (mph)	Frequency
54	2
55	3
56	2
57	4
58	5
59	12
60	6
61	4
62	3
63	2
64	2

Example 5.3

Suppose that we record speeds of all the vehicles that we observe and count at a specific point on the highway (Figure 5.10).

We can determine speeds with a certain level of precision by using different equipment (radar, microwave, inductive loops). There are two ways to express speeds on a highway: (a) the time-mean speed; (b) the space-mean speed.

The time-mean speed $\overline{u_t}$ is defined in the following way:

$$\overline{u_t} = \frac{1}{n} \sum_{i=1}^{n} u_i$$

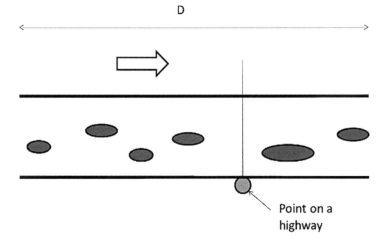

Figure 5.10 Computation of the time-mean speed and space-mean speed

where

    n – number of recorded speeds

    u_i – speed of the i-th recorded vehicle

The time-mean speed represents the average speed of all n vehicles passing a point on a highway.

The space-mean speed is the average speed that has been used in the majority of traffic models. Let us notice the section of the highway whose length equals D (Figure 5.10). We denote by t_i the amount of time needed by the i-th vehicle to travel along this highway section. The space-mean speed $\overline{u}_s$ is defined in the following way:

$$\overline{u}_s = \frac{D}{\frac{1}{n}\sum_{i=1}^{n} t_i} = \frac{D}{\overline{t}}$$

The expression $\frac{1}{N}\sum_{i=1}^{N} t_i$ represents the average travel time $\overline{t}$ of the observed n vehicles that travel along the studied highway section.

Median

Median is a more appropriate measure of the data center than the mean in cases when there are outliers in a data set. Let us clarify the concept of outliers. Outliers are some extremely high values or some extremely small values that could appear in an analyzed set of the data. When calculating the mean, these values can significantly affect the final result. In other words, because of the several unusually big or unusually small values, the mean may have a value that does not reflect the realistically considered set of data. In such situations, the median is recommended as a measure that better illustrates the data center.

Let us assume that we recorded the following speeds (mph) of 9 vehicles:

60, 65, 58, 59, 112, 55, 59, 63, 62

The average speed (mean) equals:

$$\text{Mean} = \frac{60 + 65 + 58 + 59 + 112 + 55 + 59 + 63 + 62}{9} = 65.88\,\text{mph}$$

As we see, eight of the nine recorded vehicles had a speed lower than the calculated mean speed. It is clear that the calculated mean speed does not accurately reflect the speeds of the recorded vehicles.

Let us rearrange the recorded speeds in increasing order, from smallest to largest:

55, 58, 59, 59, **60**, 62, 63, 65, 112

A speed that is equal to 60 mph is located exactly in the middle of the data that are arranged in increasing order. When data are arranged in increasing order, the median is the place that divides the data into half. As we see, in our case, four vehicles had a speed less than 60 mph and four vehicle speed exceeded 60 mph. In the case when the number of observations n is odd, the sample median is the single central value. In the opposite case, when the number of observations n is even, the sample median represents the arithmetic mean of the two central values.For example, imagine that we recorded the speed of one more vehicle. Let this speed be equal to 61 mph. The speeds in increasing order, from smallest to largest, read:

55, 58, 59, 59, 60, 61, 62, 63, 65,112

Two values in the middle are respectively 60 and 61. The median M (arithmetic mean of these values) is equal to $M = \dfrac{60+61}{2} = 60.5$ mph.

Taking into consideration previous examples, let us show how the median could be determined by Excel. Figure 5.11 shows the set of values from which the median should be determined.

The value of the median can be determined by the function: "=median(A1:A9)". For that function, Excel returns a value that is equal to 60.

	A	B	C
1	60		
2	65		
3	58		
4	59		
5	112		
6	55		
7	59		
8	63		
9	62		
10			
11			

Figure 5.11 Data in Excel sheet

Mode

Let us consider n observations. The mode represents the value with the highest frequency. We denote mode by Mo. Clearly, mode does not exist in the specific case, when all observations take place with the same frequency. The mode is not unique when two or more observations happen with the same frequency.

Measures of variability

Range

In every sample, there are deviations of the individual values from the central value. The sample range is one of the basic variability measures. The sample range r represents the difference between the biggest and smallest observations, i.e.:

$$r = \max(x_i) - \min(x_i)$$

Example

The sample is composed of 10 observations of the number of passengers that wait for the bus at the bus station. These 10 observations are: 10, 12, 9, 8, 11, 9, 9, 10, 7, 11. The numbers of passengers in increasing order, from smallest to largest, read:

7, 8, 9, 9, 9, 10, 10, 11, 11, 12

The sample range r equals:

$$r = \max(x_i) - \min(x_i) = 12 - 7 = 5$$

Standard Deviation

Mean and median help us to determine where the center of the data is. On the other hand, two sets of data can have, for example, the same mean, but, at the same time, can have very different deviations of the individual values from the center. For example, a data set of 15, 20, 25 and a set 3, 20, 37 have the same values of an arithmetic mean $\bar{x} = \dfrac{15 + 20 + 25}{3} = \dfrac{3 + 20 + 37}{3} = 20.$

The deviations of the values from the arithmetic mean in these two sets are respectively (−5, 0, 5) and (−17, 0, 17). Data variability can be measured in different ways. Sample standard deviation is the method most commonly used to measure the variability. The sample standard deviation s is defined in the following way:

$$s = \sqrt{\frac{\sum_{i=1}^{n}(x_i - \bar{x})^2}{n-1}}$$

where:

$\bar{x}$ – arithmetic mean

n – number of elements in the sample

The sample standard deviation shows approximately the average distance from any point in the data set to the center. It has been shown that the sample variance, s^2, for the data set that is given in the frequency table, equals:

$$s^2 = \frac{\sum_{i=1}^{n}x_i^2 - \frac{\left(\sum_{i=1}^{n}x_i\right)^2}{n}}{n-1}$$

Example 5.4

The sample is composed of 10 recorded numbers of passengers on a flight that departs every Monday from city A to city B at 8:30 a.m. The numbers of passengers are:

280, 300, 260, 290, 250, 275, 292, 268, 254, 292

The average (mean) number of passengers $\bar{x}$ on the flight equals:

$$\bar{x} = \frac{280 + 300 + 260 + 290 + 250 + 274 + 292 + 268 + 254 + 292}{10} = 276$$

The calculation of terms for the standard deviation is shown in Table 5.10.

The sample standard deviation is:

$$s = \sqrt{\frac{\sum_{i=1}^{n}(x_i - \bar{x})^2}{n-1}} = \sqrt{\frac{2,784}{10-1}} = \sqrt{\frac{2,784}{9}} = 17.59 \approx 18 \text{ passengers}$$

Let us demonstrate how the sample mean and the sample standard deviation can be calculated in Excel. Let us note the cells where the values of passenger numbers in the aircrafts are written (from A2 to A11) in Figure 5.12. Mean, or average, number of passengers, can be calculated by the Excel function: "=average(A2:A11)". The sample standard deviation is determined by the Excel function: "=stdev.s(A2:A11)".

Table 5.10 Calculation of terms for standard deviation

Aircraft	Number of passengers in aircraft x_i	$x_i - \bar{x}$	$(x_i - \bar{x})^2$
1	280	4	16
2	300	24	576
3	260	−16	256
4	290	14	196
5	250	−26	676
6	274	−2	4
7	292	16	256
8	268	−8	64
9	254	−22	484
10	292	16	256
Total 2,760 0 2,784			

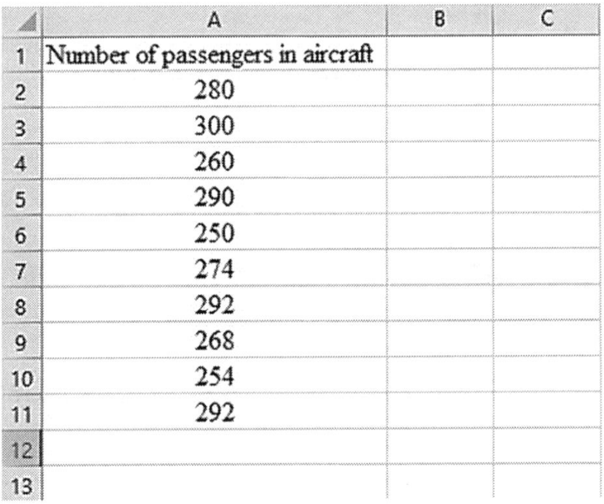

▲	A	B	C
1	Number of passengers in aircraft		
2	280		
3	300		
4	260		
5	290		
6	250		
7	274		
8	292		
9	268		
10	254		
11	292		
12			
13			

Figure 5.12 Input data in Excel

Percentiles and boxplots

The recorded travel times between point A and point B (in minutes) of 100 drivers are shown in Table 5.11.

A percentile is defined as the percentage of individuals in the data set that are below or equal where a specific number is located. In other words, the percentile p is that data value such that p percent of the data are less than or equal to it and $1-p$ percent are greater than it.

Let us note, for example, a specific driver who traveled for 95 minutes. The drivers who traveled 90, 90, 90, 91, 91, 91, 92, 92, 92, 94, and 94 minutes were driving faster and spent less time on the road. We see that there

Table 5.11 The recorded travel times between point A and point B (in minutes) of 100 drivers

108	118	115	113	107
94	108	111	102	119
101	98	97	95	112
102	101	99	99	92
118	100	114	98	119
99	120	115	120	103
109	102	120	95	100
111	120	91	97	91
110	105	90	112	97
120	102	107	113	116
94	95	112	97	103
91	116	113	118	93
106	101	92	100	117
119	102	107	98	109
111	102	117	104	108
116	108	90	116	118
96	97	106	110	109
119	92	98	116	115
112	117	96	109	111
90	115	98	104	104

are 11 drivers that were faster than our driver (11% of the total number of drivers). This means that 89% of drivers scored lower than or equal to the observed driver. In other words, the driving score of the observed driver is at the 89th percentile.

The median represents the 50th percentile, since the median is the point in the data where 50% of the data is lower than that point, and 50% of the data is higher than that point. Frequently, the following five descriptive statistics are used to describe the data set: (1) the smallest number in the data set; (2) the 25th percentile (first quartile); (3) the median (50th percentile, or second quartile); (4) the 75th percentile (third quartile); and (5) the largest number in the data set.

A boxplot represents a one-dimensional graph of numerical data based on the five descriptive statistics. Let us assume that the following are the five descriptive statistics for the analyzed data set:

The smallest number in the data set	10
The 25th percentile (first quartile)	25
The median (50th percentile, or second quartile)	40
The 75th percentile (third quartile)	55
The largest number in the data set	70

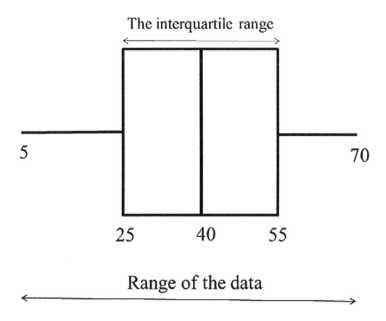

Figure 5.13 Boxplot of the analyzed data set

The corresponding boxplot is shown in Figure 5.13.

The length of the line segment on the box plot represents the difference between the largest and the smallest data value. This difference is the range of the data. The length of the box itself represents the difference between the third quartile and the first quartile. This difference is called the interquartile range.

5.3 INFERENTIAL STATISTICS

In traffic engineering, we carry out various measurements and collect various data. Based on the study and analysis of the available data, we often generate different assumptions, or hypotheses. For example, after studying the data, it is possible to set the hypothesis that the level of traffic congestion is the same in the two observed cities. It is also, for example, possible to assume that drivers younger than 23 are more likely to make traffic violations than the other drivers. These and similar hypotheses need to be tested. In other words, it is often necessary to answer with "yes" or "no" on the generated hypothesis. Hypothesis tests, confidence intervals, and regression analysis are the major techniques in inferential statistics.

5.3.1 Sampling distributions

It has already been pointed out that, based on sample characteristics, we could derive various conclusions about the population. Accordingly, the sample must appropriately represent the population. Let us assume that we want to determine the average parking duration for vehicles inside a specific parking lot. We use X to denote the random variable that represents the parking duration for a vehicle inside the parking lot. If we record parking duration for every vehicle inside the parking lot, we would know the distribution of X, as well as the population mean by μ and the population standard deviation σ. This approach could be costly and time consuming.

Instead of recording parking duration for the whole population, let us generate a few different samples. Let every generated sample contain information about parking duration in the case of three randomly chosen cars. Based on the first sample (Figure 5.14(a)), we conclude that the average parking duration for a vehicle inside parking lot equals: $\dfrac{2+4+6}{3} = 4$ hours.

Based on the second sample (Figure 5.14(b)), we conclude that the average parking duration equals: $\dfrac{1+3+5}{3} = 3$ hours, etc. As we can see, the results (in our case, the mean parking duration), vary from sample to sample.

In this way, we generated a completely new population – the population of sample means. In other words, we created a new random variable, $\overline{X}$. This random variable also has its own distribution, its own mean and its own standard deviation. The distribution of this variable is called a sampling distribution, since the data represent averages derived from generated samples. The total number of elements in the population is denoted by N, and the total number of elements in the sample is denoted by n. The total number of different samples equals $\binom{N}{n}$.

In our case, the samples are:

S_1 : $2, 4, 6$
S_2 : $1, 3, 5$

Sampling distribution of the sample means

We also denote, by μ and σ, the mean and the standard deviation of the observed characteristics of the population. In the general case, the samples $S_1, S_2, S_3, \ldots$, composed of n elements, are:

S_1 : $x_1^1, x_2^1, x_3^1, \ldots, x_n^1$

S_2 : $x_1^2, x_2^2, x_3^2, \ldots, x_n^2$

S_3 : $x_1^3, x_2^3, x_3^3, \ldots, x_n^3$

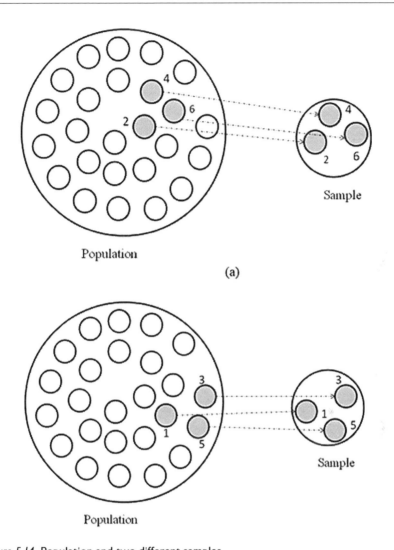

Figure 5.14 Population and two different samples

....

....

....

X_1 is the random variable that represents the parking duration time of the first-noticed car in the sample. The values of this random variable equal $x_1^1, x_1^2, x_1^3, \ldots$ X_2 is the random variable that represents the parking duration time of the second-noticed car in the sample. The values of this random variable equal $x_2^1, x_2^2, x_2^3, \ldots$, etc. The random variables $X_1, X_2, X_3, \ldots$ have the same distribution. The expected values $E(X_i)$ and the variances $\text{var}(X_i)$ of these random variables are, respectively, equal to:

$$E(X_i) = \mu_x \qquad i = 1, 2, \ldots, n$$

$$\text{var}(X_i) = \sigma_x^2 \qquad i = 1, 2, \ldots, n$$

The expected value $E(\bar{X})$ and the variances $\text{var}(\bar{X})$ of the arithmetic mean $\bar{X}$ are, respectively, equal to:

$$E(\bar{X}) = \mu_x = E\left(\frac{X_1 + X_2 + \cdots + X_n}{n}\right) = \frac{1}{n}\sum_{i=1}^{n}E(X_i) = \frac{n \cdot \mu_x}{n} = \mu_x$$

$$\text{var}(\bar{X}) = \sigma_{\bar{x}}^2 = \text{var}\left(\frac{X_1 + X_2 + \cdots + X_n}{n}\right) = \frac{1}{n^2}\sum_{i=1}^{n}\text{var}(X_i) = \frac{n \cdot \sigma_x^2}{n^2} = \frac{\sigma_x^2}{n}$$

We conclude that the mean of all possible sample means is equal to the mean of the whole population. The standard deviation $\sigma_{\bar{x}}$ of the arithmetic mean $\bar{X}$ is called the standard error and is equal to:

$$\sigma_{\bar{x}} = \frac{\sigma_x}{\sqrt{n}}$$

The standard deviation $\sigma_{\bar{x}}$ of the arithmetic mean $\bar{X}$ (standard error) is significantly less than the standard deviation of the basic population. The standard error becomes smaller and smaller as the sample size n increases. In other words, the larger the sample size, the more accurate are our sample estimates, and the less dissimilarity there is between the samples.

The distribution of the random variable $\bar{X}$ (distribution of the sample means) is narrower than the distribution of the random variable X (Figure 5.15). Figure 5.15 shows a normal distribution of the random variable X and a corresponding normal distribution of the random variable $\bar{X}$ (distribution of sample means).

It has been shown that the distribution of the sample means can be approximated to by a normal distribution when the number of elements in the sample, n, approaches infinity. The distribution of the sample means, in cases of higher n values ($n \geq 30$), could be treated as a normal distribution $N\left(\mu_x, \frac{\sigma_x}{\sqrt{n}}\right)$. This approximation is valid also in the cases where the distribution of X is non-normal, or unknown. These results are known as the Central Limit Theorem.

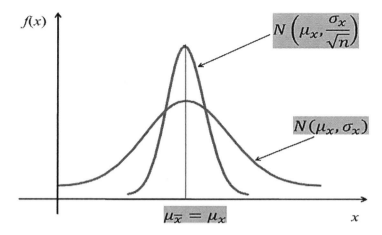

Figure 5.15 The distribution of the sample means and the distribution of the random variable *X*

Sampling distribution of the sample proportion

Let us assume that we are located at a specific point on the highway. We are interested in determining the population proportion p of heavy vehicles in traffic flow. Let us observe n vehicles in the traffic flow that appeared during a specific time period. These observed n vehicles represent one possible sample of the traffic flow. Assume that we noticed x heavy vehicles among these n vehicles. The sample proportion $\bar{p}$ of heavy vehicles in traffic flow equals:

$$\bar{p} = \frac{x}{n}$$

Consequently, the sample proportion $\bar{q}$ of passenger cars and motorcycles equals:

$$\bar{q} = \frac{n - x}{n}$$

Let us assign the random variable X_i to the *i*-th observed vehicle, defined in the following way:

$$X_i = \begin{Bmatrix} 0 & 1 \\ q & p \end{Bmatrix}$$

$$p + q = 1$$

The mean $E(X_i)$ and the variance $\text{var}(X_i)$ of the random variable X_i are, respectively, equal to:

$$E(X_i) = 0 \cdot q + 1 \cdot p = p$$

$$\text{var}(X_i) = E(X_i^2) - p^2 = p - p^2 = p \cdot q$$

We denote by X the random variable that represents the total number of heavy vehicles in the sample, i.e.:

$$X = \sum_{i=1}^{n} X_i$$

The mean $E(X)$ and the variance $\text{var}(X)$ of the random variable X are, respectively, equal:

$$E(X) = \sum_{i=1}^{n} E(X_i) = \sum_{i=1}^{n} p = n \cdot p$$

$$\text{var}(X) = \sigma_x^2 = \sum_{i=1}^{n} \text{var}(X_i) = \sum_{i=1}^{n} p \cdot q = n \cdot p \cdot q$$

The mean $E(\bar{p})$ and the variance $\text{var}(\bar{p})$ of the sample proportion $\bar{p}$ of heavy vehicles in traffic flow equals are, respectively, equal to:

$$E(\bar{p}) = E\left(\frac{X}{n}\right) = \frac{n \cdot p}{n} = p$$

$$\sigma_x^2 = \text{var}(\bar{p}) = \text{var}\left(\frac{X}{n}\right) = \frac{n \cdot p \cdot q}{n^2} = \frac{p \cdot q}{n}$$

The standard deviation of the sample proportion $\bar{p}$ equals:

$$\bar{\sigma}_p = \sqrt{\frac{p \cdot (1-p)}{n}}$$

5.3.2 Parameter estimation

We have studied in previous chapters the probability density functions that frequently appear in the transportation arena. These probability density functions are characterized by specific parameters. Thus, for example, the

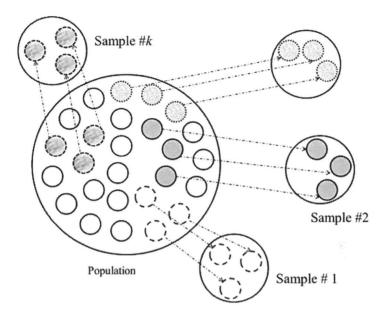

Figure 5.16 Population and various samples

population parameters of the normal distribution are μ and σ, the parameter of the Poisson distribution is λ, etc.

In order to determine the value of the population parameters, analysts generate samples (Figure 5.16). In this way, the analysts use some of the collected data to make inferences about the unknown population parameters.

The area of mathematical statistics that estimates the parameter values, based on sample statistics, is called parameter estimation. There are two possible approaches in parameter estimation. The first approach uses a single value as an estimate of the unknown population parameter. This single value is called a point estimate (Figure 5.17). The second approach specifies an interval in which the unknown population parameter lies. The issue of confidence in such interval estimate is also raised within the second approach. The interval in which the unknown population parameter lies is called the confidence interval (Figure 5.17).

Maximum likelihood method

Let us consider the following example. Four travelers have to choose between car and public transport. The travelers are independent in making their decisions. We introduce simplified assumptions that the probability of choosing a specific transportation mode is constant and does not vary from traveler to traveler. We denote by p the probability of choosing the car.

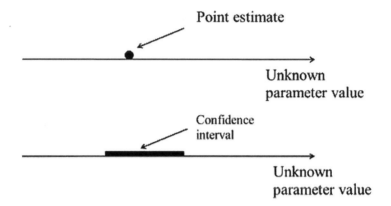

Figure 5.17 Point estimate and confidence interval

We also denote by X the random variable that represents the number of passengers that choose the car among the four passengers. The random variable X can take the following values: 0, 1, 2, 3, and 4. Let us assume that only one traveler chose public transport; it could be the first, second, third, or fourth traveler. The probability of the event that, among four travelers, only one chose public transport, equals:

$$ppp(1-p) + pp(1-p)p + p(1-p)pp + (1-p)ppp = 4p^3(1-p)$$

The probability p of choosing the car is unknown. We estimate this probability by the maximum likelihood method. This method determines parameters that maximize the likelihood that the sample was produced from the model with the chosen parameter values. In other words, the maximum likelihood method finds the values of parameters that are most likely to produce the choices detected in the sample.

The probability $4p^3(1-p)$ represents the likelihood function $L(p)$ of a random sample of size 4, i.e.:

$$L(p) = 4p^3(1-p)$$

We are looking for the p value that maximizes $L(p)$. Therefore:

$$\frac{dL(p)}{dp} = \frac{d\left(4p^3(1-p)\right)}{dp}$$

$$\frac{dL(p)}{dp} = 12p^2 - 16p^3$$

Equating $\dfrac{dL(p)}{dp}$ to zero and solving for p yields:

$$p = \bar{p} = \frac{3}{4}$$

where:

$\bar{p}$ – the sample proportion of car drivers

It looks intuitively quite rational to use $\bar{p}$ as an estimate of the probability p. The maximum likelihood method is used to estimate parameters of the discrete random variables, as well as for the estimation of the parameters of continuous distributions.

Example 5.5

The random variable X has a normal distribution, i.e. $X \sim N(\mu, \sigma)$. The likelihood function of a random sample of size n is:

$$L = \prod_{i=1}^{n} \frac{1}{\sigma\sqrt{2 \cdot \pi}} e^{-\frac{(x_i - \mu)^2}{2 \cdot \sigma^2}}$$

$$L = \left(\frac{1}{\sqrt{2 \cdot \pi}}\right)^n \left(\frac{1}{\sigma}\right)^n e^{-\frac{1}{2 \cdot \sigma^2}\sum_{i=1}^{n}(x_i - \mu)^2}$$

The $\bar{\mu}$ that maximizes $L(\mu)$ also maximizes $\ln L(\mu)$. The $\ln L(\mu)$ equals:

$$\ln L = -n \cdot \ln\sqrt{2\pi} - n \cdot \ln\sigma - \frac{1}{2 \cdot \sigma^2}\sum_{i=1}^{n}(x_i - \mu)^2$$

We are looking for the μ value that maximizes $L(p)$. Therefore:

$$\frac{d\ln L}{\mu} = \frac{1}{\sigma^2}\sum_{i=1}^{n}(x_i - \mu)^2 = 0$$

Finally, we have:

$$\bar{\mu} = \frac{1}{n}\sum_{i=1}^{n}x_i = \bar{x}$$

We can conclude that the sample mean is the maximum likelihood estimator of μ.

We are also interested to estimate the standard deviation σ. Therefore:

$$\frac{d\ln L}{\sigma} = -\frac{n}{\sigma} + \frac{1}{\sigma^3}\sum_{i=1}^{n}(x_i - \mu)^2 = 0$$

Finally, we have:

$$\bar{\sigma}^2 = \frac{1}{n}\sum_{i=1}^{n}(x_i - \bar{x})^2$$

i.e.:

$$\bar{\sigma} = \sqrt{\frac{1}{n}\sum_{i=1}^{n}(x_i - \bar{x})^2}$$

Similarly, we can conclude that the sample standard deviation is the maximum likelihood estimator of σ.

5.3.3 Confidence intervals

Let us again assume that we are interested to determine the average parking duration, μ, for vehicles inside a specific parking lot. We denote by X to the random variable that represents the parking duration for a vehicle inside the parking lot. If we record parking duration for every vehicle, we would know the distribution of X, as well as the mean, μ, and the standard deviation, σ. This approach could be costly and time consuming.

We can also generate a few different samples (Figure 5.16). Let every sample contain n elements: $x_1, x_2, \dots, x_n$. We could estimate the average parking duration that is based on the random sample of n vehicles. Since various samples generate different average parking durations, we conclude that there is variability in our estimate. We introduce into the analysis the margin of errors (the amount of "plus or minus"). This margin of errors measures the variability of the parameter estimate. The greater the sample size, the smaller the margin of error.

For the chosen α value, we want to determine U_1 and U_2, such that the probability of the event that the unknown parameter lies within the interval $[U_1, U_2]$ equals $1 - \alpha$. The probability that the parameter lies outside of the interval $[U_1, U_2]$ is equal to α. We choose a probability $1 - \alpha$ that is close to 1. The probability $1 - \alpha$ is called the confidence level of a confidence interval. Practically, the confidence level of a confidence interval represents the percentage of the time our result would be accurate if we generated many random samples. Usual confidence levels are 95% or 99%.

The interval $[u_1, u_2]$ represents the confidence interval of the unknown parameter β if the true value of β lies within the interval $[u_1, u_2]$ with the probability equal to $1 - \alpha$. The confidence interval should be understood as an interval of likely values for the unknown parameter.

More formally, we could write the following:

$$P(u_1 \leq \beta \leq u_2) = 1 - \alpha$$

For example, we can claim that the average parking duration is 210 minutes plus or minus 15 minutes. This actually means that the average parking duration is somewhere between 195 minutes and 225 minutes.

Confidence interval for the mean with a known variance

We consider the population that has an unknown mean μ and a known variance σ^2; the population standard deviation could be known, for example, from previous studies. We generate the sample $(x_1, x_2, ..., x_n)$. It has been shown that the distribution of the sample means can be approximated by a normal distribution when the number of elements in the sample n approaches infinity. The distribution of the sample means, where the number of n values is high $(n \geq 30)$, could be treated as a normal distribution $N\left(\mu, \dfrac{\sigma}{\sqrt{n}}\right)$. Consequently:

$$P\left(-z^* < \frac{\bar{x} - \mu}{\dfrac{\sigma}{\sqrt{n}}} < z^*\right) = \Phi\left(z^*\right) - \Phi\left(-z^*\right) = 2 \cdot \Phi\left(z^*\right) = 1 - \alpha$$

$$P\left(\bar{x} - z^* < \frac{\mu}{\dfrac{\sigma}{\sqrt{n}}} < \bar{x} + z^*\right) = 2 \cdot \Phi\left(z^*\right) = 1 - \alpha$$

$$P\left(\bar{x} - z^* \cdot \frac{\sigma}{\sqrt{n}} < \mu < \bar{x} + z^* \cdot \frac{\sigma}{\sqrt{n}}\right) = 2 \cdot \Phi\left(z^*\right) = 1 - \alpha$$

In words, we expect that the population mean lies in the interval $\left(\bar{x} - z^* \cdot \dfrac{\sigma}{\sqrt{n}}, \bar{x} + z^* \cdot \dfrac{\sigma}{\sqrt{n}}\right)$ with the probability equal to $1 - \alpha$. For example, the interval $\left(\bar{x} - 1.96 \cdot \dfrac{\sigma}{\sqrt{n}}, \bar{x} + 1.96 \cdot \dfrac{\sigma}{\sqrt{n}}\right)$ is the 95% confidence interval estimator of the population mean. Table 5.12 contains some of the frequently used confidence levels, and the related k-values.

Example 5.6

The sample contains recorder speeds of 100 vehicles. The sample mean and the variance equal:

$$\bar{x} = 64.25 \text{ mph}$$

Table 5.12 Frequently used confidence levels, and the related z*-values

Confidence level $1-\alpha$	0.8	0.9	0.95	0.96	0.98	0.99
Value from the standard normal distribution z*	1.28	1.645	1.96	2.05	2.33	2.58

$$\sigma = 3.25\,\text{mph}$$

Find the 95% confidence interval and the 99% confidence interval estimators of the population mean.

SOLUTION:

The 95% confidence interval estimator is:

$$\left(\bar{x}-1.96\cdot\frac{\sigma}{\sqrt{n}}, \bar{x}+1.96\cdot\frac{\sigma}{\sqrt{n}}\right)=\left(64.25-1.96\cdot\frac{3.25}{\sqrt{100}}, 64.25+1.96\cdot\frac{3.25}{\sqrt{100}}\right)$$

$$=\left(63.613, 64.887\right)$$

Similarly, the 99% confidence interval estimator is:

$$\left(\bar{x}-2.58\cdot\frac{\sigma}{\sqrt{n}}, \bar{x}+2.58\cdot\frac{\sigma}{\sqrt{n}}\right)=\left(64.25-2.58\cdot\frac{3.25}{\sqrt{100}}, 64.25+2.58\cdot\frac{3.25}{\sqrt{100}}\right)$$

$$=\left(63.4115, 65.0885\right)$$

Let us show how the confidence interval can be determined by Excel in the case of previous examples. The value of $z^{*}\cdot\frac{\sigma}{\sqrt{n}}$ can be determined by the Excel function "CONFIDENCE.NORM" which has three parameters: the first is the probability α, the second is a standard deviation, and the third is the size of a sample. Figure 5.18 shows an example of how a confidence interval can be determined when $\alpha=0.05$, and Figure 5.19 shows an example when $\alpha=0.01$. In both cases, the Excel functions in the cells B6 and B7 are:

- Cell B6: "=B1-CONFIDENCE.NORM(B3,B2,B4)"
- Cell B7: "=B1+CONFIDENCE.NORM(B3,B2,B4)"

Confidence interval for a population proportion

We could be interested, for example, to determine the proportion of drivers who do not wear seat belts, or the proportion of air passengers that have a fear of flying. In situations like these, we try to estimate a population

▲	A	B	C	D	E
1	$\bar{x} =$	64.25			
2	$\sigma =$	3.25			
3	$\alpha =$	0.05			
4	$n =$	100			
5					
6	The left boundary	63.613			
7	The right boundary	64.887			
8					
9					
10					

Figure 5.18 Confidence interval when $\alpha = 0.05$

▲	A	B	C	D	E
1	$\bar{x} =$	64.25			
2	$\sigma =$	3.25			
3	$\alpha =$	0.01			
4	$n =$	100			
5					
6	The left boundary	63.4129			
7	The right boundary	65.0871			
8					
9					
10					

Figure 5.19 Confidence interval when $\alpha = 0.01$

proportion based on a sample proportion. As with the population mean estimation procedure, we should include in the analysis a margin of error.

It has been shown that the following is the confidence interval for a population proportion p:

$$\left(\bar{p} - z^* \cdot \sqrt{\frac{\bar{p} \cdot (1 - \bar{p})}{n}}, \quad \bar{p} + z^* \cdot \sqrt{\frac{\bar{p} \cdot (1 - \bar{p})}{n}} \right)$$

where:

$\bar{p}$ – the sample proportion

z^* – the appropriate value from the standard normal distribution for the desired level of confidence

n – the sample size

The population proportion is frequently estimated in situations when the feature being observed is categorical. For example, travelers' behavior is studied by using various surveys (do/do not ride public transport, do/do not use toll route, do/do not wear a seatbelt while driving, etc.).

Example 5.7

An analysis of 100 fatal crashes showed that 17% of all fatal crashes involve drivers older than or equal to 65 years. Estimate, with the 95% confidence interval, the percentage of potential fatal crashes caused by older drivers.

SOLUTION:

The left boundary of the interval is:

$$\bar{p} - z^* \cdot \sqrt{\frac{\bar{p} \cdot (1 - \bar{p})}{n}} = 0.17 - 1.96 \cdot \sqrt{\frac{0.17 \cdot 0.83}{100}} = 0.17 - 1.96 \cdot 0.038$$

$$= 0.1 = 10\%$$

The right boundary of the interval is:

$$\bar{p} + z^* \cdot \sqrt{\frac{\bar{p} \cdot (1 - \bar{p})}{n}} = 0.17 + 1.96 \cdot \sqrt{\frac{0.17 \cdot 0.83}{100}} = 0.17 + 1.96 \cdot 0.038$$

$$= 0.24 = 24\%$$

We can conclude that the percentage of potential fatal crashes caused by older drivers is between 10% and 24%, with a confidence level of 95%.

Figure 5.20 shows how the confidence interval for this example can be determined in Excel. The input data for p, α, and n are given in the cells B1, B2, and B3. Cells B5, B7, and B8 have the following excel functions:

◢	A	B	C	D	E
1	p =	0.17			
2	α =	0.05			
3	n =	100			
4					
5	z* =	1.95996			
6					
7	The left boundary	0.09638			
8	The right boundary	0.24362			
9					
10					

Figure 5.20 Confidence interval for a population proportion

- Cell B5: "=NORM.S.INV(1-B2/2)"
- Cell B7: "=B1-B5*SQRT(B1*(1-B1)/B3)"
- Cell B8: "=B1+B5*SQRT(B1*(1-B1)/B3)"

Confidence interval for the difference of two means

There are situations in transportation analysis where we want to compare two populations. For example, we can compare average travel time by car, or average travel time by public transport in city A versus city B. When a characteristic that is being observed and compared is numerical, we are interested in the difference in the means between the two specified populations. In the first step, we take a sample from each population and calculate the sample means, $\bar{x}_1$ and $\bar{x}_2$. In the next step, we calculate the difference between the sample means, $\bar{x}_1 - \bar{x}_2$. We also include in the analysis a margin of error.

We can write the following:

$$P\left(\bar{x}_1 - \bar{x}_2 - z^* \cdot \sqrt{\frac{\sigma_1^2}{n_1} + \frac{\sigma_2^2}{n_2}} < \mu_1 - \mu_2 < \bar{x}_1 - \bar{x}_2 + z^* \cdot \sqrt{\frac{\sigma_1^2}{n_1} + \frac{\sigma_2^2}{n_2}} \right)$$

$$= 2 \cdot \Phi\left(z^*\right) = 1 - \alpha$$

where:

$\bar{x}_1, \bar{x}_2$ – sample means
n_1, n_2 – sample sizes
σ_1, σ_2 – population standard deviations
z^* – the appropriate value from the standard normal distribution for the desired level of confidence

In other words, the following is the confidence interval for the difference of two means:

$$\left(\bar{x}_1 - \bar{x}_2 - z^* \cdot \sqrt{\frac{\sigma_1^2}{n_1} + \frac{\sigma_2^2}{n_2}}, \bar{x}_1 - \bar{x}_2 + z^* \cdot \sqrt{\frac{\sigma_1^2}{n_1} + \frac{\sigma_2^2}{n_2}} \right)$$

Example 5.8

Frequently, tows wait to get through the lock chamber. The lock chambers at location 1 and location 2 are observed. There are 100 observations ($n_1 = 100$) from the first location and 120 observations ($n_2 = 120$) from the second location. The average delay times and standard deviations equal:

$\bar{x}_1 = 3.2$ hours

$$\bar{x}_2 = 2.8 \text{ hours}$$

$$\sigma_1 = 0.5 \text{ hours}$$

$$\sigma_2 = 0.6 \text{ hours}$$

Compare the average waiting times at the two locations, using a 95% confidence interval.

SOLUTION:

The left boundary of the interval is:

$$\bar{x}_1 - \bar{x}_2 - z^* \cdot \sqrt{\frac{\sigma_1^2}{n_1} + \frac{\sigma_2^2}{n_2}} = 3.2 - 2.8 - 1.96 \cdot \sqrt{\frac{0.5^2}{100} + \frac{0.6^2}{120}}$$

$$= 0.4 - 1.96 \cdot 0.074 = 0.25$$

The right boundary of the interval is:

$$\bar{x}_1 - \bar{x}_2 + z^* \cdot \sqrt{\frac{\sigma_1^2}{n_1} + \frac{\sigma_2^2}{n_2}} = 3.2 - 2.8 + 1.96 \cdot \sqrt{\frac{0.5^2}{100} + \frac{0.6^2}{120}}$$

$$= 0.4 + 1.96 \cdot 0.074 = 0.55$$

We can conclude that the average waiting time at the first location is longer than that at the second location by between 0.25 and 0.55 hours, with a 95% level of confidence.

Figure 5.21 shows how this example can be solved in Excel. The input data are given in the cells from B1 to B7. The cells B9, B11, and B12 have the following Excel functions:

- Cell B9: "=NORM.S.INV(1-B7/2)"
- Cell B11: "=B1-B2-B9*SQRT((B3/B5)+(B4/B6))"
- Cell B12: "=B1-B2+B9*SQRT((B3/B5)+(B4/B6))"

Confidence interval for the difference between two proportions

There are also situations in transportation studies when we want to compare two proportions. For example, we can compare proportions (percentages) of public transport riders, or percentages of solo drivers in city A versus city B. When the characteristic that is being observed and compared is categorical, we are interested in the difference in the proportions for the two specified populations. In the first step, we take a sample from each population and calculate sample proportions $\bar{p}_1$ and $\bar{p}_2$. In the next step, we calculate the difference between the sample means, $\bar{p}_1 - \bar{p}_2$. We also include, in the analysis, a margin of error. We can write the following:

	A	B	C	D	E
1	$\overline{x_1} =$	3.2			
2	$\overline{x_2} =$	2.8			
3	$\sigma_1^2 =$	0.25			
4	$\sigma_2^2 =$	0.36			
5	$n_1 =$	100			
6	$n_2 =$	120			
7	$\alpha =$	0.05			
8					
9	$z^* =$	1.95996			
10					
11	The left boundary	0.25465			
12	The right boundary	0.54535			
13					
14					

Figure 5.21 Solution of example in Excel

$$P\left(\begin{array}{c} \bar{p}_1 - \bar{p}_2 - z^* \cdot \sqrt{\dfrac{\bar{p}_1 \cdot (1-\bar{p}_1)}{n_1} + \dfrac{\bar{p}_2 \cdot (1-\bar{p}_2)}{n_2}} \\[3mm] < p_1 - p_2 < \bar{p}_1 - \bar{p}_2 + z^* \cdot \sqrt{\dfrac{\bar{p}_1 \cdot (1-\bar{p}_1)}{n_1} + \dfrac{\bar{p}_2 \cdot (1-\bar{p}_2)}{n_2}} \end{array}\right) = 2 \cdot \Phi\left(z^*\right) = 1 - \alpha$$

where:

$\bar{p}_1, \bar{p}_2$ – sample proportions

n_1, n_2 – sample sizes

z^* – the appropriate value from the standard normal distribution for the desired level of confidence

The following is the confidence interval for the difference between two means:

$$\left(\bar{p}_1 - \bar{p}_2 - z^* \cdot \sqrt{\frac{\bar{p}_1 \cdot (1-\bar{p}_1)}{n_1} + \frac{\bar{p}_2 \cdot (1-\bar{p}_2)}{n_2}},\right.$$

$$\left.\bar{p}_1 - \bar{p}_2 + z^* \cdot \sqrt{\frac{\bar{p}_1 \cdot (1-\bar{p}_1)}{n_1} + \frac{\bar{p}_2 \cdot (1-\bar{p}_2)}{n_2}}\right)$$

Example 5.9

Cities A and B are connected by air and by train. A survey of 120 randomly chosen air passengers indicated 80 business travelers. Among 100 randomly chosen train passengers surveyed, 13 were business passengers. We want to estimate, with 95% confidence, the difference between the proportions of air business passengers versus train business passengers.

SOLUTION:

The sample proportions and the sample sizes equal:

$$\bar{p}_1 = \frac{80}{120} = 0.67 \quad \bar{p}_2 = \frac{13}{100} = 0.13 \quad n_1 = 120 \quad n_2 = 100$$

The left-hand boundary of the interval is:

$$\bar{p}_1 - \bar{p}_2 - z^* \cdot \sqrt{\frac{\bar{p}_1 \cdot (1-\bar{p}_1)}{n_1} + \frac{\bar{p}_2 \cdot (1-\bar{p}_2)}{n_2}}$$

$$= 0.67 - 0.13 - 1.96 \cdot \sqrt{\frac{0.67 \cdot 0.33}{120} + \frac{0.13 \cdot 0.87}{100}}$$

$$= 0.54 - 1.96 \cdot 0.055 = 0.43$$

The right-hand boundary of the interval is:

$$\bar{p}_1 - \bar{p}_2 + z^* \cdot \sqrt{\frac{\bar{p}_1 \cdot (1-\bar{p}_1)}{n_1} + \frac{\bar{p}_2 \cdot (1-\bar{p}_2)}{n_2}}$$

$$= 0.67 - 0.13 + 1.96 \cdot \sqrt{\frac{0.67 \cdot 0.33}{120} + \frac{0.13 \cdot 0.87}{100}}$$

$$= 0.54 + 1.96 \cdot 0.055 = 0.65$$

We can conclude that air travel has a higher percentage of business passengers than does train travel. The difference in percentage of business travelers is between 43% and 65%, with a 95% level of confidence.

The solution for this problem, obtained in Excel, is given in Figure 5.22. The input data are given in the cells from B1 to B5. The cells B7, B9, and B10 have the following Excel functions:

- Cell B7: "=NORM.S.INV(1-B5/2)"
- Cell B9: "=B1-B2-B7*SQRT((B1*(1-B1)/B3)+(B2*(1-B2)/B4))"
- Cell B10: "=B1-B2+B7*SQRT((B1*(1-B1)/B3)+(B2*(1-B2)/B4))"

5.3.4 Testing of a statistical hypothesis

Citizens, dissatisfied with public transport, can claim that over 90% of bus departures are not on time, or that, in 70% –80% of cases during peak hours, there is a bunching of buses. Similarly, city authorities can claim that

◢	A	B	C	D	E
1	$\overline{p_1} =$	0.67			
2	$\overline{p_2} =$	0.13			
3	$n_1 =$	120			
4	$n_2 =$	100			
5	$\alpha =$	0.05			
6					
7	$z^* =$	1.95996			
8					
9	The left boundary	0.43312			
10	The right boundary	0.64688			
11					
12					

Figure 5.22 Confidence interval made in Excel

the construction of a new bridge reduced the average travel time between the two parts of the city by 10 minutes, etc. How can the truthfulness of these claims be assessed? In mathematical statistics, all claims are checked. Statistical procedures for checking claims are called statistical tests.

There are parametric and nonparametric statistical tests. We use non-parametric statistical tests to check the assumption that the population data has a specific distribution. For example, let us assume that the measured time intervals between the appearances of successive vehicles (headways) are, for example, 6 s, 11 s, 13 s, 16 s, etc. After studying these data, we can generate the hypothesis that the time interval between vehicle arrivals is a random variable that has an exponential distribution.

On the other hand, we use parametric tests to check the claim about a population parameter value (the mean, for example).

We can test hypotheses about numerical variables, as well as hypotheses about categorical variables. After the statistical test is performed, we cannot, with absolute confidence, claim that the hypothesis is true or false. On the other hand, these tests make it possible, with a probability close to 1, to make a judgment on the accuracy of the hypothesis. The generated hypothesis is denoted by H_0. We call this hypothesis a null hypothesis. An alternative hypothesis, different from the null hypothesis, is called the alternative hypothesis (the term research hypothesis is also used). We denote the alternative hypothesis by H_1. For example the null hypothesis $H_0(p_t=0.1)$ could claim that the probability p_t of a truck appearing in the traffic flow to equal 0.1. The alternative hypothesis could be $H_1(p_t=0.2)$, or $H_1(p_t>0.1)$, or $H_1(p_t\neq0.1)$, or $H_1(p_t=0.3)$, etc.

Generally, we consider that the null hypothesis H_0 is true unless our data and statistics prove otherwise. Once we have enough proof against the null hypothesis H_0, we reject H_0 and accept H_1. Many statisticians compare hypothesis testing with jury trials. The H_0 could be treated as the "not-guilty" verdict, and H_1 as the "guilty" verdict. Like in the case of jury trials, the "burden of proof" is on the analyst.

Testing a hypothesis about the population mean

Let us assume that we want to test the hypothesis about the population mean. For example, we want to test the hypothesis that the population mean speed on an observed highway section is equal to 62 mph, i.e. that $H_0(\mu=62)$. The alternative hypothesis could be $H_1(\mu>62)$, $H_1(\mu<62)$, or $H_1(\mu\neq62)$.

We showed in previous sections that the distribution of sample means, in cases with high n values ($n\geq30$), could be treated as a normal distribution $N\left(\mu,\dfrac{\sigma}{\sqrt{n}}\right)$. This result is known as the Central Limit Theorem. The samples, sample means, and the distribution of the sample means are shown in Figure 5.23.

The standard normal distribution Z is used when testing the hypothesis. We switch our sample statistic (mean) to a test statistic by altering it to a standard score, i.e.:

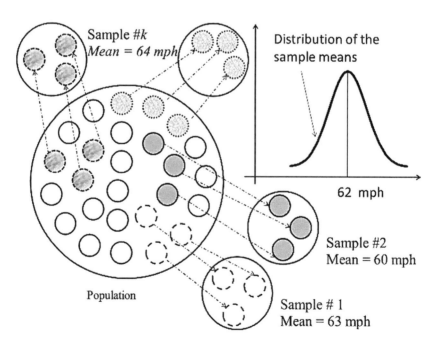

Figure 5.23 The samples, sample means, and the distribution of the sample means

$$Z = \frac{\bar{X} - \mu}{\frac{\sigma}{\sqrt{n}}} \sim N(0,1)$$

We see that we obtained a standard score for a sample mean by subtracting the mean from our statistic and dividing the result obtained by the standard error. The random variable Z is normally distributed (Figure 5.24), with parameters 0 and 1, i.e. $Z \sim N(0,1)$. The standard score measures the distance between the sample results and the declared population value, expressed by the number of standard errors.

In a general case, we want to test the hypothesis that the population mean is equal to the specific value, a, i.e. that $H_0(\mu = a)$. It is logical to assume that, if our test statistic is relatively close to 0, there is no need to reject the hypothesis $H_0(\mu = a)$. In the case when the test statistic is far from 0, we conclude that the results of the sample do not confirm the hypothesis. In this case we should reject the hypothesis $H_0(\mu = a)$. We introduce into the analysis the p-value (Figure 5.24).

The p-value measures how likely it was that we obtained our sample results if the null hypothesis were true. Obviously, the smaller the p-value is, the stronger is the evidence against the null hypothesis, H_0. The value of p should be determined in the following way:

1. Determine the probability of the standard normal distribution, according to the value of the test statistics (Table A.1 in the appendix), i.e.: $P(Z < z)$, where z is the value of the test statistic.

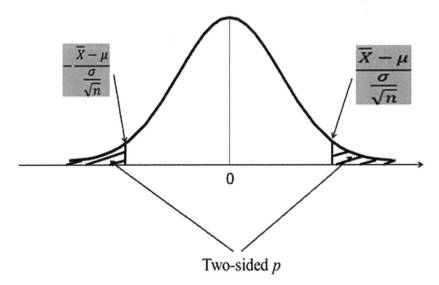

Two-sided p

Figure 5.24 The p-value

Table 5.13 Recommended conclusions related to p-values

p-value	Conclusion
$p > 0.10$	Non-significant evidence against H_0
$0.05 < p < 0.10$	Marginally significant evidence against H_0
$0.01 < p < 0.05$	Significant evidence against H_0
$p \leq 0.01$	Highly significant evidence against H_0

2. If $z > 0$ then $p_1 = 1 - P(Z < z)$, otherwise $p_1 = P(Z < z)$.
3. If we have the following types of hypothesis:
   a. $H_0(\mu = \mu_0)$ and $H_1(\mu \neq \mu_0)$ then $p = 2 \cdot p_1$
   b. $H_0(\mu = \mu_0)$ and $\left(H_1 \left(\mu > \mu_0 \right) \text{ or } H_1(\mu < \mu_0) \right)$ then $p = p_1$

Table 5.13 shows the recommended conclusions related to the p-values.

We denote by α the probability of incorrectly rejecting the null hypothesis H_0. The α value represents the threshold that enables the analyst to reject or retain the null hypothesis. The usual values for α are 0.10, 0.05, etc.

We reject the null hypothesis, H_0, when $p \leq \alpha$. In the opposite case, when $p > \alpha$ we retain the null hypothesis, H_0.

Example 5.10

We want to test the hypothesis that the population mean speed on a highway section is equal to 62 mph, i.e. that $H_0(\mu = 62)$. The sample contains $n = 100$ measured speeds with the mean value $\bar{x} = 63$ mph and the standard deviation $s = 3$ mph. The alternative hypothesis is $H_1(\mu \neq 62)$. The probability of incorrectly rejecting the null hypothesis is $\alpha = 0.05$.

SOLUTION:

Our test statistic is:

$$\frac{\bar{x} - a}{\frac{\sigma}{\sqrt{n}}} = \frac{63 - 62}{\frac{3}{\sqrt{100}}} = 3.33$$

In words, the sample mean is 3.33 standard errors higher than the declared population mean. From Table A.1, we read that $P(Z < 3.33) = 0.9996$.

This value can be also obtained by the Excel function: "=NORM.S.DIST(3.33,TRUE)". The value of p is: $p = 2 \cdot \left(1 - P(Z < 3.33) \right) = 2 \cdot (1 - 0.9996) = 2 \cdot 0.0004 = 0.0008$. Because

$$p = 0.0008 < 0.01,$$

we have highly significant evidence against H_0, and we reject this hypothesis.

Testing one population proportions

Let us assume that we want to test the hypothesis about the population proportion. For example, we want to test the hypothesis that the proportion of heavy vehicles in traffic flow on an observed highway section is equal to p_0, i.e. that $H_0(p=p_0)$ is accepted. The alternative hypothesis could be $H_0(p>p_0)$, $H_1(p<p_0)$, or $H_1(p\neq p_0)$. We take a sample, composed of n elements $(n \geq 30)$. Let us assume that there are m heavy vehicles among n vehicles in the sample. The proportion $\bar{p}$ of heavy vehicles in the sample equals:

$$\bar{p} = \frac{m}{n}$$

The random variable $\bar{p}$ has the following normal distribution:

$$\bar{p} \sim N\left(p, \sqrt{\frac{p \cdot q}{n}}\right) \approx N\left(\bar{p}, \sqrt{\frac{\bar{p} \cdot \bar{q}}{n}}\right)$$

The standardized random variable:

$$\frac{\bar{p} - p}{\sqrt{\dfrac{\bar{p} \cdot \bar{q}}{n}}} = \frac{\bar{p} - p_0}{\sqrt{\dfrac{\bar{p} \cdot \bar{q}}{n}}}$$

is normally distributed with parameters 0 and 1, i.e. $N(0,1)$.

Example 5.11

Based on previous data, transportation planners claim that 35% of the total population of passengers between city A and city B are air passengers. The sample, composed of $n=200$ passengers, showed that $m=64$ of them were air passengers. We denote by p the proportion of air passengers in the population. We want to test the null hypothesis $H_0(p=0.35)$, against the alternative hypothesis $H_1(p<0.35)$.

SOLUTION:

The sample proportion is:

$$\bar{p} = \frac{m}{n} = \frac{64}{200} = 0.32$$

The test statistic is:

$$\frac{\bar{p}-p_0}{\sqrt{\dfrac{p\cdot q}{n}}} = \frac{0.32-0.35}{\sqrt{\dfrac{0.32\cdot 0.68}{200}}} = -0.91$$

In words, the sample taken generated a result that is 0.91 standard errors below the declared value for the population of passengers. From Table A.1, we read that the $P(Z < -0.91)$ is equal to 0.1814 (one can use the Excel function "=NORM.S.DIST(-0.91,TRUE)"). Since $p = P(Z < -0.91) = 0.1814$ and:

$$p = 0.1814 > 0.10$$

we do not have enough evidence to reject the null hypothesis $H_0(p=0.35)$. The statement that there are 35% of the total population of passengers between city A and city B are air passengers is accepted.

Comparing two population means

Let us assume that we want to compare two population means. In this case, we need to take two distinct samples, one from each population. We denote the sample means by $\bar{x}_1$ and $\bar{x}_2$, respectively, the standard errors by s_1 and s_2, respectively and the sample sizes by n_1 and n_1, respectively.

In this test, the null hypothesis $H_0(\mu_1 - \mu_2 = 0)$ states that the two population means are equal, i.e. that the difference between these two means is equal to zero.

The test statistic is:

$$\frac{\left(\bar{x}_1 - \bar{x}_2\right) - 0}{\sqrt{\dfrac{s_1^{\,2}}{n_1} + \dfrac{s_2^{\,2}}{n_2}}}$$

Example 5.12

Two independent samples are taken in the catchment areas of two bus stations. Each sample contained $n_1 = n_2 = 100$ surveyed inhabitants. The distances between the homes and the bus stations were measured. The average walking distance in the case of the first bus station was equal to $\bar{x}_1 = 285$ m, with a standard deviation of $s_1 = 15$ m. The average walking distance in the case of the second bus station was equal to $\bar{x}_2 = 290$ m, with a standard deviation of $s_1 = 10$ m. Compare the two population means, i.e. $H_0(\mu_1 - \mu_2 = 0)$, against the alternative hypothesis $H_1(\mu_1 - \mu_2 \neq 0)$.

SOLUTION:

The test statistics is:

$$\frac{\left(\overline{x_1} - \overline{x_2}\right) - 0}{\sqrt{\dfrac{s_1^2}{n_1} + \dfrac{s_2^2}{n_2}}} = \frac{(285 - 290) - 0}{\sqrt{\dfrac{15^2}{100} + \dfrac{10^2}{100}}} = -2.77$$

The sample taken generated a result that is –2.77 standard errors below the declared value for the population of transit riders. From Table A.1, we read that $P(Z < -2.77)$ is equal to 0.028 (it can be obtained by the Excel function "=NORM.S.DIST(-2.77,TRUE)". The value of p is obtained as: $p = 2 \cdot P(Z < -2.77) = 2 \cdot 0.028 = 0.056$. Since

$$0.05 < p = 0.056 < 0.10,$$

there is marginally significant evidence against $H_0\left(\mu_1 - \mu_2 = 0\right)$.

The chi-squared test

Let us assume that we took a random sample of size n from the population whose probability distribution is unknown. These n observations are put together in a frequency histogram that has k class intervals. The empirical distribution obtained could be less or more similar to particular theoretical distributions. We frequently want to test the hypothesis that a given empirical distribution is an approximation to a specific theoretical distribution. For example, after studying the data, we could set up the hypothesis that the population approximates to a Poisson distribution. This hypothesis should then be tested.

The χ^2 test (chi-squared test) is based on the fact that the sampling distribution of the test statistic is a chi-squared distribution, should the null hypothesis be confirmed.

Let us briefly explain the chi-squared distribution. If $Z_1, Z_2, \ldots, Z_n$ are independent standard normal random variables, then the random variable X that is defined as:

$$X = Z_1^2 + Z_2^2 + \cdots + Z_n^2$$

Has a chi-square distribution with n degrees of freedom, i.e. $X \sim \chi_n^2$.

Let us briefly describe the concept of degree of freedom. Let us assume that you could walk, ride a bicycle, drive a car, take a tram or bus to go from home to your office. Imagine that you want to use different modes of transportation every working day of the week. On Monday you can opt for any of the five transportation modes. On Tuesday, you can choose from the four remaining modes, on Wednesday you can choose from three remaining transportation modes, etc. Finally, on Friday, you must choose the one remaining transportation mode. Obviously, we have $5 - 1 = 4$ working days in which we are free to choose. In other words, we have four working days in which transportation mode could vary. Similarly, in statistics, the number of degrees of freedom represents the number of values (observations) in the calculation of a statistic that are free to vary. Let us clarify this definition by considering the following example. Let us assume that we have a sample of 10 vehicle speeds (mph) and that we want to test the hypothesis about the population mean speed. The assumption is that the average speed is 63 mph. The sum of the 10 speeds must be equal to mean $\cdot 10 = 63 \cdot 10 = 630$. The first value in the data set is free to vary, the second value in the data set is free to vary, etc. Obviously, in order to have a mean equal to 63, the 10th value cannot vary and must take a specific value. In this case we have $10 - 1 = 9$ degrees of freedom.

The number of degrees of freedom is calculated as:

$$k - p - 1$$

where:

p – the number of parameters of the assumed distribution estimated by sample statistics

k – the number of class intervals

When X is a chi-square random variable with n degrees of freedom, then for any $\alpha \in (0,1)$, the quantity χ_n^2 is defined in such a way, that the following is satisfied (Figure 5.25):

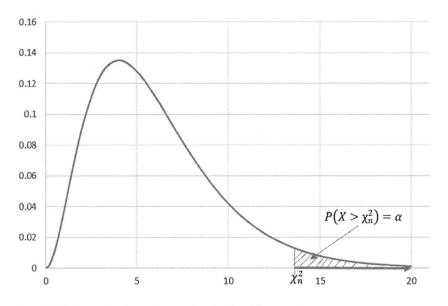

Figure 5.25 Example of rejection region for the chi-squared test

$$P\left(X > \chi_n^2\right) = \alpha$$

Table A.2 contains the values of χ_n^2 for various values of n and α.

We use the chi-squared test to conclude whether there is a significant difference between the expected frequencies and the observed frequencies. We denote by O_i the observed frequency in the i-th class interval. From the assumed probability distribution, we compute the expected frequency in the i-th class interval. We denote this frequency by E_i. The test statistic is:

$$\chi^2 = \sum_{i=1}^{k} \frac{\left(O_i - E_i\right)^2}{E_i}$$

where:

χ^2 – the test statistic that measures the difference between the expected frequencies and the observed frequencies

O_i – the observed frequency in the i-th class interval

E_i – the expected frequency in the i-th class interval

We reject the null hypothesis (that the population has the assumed chi-squared distribution) when the test statistic $\chi^2 > \chi_{\alpha,\,k-p-1}^2$.

Example 5.13

The bus travel times between terminal A and terminal B are recorded 160 times between 8:00 a.m. and 9:00 a.m. The obtained results are given in Table 5.14.

It has been assumed that the bus travel time between terminal A and terminal B has a uniform distribution (null hypothesis). The χ^2-test proves this hypothesis.

Table 5.14 Bus travel times between terminal A and terminal B

Bus travel time t_i	Observed frequency O_i	Expected frequency E_i	$O_i - E_i$	$\dfrac{\left(O_i - E_i\right)^2}{E_i}$
27	22	20	2	0.2
28	18	20	−2	0.2
29	23	20	3	0.45
30	17	20	−3	0.45
31	19	20	−1	0.05
32	20	20	0	0
33	21	20	1	0.05
34	20	20	0	0
Total	160			$\chi^2 = 1.4$

SOLUTION:

The number of class intervals is $k = 10$. We do not need to calculate any parameter of the uniform distribution. In other words, the number of parameters p of the assumed distribution estimated by the sample statistics is equal to 0. The number of degrees of freedom is equal to:

$$k - p - 1 = 8 - 0 - 1 = 7$$

We find from the Table in the Appendix, that contains values for Chi-Square distribution (this value can be also obtained from the Excel function: =CHISQ.INV.RT(0.05,7)") that:

$$\chi^2_{0.05\ n} = \chi^2_{0.05\ 7} = 14.0671$$

Since

$$1.4 = \chi^2 < \chi^2_{0.05\ 7} = 14.0671$$

we cannot reject the null hypothesis. We can claim that the bus travel time between terminal A and terminal B are uniformly distributed.

Example 5.14

The queue length at a traffic signal was recorded 95 times. The obtained results are given in Table 5.15.

The first three columns of Table 5.15 contains information on the number of vehicles in a queue (class interval), the means of the class intervals, and the observed frequencies O_i. The sample mean $\bar{x}$, and the standard deviation s are:

$$\bar{x} = 13.85$$

$$s = 3.71$$

We use these values as approximations of the population mean μ and standard deviation σ. Let us calculate the probabilities related to the class intervals. For example, the probability $P(12 < X < 14)$ represents the probability of the event that the random variable $X \sim N(13.85, 3.71)$ has the value from the class interval 12–14. This probability equals:

$$P(12 < X < 14) = P\left(\frac{12 - 13.85}{3.71} < \frac{X - 13.85}{3.71} < \frac{14 - 13.85}{3.71}\right)$$

$$= P\left(-0.4986 < \frac{X - 13.85}{3.71} < 0.0404\right) = 0.2075$$

Table 5.15 The queue length at a traffic signal

Number of vehicles in a queue (class interval)	Mean of the class interval x_i	Observed frequency O_i		Expected frequency E_i		$O_i - E_i$	$\dfrac{(O_i - E_i)^2}{E_i}$
4–6	5	2	11	1,23	14.84	−3.84	0.99
6–8	7	4		3.96			
8–10	9	5		9.65			
10–12	11	17		14.57		2.03	0.28
12–14	13	23		19.71		3.29	0.55
14–16	15	20		19.28		0.72	0.03
16–18	17	13		13.99		−0.99	0.07
18–20	19	5	11	12.70	17.42	−6.42	0.37
20–22	21	3		3.22			
22–24	23	2		1.03			
26–28	25	1		0.47			
Total		95					$\chi^2 = 2.29$

The expected frequency E_s for the fifth class interval (12–14) equals:

$$E_5 = 95 \cdot 0.2075 = 19.71$$

The other expected frequencies are calculated in the same way and are shown in Table 5.15. The number of parameters of the assumed distribution p, estimated by sample statistics, is equal to 2 (we estimated μ and σ). The number of degrees of freedom is:

$$k - p - 1 = 6 - 2 - 1 = 3$$

We find from Table A.2 (it can be also be obtained from the Excel function "=CHISQ.INV.RT(0.05,3)") that:
  Significantly since

$$2.29 = \chi^2 < \chi^2_{0.05\ 3} = 7.815$$

we cannot reject the null hypothesis. We can claim that the number of vehicles in a queue at the traffic signal is normally distributed.

5.4 CORRELATION AND REGRESSION

Traffic engineers frequently face the question about the relationship between two or more variables. By measuring the correlation coefficient, analysts determine the strength of this relationship. In many situations, we analyze various "If/Then" scenarios. In other words, we are often interested in answering the question of what will happen "then" to a value of a dependent variable "if" the values of the independent variable change in the future.

Let us consider the problem of ridership prediction in public transport. Several factors influence the ridership in public transport. We might be interested, for example, in the relationship between the ridership in public transport and a service frequency. We denote ridership by y, to represent the dependent variable, and the bus frequency by x, to represent the independent variable. Regression analysis mathematically explains the relationship between two or more variables. Regression analysis helps us to determine how well, for example, an independent variable (bus frequency) predicts a dependent variable (ridership). The following are the most common relationships between a dependent variable (y) and one independent variable (x):

$y = a_0 + a_1 \cdot x$	(Straight line)
$y = a_0 + a_1 \cdot x + a_2 \cdot x^2$	(Parabola)
$y = a_0 + a_1 \cdot x + a_2 \cdot x^2 + a_3 \cdot x^3$	(Cubical parabola)
$\cdots\cdots\cdots\cdots\cdots$	
$y = a_0 + a_1 \cdot x + a_2 \cdot x^2 + a_3 \cdot x^3 + \ldots + a_N \cdot x^n$	(Parabola of the n-th power)
$\dfrac{1}{y} = a_0 + a_1 \cdot x$	(Hyperbola)
$y = a \cdot b^x$	(Exponential curve)
$y = a \cdot x^b$	(Geometric curve)
$y = a \cdot b^x + c$	(Modified exponential curve)
$y = \dfrac{1}{a \cdot b^x + c}$	(Logistic curve)

A simple regression equation contains only one independent variable and reads:

$$y = a_0 + a_1 \cdot x$$

A linear regression equation, that has a few independent variables, reads:

$$y = a_0 + a_1 \cdot x_1 + a_2 \cdot x_2 + a_3 \cdot x_3 + \cdots + a_N \cdot x_n$$

where:

y – dependent variable

$x_1, x_2, x_3, \ldots, x_n$ – independent (explanatory) variables

$a_0, a_1, a_2, a_3, \ldots, a_n$ – regression coefficients

In the next paragraph, we will explain the way in which regression coefficients are computed. Every regression coefficient is associated with one independent variable. They represent the strength, as well as the type of relationship, between the independent variable and the dependent variable. A high value of the specific regression coefficient indicates a strong relationship between the corresponding independent variable and the dependent variable. Let us assume that the last relation represents the relationship between the ridership in public transport and several independent (explanatory) variables. Let x_1 represent the bus frequency. It is logical to assume that, the higher the bus frequency, the higher the ridership value. In this case the sign of the coefficient a_1 should be positive. Let x_2 represents the bus fare. It is also logical to assume that, the higher the bus fare, the lower the ridership value. In this case the sign of the coefficient a_2 should be negative.

Developing a regression model is an iterative process that could be time consuming. An analyst searches for the independent variables that could successfully predict dependent variable values. In a few iterations, an analyst eliminates and/or adds independent variables, until generating the best regression model.

5.4.1 Estimation of the regression coefficients by the least squares method

Let us consider a simple regression equation. For example, let us assume that public transport ridership (y) linearly depends on service frequency (x), i.e. $y = a \cdot x + b$ (Figure 5.26). We perform the model calibration, i.e. the estimation of the parameters a and b, by the least squares method.

The parameters a and b are calculated to best fit a data set. In other words, using the least squares method, we try to minimize the distance between the straight line and the points. Let us assume that the data set consists of n data pairs (x_i, y_i), $i = 1, 2, 3, \ldots, n$ where x_i is an independent variable (frequency) and y_i is a dependent variable (ridership). The equation of the estimated line is:

$$Y = a \cdot X + b$$

The difference e_i between the actual value of the transportation demand y_i and the value of the transportation demand predicted by the linear model Y_i (the i-th error) equals:

$$e_i = y_i - Y_i$$

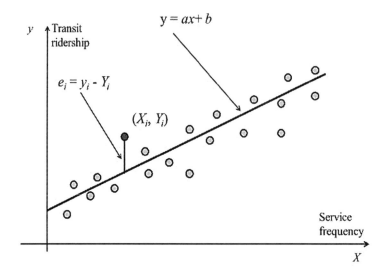

Figure 5.26 The estimation of the parameters a, and b by the least squares method

where:
  e_i is the *i*-th error
  y_i is the *i*-th actual value
  Y_I is the *i*-th predicted value

The sum of squared errors S equals:

$$S = \sum_{i=1}^{n}\left(e_i\right)^2 = \sum_{i=1}^{n}\left(y_i - Y_i\right)^2$$

$$S = \sum_{i=1}^{n}\left(e_i\right)^2 = \sum_{i=1}^{n}\left(y_i - a \cdot X_i - b\right)^2$$

The optimal values of the parameters a and b are obtained when the sum S is minimal. The minimum value of the sum S is obtained by setting the gradient to zero, i.e.:

$$\frac{\partial S}{\partial a} = \sum_{i=1}^{n} 2 \cdot \left(y_i - a \cdot X_i - b\right) \cdot \left(-X_i\right) = 0$$

$$\frac{\partial S}{\partial b} = \sum_{i=1}^{n} 2 \cdot \left(y_i - a \cdot X_i - b\right) \cdot 1 = 0$$

After solving the previous equations, the following coefficients of the esti-mated straight line are obtained:

$$a = \frac{\sum_{i=1}^{n}(x_i - \bar{x})\cdot(y_i - \bar{y})}{\sum_{i=1}^{n}(x_i - \bar{x})^2}$$

$$b = \frac{\sum_{i=1}^{n}y_i}{n} - a\cdot\frac{\sum_{i=1}^{n}x_i}{n}$$

where:

$$\bar{x} = \frac{\sum_{i=1}^{n}x_i}{n}$$

$$\bar{y} = \frac{\sum_{i=1}^{n}y_i}{n}$$

Example 5.15

Table 5.16 shows numbers of passengers on U.S. airlines from 2003 to 2019. If we suppose that there is a linear trend, then determine coefficients a and b.

SOLUTION:

When we have a linear trend, the equation of the line-of-best-fit is the following:

$$y = a\cdot x + b$$

where we have to determine the coefficients a and b, according to val-ues of x and y that are given in Table 5.17. The values for x are years, while the numbers of passengers are values for y. The values of $\bar{x}$ and $\bar{y}$ are given by:

$$\bar{x} = \frac{\sum_{i=1}^{n}x_i}{n} = \frac{34,187}{17} = 2011$$

$$\bar{y} = \frac{\sum_{i=1}^{n}y_i}{n} = \frac{13,032}{17} = 766.59$$

Table 5.16 Number of passengers on U.S. airlines from 2003 to 2019

Years	Number of passengers (in millions)
2003	647
2004	704
2005	739
2006	745
2007	770
2008	743
2009	704
2010	720
2011	731
2012	737
2013	743
2014	763
2015	798
2016	824
2017	849
2018	889
2019	926

Table 5.17 shows the values used for calculation of coefficients a and b. The coefficients a and b are given by:

$$a = \frac{\sum_{i=1}^{n}(x_i - \bar{x})\cdot(y_i - \bar{y})}{\sum_{i=1}^{n}(x_i - \bar{x})^2} = \frac{4,849}{408} = 11.88$$

$$b = \frac{\sum_{i=1}^{n} y_i}{n} - a \cdot \frac{\sum_{i=1}^{n} x_i}{n} = \bar{y} - a \cdot \bar{x} = 766.59 - 11.88 \cdot 2,011 = -23,134$$

Now, we can write the following:

$$y = 11.88 \cdot x - 23,134$$

Values for a and b can be obtained in Excel in the way shown in Figures 5.27, 5.28, and 5.29 To obtain Figure 5.27 we have to select Input Data and insert Scatter Chart. Make a right click on one of the points in the chart and select the option Add Trendline (see Figure 5.28). In the section Format Trendline, Linear and Display Equation should be selected on the chart (see Figure 5.29). From the equation in Figure 5.29, we can see the values for a and b.

Table 5.17 The necessary values for calculation of coefficients a and b

x_i	y_i	$x_i - \overline{x}$	$y_i - \overline{y}$	$(x_i - \overline{x}) \cdot (y_i - \overline{y})$	$(x_i - \overline{x})^2$
2003	647	−8	−119.59	956.706	64
2004	704	−7	−62.588	438.118	49
2005	739	−6	−27.588	165.529	36
2006	745	−5	−21.588	107.941	25
2007	770	−4	3.41176	−13.647	16
2008	743	−3	−23.588	70.7647	9
2009	704	−2	−62.588	125.176	4
2010	720	−1	−46.588	46.5882	1
2011	731	0	−35.588	0	0
2012	737	1	−29.588	−29.588	1
2013	743	2	−23.588	−47.176	4
2014	763	3	−3.5882	−10.765	9
2015	798	4	31.4118	125.647	16
2016	824	5	57.4118	287.059	25
2017	849	6	82.4118	494.471	36
2018	889	7	122.412	856.882	49
2019	926	8	159.412	1275.29	64
			Total:	4,849	408

5.4.2 Correlation coefficient

Statisticians use a correlation coefficient to determine the strength of the relationship between the two variables. If the values of the second variable increase as the values of the first variable increase, we are talking about a positive relationship. Thus, for example, there is a positive relationship

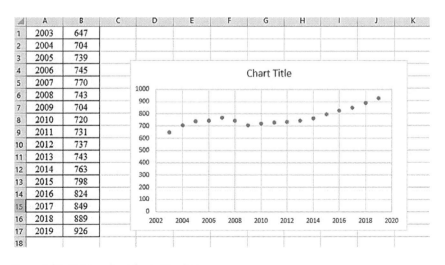

Figure 5.27 Making chart from the data

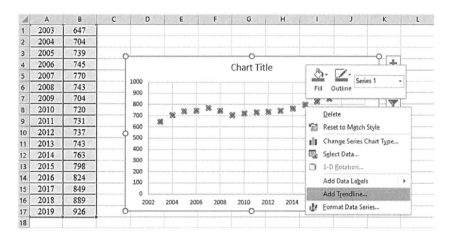

Figure 5.28 Adding the trendline

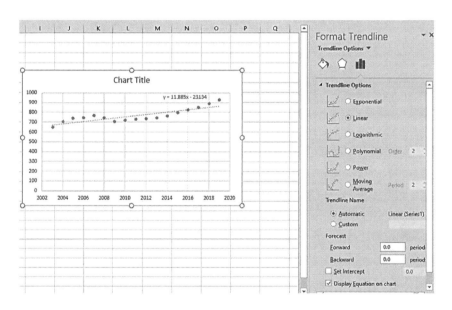

Figure 5.29 Displaying trendline and equation on the chart

between the number of inhabitants of a city and the number of registered motor vehicles. Cities with a larger number of residents, as a rule, have a higher number of registered motor vehicles and *vice versa*.

In cases when the values of one variable increase when the corresponding values of the second variable decrease, we talk about a negative relationship. Experiences from several cities show that the number of vehicles entering the downtown (business)district decreases when the congestion

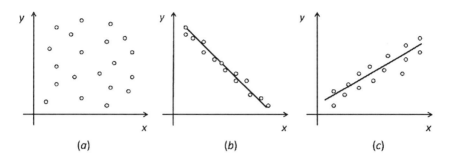

Figure 5.30 Scatter diagram

pricing fee increases. In this case, there is a negative relationship between the number of vehicles entering the central business district and the congestion pricing fee.

The relationship (positive or negative) between the two variables can be stronger or weaker. The scatter diagram (Figure 5.30) helps us to get a certain impression about the strength of the relationship.

By visual inspection of the scatter diagram, we can easily discover that, in case (*a*), there is no relationship between the variable x and the variable y. In case (*b*), a strong negative relationship is observed. In case (*c*), the relationship is positive, but less strong than in the case (*b*).

There are several correlation measures reported in the literature. The most widely used one is the Pearson correlation coefficient that is defined in the following way:

$$r = \frac{\sum_{i=1}^{n}(x_i - \bar{x}) \cdot (y_i - \bar{y})}{\sqrt{\sum_{i=1}^{n}(x_i - \bar{x})^2}\sqrt{\sum_{i=1}^{n}(y_i - \bar{y})^2}}$$

The correlation coefficient r takes the values from the interval $[-1,1]$. This coefficient is without dimensionality and does not depend on the dimensions of variables x and y. The closer the correlation coefficient to one, the stronger the linear relationship between the independent variable x and the dependent variable y. The closer the correlation coefficient is to zero, the weaker the linear relationship is between the independent variable x and the dependent variable y.

The coefficient r^2 is called the coefficient of determination.

Example 5.16

The values of independent variable x and dependent variable y are shown in Table 5.18. Determine the correlation coefficient between these variables.

Table 5.18 The values of the independent
variable x and dependent variable
y used to calculate the
correlation coefficient

x	y
2	3
6	6
8	12
12	12
16	15
18	21
22	24
28	27

The arithmetic means of the variables x and y are respectively equal:

$$\bar{x} = \frac{2+6+8+12+16+18+22+28}{8} = \frac{112}{8} = 14$$

$$\bar{y} = \frac{3+6+12+12+15+21+24+27}{8} = \frac{120}{8} = 15$$

The correlation coefficient (Table 5.19) reads:

$$r = \frac{\sum_{i=1}^{n}(x_i - \bar{x})\cdot(y_i - \bar{y})}{\sqrt{\sum_{i=1}^{n}(x_i - \bar{x})^2 \sum_{i=1}^{n}(y_i - \bar{y})^2}} = \frac{504}{\sqrt{528\cdot505}} = 0.976$$

Since the correlation coefficient is close to one, we conclude that there is a strong linear relationship between the independent variable x and the dependent variable y.

Let us show how the correlation coefficient can be determined by Excel. For that purpose, we use the same example that is given above. Values of x and y are given in Figure 5.31. We can determine the correlation coefficient by the following Excel function "=CORREL(A2:A9,B2:B9)". By using this function, we obtain the value of 0.977.

The second way is to use the tool from Data Analysis. We take Data Analysis from the Data Menu (see Figure 5.32).

We choose "correlation" from the list of tools. In the Input Range field, we select cells with input data (see Figure 5.33). We select new workbook for output options.

The results obtained are given in Figure 5.34. We note from the Figure that the correlation between Column 2 (dependent variable y) and Column 1 (independent variable x) is 0.977.

Table 5.19 Calculation of the correlation coefficient

x_i	y_i	$x_i - \bar{x}$	$y_i - \bar{y}$	$(x_i - \bar{x})^2$	$(x_i - \bar{x}) \cdot (y_i - \bar{y})$	$(y_i - \bar{y})^2$
2	3	-12	-12	144	144	144
6	6	-8	-9	64	72	81
8	12	-6	-3	36	18	9
12	12	-2	-3	4	6	9
16	15	2	0	4	0	0
18	21	4	6	16	24	36
22	24	8	9	64	72	81
28	27	14	12	196	168	144
				$\sum_{i=1}^{n}(x_i - \bar{x})^2 = 528$	$\sum_{i=1}^{n}(x_i - \bar{x}) \cdot (y_i - \bar{y}) = 504$	$\sum_{i=1}^{n}(y_i - \bar{y})^2 = 505$

	A	B	C
1	x	y	
2	2	3	
3	6	6	
4	8	12	
5	12	12	
6	16	15	
7	18	21	
8	22	24	
9	28	27	
10			
11			

Figure 5.31 Input data

5.4.3 Multiple linear regression

In many cases, the dependent variable depends on more than one independent variable. For example, when estimating transportation demand, analysts usually choose for the dependent variable the number of trips, the number of passengers, the number of operations, or the number of passenger kilometers. Independent variables are chosen from the set of socioeconomic characteristics and characteristics of the transportation system. Multiple regression analysis has been widely used to predict transportation demand. In many cases, nonlinear regression (which does not assume a linear relationship between the variables) is also used for prediction. The typical independent variables are population, income, employment, volume of trade, service frequencies, total travel times, fares, etc.

The following is the form of a multiple linear regression:

$$y = a_0 + a_1 \cdot x_1 + a_2 \cdot x_2 + \cdots + a_n \cdot x_n$$

where:
   y – dependent variable

Figure 5.32 Data Analysis in Data Menu

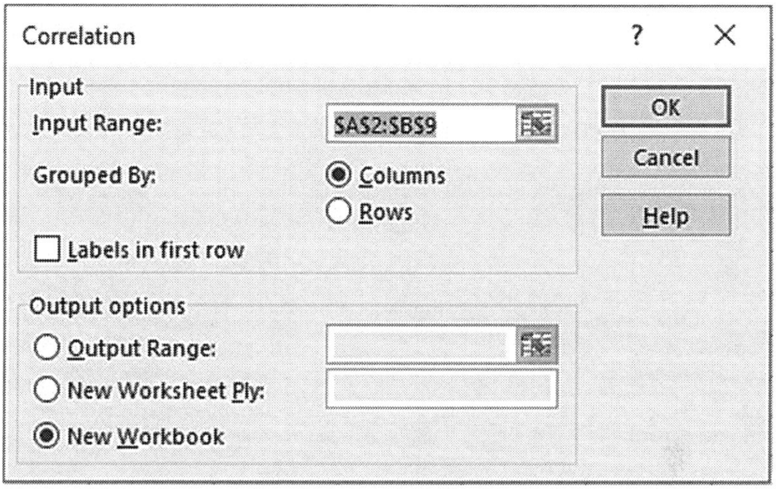

Figure 5.33 Correlation frame

$x_1, x_2, x_3, \ldots, x_n$ – independent (explanatory) variables
$a_0, a_1, a_2, a_3, \ldots, a_n$ – regression coefficients

The simplest case of a multiple linear regression is the case with one dependent and two independent variables:

$$y = a_0 + a_1 \cdot x_1 + a_2 \cdot x_2$$

The dependent variables are sometimes called outcome variables or regressands. The independent variables are often called the predictor variables or regressors.

Let us determine the regression coefficients in a case with one dependent and two independent variables. We use the least squares method.

	A	B	C	D
1		*Column 1*	*Column 2*	
2	Column 1	1		
3	Column 2	0.977008	1	
4				
5				
6				

Figure 5.34 Results of correlations

The sum S of the squared differences between the y_i and their estimated values equals:

$$S = \frac{1}{n} \sum_{i=1}^{n} \left(a_0 + a_1 \cdot x_1 + a_2 \cdot x_2 - y_i \right)^2$$

We take partial derivatives of the sum S with respect to a_0, a_1, and a_2. After equating partial derivatives to 0, the following set of equations is obtained:

$$\sum_{i=1}^{n} y_i = a_0 \cdot n + a_1 \cdot \sum_{i=1}^{n} x_{1i} + a_2 \cdot \sum_{i=1}^{n} x_{2i}$$

$$\sum_{i=1}^{n} y_i \cdot x_{1i} = a_0 \cdot \sum_{i=1}^{n} x_{1i} + a_1 \cdot \sum_{i=1}^{n} x_{1i}^2 + a_2 \cdot \sum_{i=1}^{n} x_{1i} \cdot x_{2i}$$

$$\sum_{i=1}^{n} y_i \cdot x_{2i} = a_0 \cdot \sum_{i=1}^{n} x_{2i} + a_1 \cdot \sum_{i=1}^{n} x_{1i} \cdot x_{2i} + a_2 \cdot \sum_{i=1}^{n} x_{2i}^2$$

The regression coefficients a_0, a_1, and a_2 are obtained after solving this system of equations.

Example 5.17

Models where transportation demand is a function of socio-economic characteristics and the characteristics of the transportation system can be written in the following general form:

$$D = a \cdot \prod_{i=1}^{m} S_i^{b_i} \cdot \prod_{j=1}^{n} T_j^{c_j}$$

where:

   m – the total number of socio-economic characteristics
   n – the total number of transportation system characteristics
   D – the number of passengers
   S_i – the value of the i-th socio-economic characteristic
   T_j – the value of the j-th transportation system characteristic
   a, b_i, c_j – coefficients to be estimated

Let us introduce the simplified assumption that the yearly number of air passengers D from one city depends on the city's total population s_1, the city's average income per capita s_2, and the yearly number of offered seats in the aircraft departing the city T_1. In other words:

$$D = a \cdot S_1^{b_1} \cdot S_2^{b_2} \cdot T_1^{c_1}$$

The number of passengers D is not a linear function of independent variables s_1, s_2, and T_1. We transform our problem to linearity. On taking logarithms, we obtain the following relation:

$$\log D = \log\left(a \cdot S_1^{b_1} \cdot S_2^{b_2} \cdot T_1^{c_1}\right)$$

$$\log D = \log a + b_1 \cdot \log S_1 + b_2 \cdot \log S_2 + c_1 \cdot \log T_1$$

After substitution:

$$y = \log D$$

$$a_0 = \log a$$

$$x_1 = \log S_1$$

$$x_2 = \log S_2$$

$$x_3 = \log T_1$$

we obtain the following multiple linear regression model:

$$y = a_0 + b_1 \cdot x_1 + b_2 \cdot x_2 + c_1 \cdot x_3$$

We can now estimate the regression coefficients a_0, b_1, b_2, and c_1 by the least squares method. Once the regression coefficients have been determined, the number of passengers D can be predicted by using the formula $D = a \cdot S_1^{b_1} \cdot S_2^{b_2} \cdot T_1^{c_1}$.

5.5 DISCRETE OUTCOME MODELS

Frequently, within the transport analysis, discrete or nominal data appear as the model outcome. We use nominal scales for labeling variables, which have no quantitative value. The nominal scales that appear in transportation analysis are mutually exclusive. Some authors suggest that nominal scales should be called "labels", or "names". Thus, for example, within the surveys related to mode of transport choice in a city, the following question may appear: From home to work you: (a) walk; (b) use bicycle; (c) use car; (d) ride public transport. In some analyses and surveys, ordinal discrete data appear. The typical is the following question. How many times per week do you use public transport: (1) I do not use public transport; (2) once per week; (2) two times per week,... etc.

In addition to appearing in mode choice problems, discrete outcomes also appear in many other transportation problems. Trip makers decide whether or not to make a certain trip. They choose the destination and the mode of transportation, choose the departure time, choose the carrier and route. Drivers choose the route within the city, while air passengers make the choice of airline carrier, and the class of transportation (business, first, and tourist class in airplanes). All these discrete outcomes are the result of the specific behavior of passengers, and drivers.

The discrete outcome (the dependent variable which is discrete) represents choice (mode choice, route choice), or category (type of traffic accident). Probit and logit models are the best-known and most widely used discrete outcome models.

5.5.1 Basics of discrete choice models

Discrete choice models begin with the passenger as the trip's decision maker. A decision maker is faced with the problem of choice of one alternative from a finite set of mutually exclusive alternatives. Discrete choice models are disaggregated by their nature. This means that these models try to describe, analyze, and predict choice behavior of the individual passengers or other decision makers (organization, household, shipper, etc.). Discrete choice models are able to provide an explanation as to why individual decision makers make specific choices in a specific situation. These models are also capable of predicting changes in choice behavior due to changes in a decision maker's characteristics and changes in alternatives' attributes. For example, in many situations, it is valuable to analyze at what limit the probability of choosing a transportation mode will decline, e.g. if the expense for that mode increased by a specific amount.

In addition to the frequency, travel time, and cost of different modes of transportation, many other factors are present, the effect of which on the decision-making process cannot be quantified without pooling the passenger population. Therefore, without surveys, it is impossible to determine the effect of comfort, the passenger's feeling of safety during the trip, or schedule reliability on transportation mode choice.

The majority of models applied to travel behavior analysis are based on utility theory. Utility denotes the measure of the alternative attractiveness. Utility depends on the individual passenger or driver who makes the choice. Utility maximization discrete choice models assume that the decision maker always selects the alternative having the maximum utility (Figure 5.34). These deterministic choice models have been widely used in transportation engineering. Deterministic choice models assume that the process of choosing a mode of transportation, for example, is basically deterministic, and that the individual permanently selects the mode of transportation in the same way. In other words, it is assumed that if the individual chose the bus once when traveling from city A to city B, he/she will also choose the bus in subsequent situations. These models also assume that passengers with similar characteristics will choose in an identical way when confronted with the same set of alternatives.

Figure 5.35 shows deterministic choice in the case of two alternatives. The utility of Alternative 1 is plotted along the horizontal axis, while the utility of Alternative 2 is plotted along the vertical axis. The utilities of specific passengers are shown within the utility space. The points that have equal utilities for the two considered alternatives lie on the 45° line. All six passengers above the 45° line have higher utility for Alternative 2 than for Alternative 1. These six passengers select Alternative 2. In the same way, four passengers below the 45° line, that have greater utility for Alternative 1, select Alternative 1.

Many real-life examples showed that the assumption about choosing the alternative having the greatest utility has not always been accurate. This triggered the development of the choice models that also incorporate a probabilistic aspect. Stochastic choice models assume that the choice process is subject to many random effects that cannot be exactly perceived. Stochastic models more reasonably explain the process of passenger choice than do the deterministic models.

Let us note passenger q who travels between two cities. We denote by A_i to one of the possible alternatives to make this trip. We also denote by choice set C_q to represent a set of considered alternatives. To every alternative, we join function V_{iq}, called the choice function. The choice function

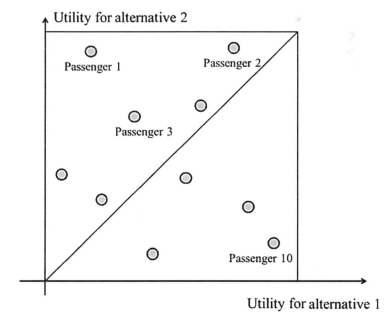

Figure 5.35 Deterministic choice

is a function of the alternative attributes and the characteristics of the passenger. The choice function V_{iq} is mostly of the form:

$$V_{iq} = \sum_j a_i \cdot x_{iqj}$$

where:

x_{iqj} – variables affecting the choice of mode that relate to alternative A_i (total travel time, total travel cost, number of transfers, walking distance, schedule reliability, etc.)

a_i – parameters estimated when calibrating the choice model that indicate the effect of variables x_{iqj} on the choice of alternative A_i

We represent the utility of each alternative as the weighted sum of the alternative attributes.

The choice process is subject to many random effects. Passengers often do not have complete information on the frequency of competitive modes of transportation, on the departure schedules, probability of getting a vacant seat at a specific time of day, or the probability of finding a parking space. In some cases, passengers do not know precisely the cost of some of the modes. In the same vein, some passengers do not opt for a certain alternative even we think it would be very logical to do so because of its advantages compared to other alternatives.

We denote by U_{iq} to the utility that passenger q connects with alternative A_i. The utility U_{iq} is given by:

$$U_{iq} = V_{iq} + \varepsilon_{iq}$$

where

V_{iq} is the deterministic part of the utility
ε_{iq} is the random term

This random term practically represents the piece of the utility unidentified by the analyst. The total error ε, that is the sum of errors from many sources (inadequate information, absence of some modal attributes, exclusion of some characteristics of the passenger, etc.), is represented by a random variable. For example, two passengers, with the same or similar attributes, can make different choices. In addition, some passengers and/ or drivers sometimes behave irrationally and do not always choose the best alternative. For a different probability density function of the random term, different stochastic choice models are obtained. The widespread assumption for the error distribution is that errors are distributed normally. This assumption leads to the multinomial probit model (MNP). We do not cover MNP in this book. It has been shown that this model has some properties that make it complicated to apply in choice analysis. This caused

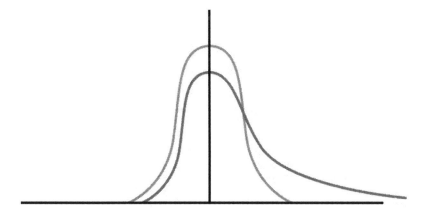

Figure 5.36 Probability density function for Gumbel and normal Distributions (same mean and variance)

researchers to use the assumption about the Gumbel distribution of error term (Figure 5.36).

The Gumbel distribution is a very good approximation of the normal distribution. Additionally, the Gumbel distribution creates a closed-form probabilistic choice model.

5.5.2 Logit model

One of the best-known stochastic choice models is the *logit* model. The logit model, was first introduced in the case of binary choice. The *multi-nomial logit* represents the generalization of the original model to more than two alternatives. The family of logit models was extensively used in various transportation studies. The logit model is based on the assumption that random variables representing random terms are independent and distributed by *Gumbel*'s probability density function. It has been shown, in this case, that the probability that the alternative A_i will be chosen, equals:

$$P(i) = \frac{e^{V_i}}{\sum_k e^{V_k}}$$

The upper equation is known as the *logit* model. The standard utility function of the logit model is linear. In some cases, the choice function is not linear so the least squares method or the multiple regression technique can only be used to estimate the parameters of the choice function in special cases. The maximum likelihood method is used to estimate parameters of choice functions. Choice models can be based on disaggregated or aggre-gated data. Choice models based on disaggregated data presume that each passenger evaluates the advantages and disadvantages of each alternative differently, so that the passenger population must be polled before estimating the parameters of the choice function. In this case, each polled passenger corresponds to a different choice function value. For choice models, based on aggregate data, all passengers correspond to the same choice function value, since it is assumed that the variables in the choice function are equal for all passengers. Therefore, the probability of choosing a certain alternative for aggregated data is equal for all passengers.

Example 5.18

Let us consider the following example. Passengers traveling between node A and node B can use private cars, or public transport (Figure 5.37).

A modal split has been calibrated using the maximum likelihood technique (an advanced statistical method). The following equations that describe utilities have been obtained:

$$U_{\text{Car}} = 2.4 - 0.2 \cdot C - 0.02 \cdot T$$

$$U_{\text{Publictransit}} = 0.4 - 0.2 \cdot C - 0.02 \cdot T$$

where:
  C – the out-of-pocket cost [$]
  T – the travel time [minutes]

Calibrated logit models help us to answer the following questions: How many trips between A and B will be by car? How many trips between A and B will be by public transport? The following are characteristics of the competitive modes:

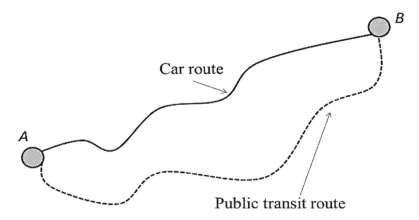

Figure 5.37 Logit model: Passengers traveling between node A and node B can use private cars or public transport

Travel time (transit) = 1 hour = 60 minutes
Travel cost (transit) = $2
Travel time (auto) = 40 minutes
Travel cost (car) = $8.00 (including parking)

The utilities U_{Car} and $U_{Publictransit}$ equal:

$$U_{Car} = 2.4 - 0.2 \cdot 8 - 0.02 \cdot 40 = 2.4 - 1.6 - 0.8 = 0$$

$$U_{Publictransit} = 0.4 - 0.2 \cdot 2 - 0.02 \cdot 60 = 0.4 - 0.4 - 1.2 = -1.2$$

The estimated probabilities of travel by competitive modes are:

$$P_{car} = \frac{e^{U_{car}}}{e^{U_{car}} + e^{U_{Publictransit}}} = \frac{e^0}{e^0 + e^{-1.2}} = 0.77$$

$$P_{Publictransit} = \frac{e^{U_{Publictransit}}}{e^{U_{car}} + e^{U_{Publictransit}}} = \frac{e^{-1.2}}{e^0 + e^{-1.2}} = 0.23$$

How should we understand the results obtained? We can translate the calculated probabilities to percentages and say: (a) The probability that a traveler from A to B uses a car is 77%; (b) The probability that a passenger from A to B uses public transport is 23%. A similar interpretation is that 77% of travelers between A and B use a car, and 23% of travelers ride public transport. Why is this important? The logit model and similar models are helpful tools in the transportation planning process. The logit model allow s us to carry out sensitivity analysis. For example, imagine that the car cost increases significantly (due to a significant increase in parking fees). Let the new car cost equals $20. The new utilities U_{Car} and $U_{Publictransit}$ then equal:

$$U_{Car} = 2.4 - 0.2 \cdot 20 - 0.02 \cdot 40 = 2.4 - 4 - 0.8 = -2.4$$

$$U_{Publictransit} = 0.4 - 0.2 \cdot 2 - 0.02 \cdot 60 = 0.4 - 0.4 - 1.2 = -1.2$$

The new estimated probabilities of travel by competitive modes are:

$$P_{car} = \frac{e^{U_{car}}}{e^{U_{car}} + e^{U_{Publictransit}}} = \frac{e^{-2.4}}{e^{-2.4} + e^{-1.2}} = \frac{e^{-2.4}}{e^{-2.4} + e^{-1.2}} = \frac{0.09}{0.09 + 0.3} = 0.23$$

$$P_{Publictransit} = \frac{e^{U_{Publictransit}}}{e^{U_{car}} + e^{U_{Publictransit}}} = \frac{e^{-1.2}}{e^{-2.4} + e^{-1.2}} = \frac{0.3}{0.09 + 0.3} = 0.77$$

The cost of the car is quite high and forces many drivers to switch from car and to public transport.

There are a lot of variations, modifications, and improvements of the original logit model (Multinomial Logit Model, Nested Logit Model, Cross-nested Logit model, C-Logit, etc.).

Example 5.19

Assume that passengers taking a private trip on a certain route choose between two possible modes of transportation solely on the basis of the travel cost. We assume that the following choice function has been calibrated:

$$V_i = -0.15 \cdot c_i$$

where:

V_i is the utility of the i-th alternative
c_i is the travel cost of the i-th alternative.

Data on the travel cost are given in Table 5.20.

The probabilities $P(A_1)$ and $P(A_2)$ that the competitive transportation modes will be chosen are, respectively, equal to:

$$P(A_1) = \frac{e^{-0.15 \cdot 10}}{e^{-0.15 \cdot 10} + e^{-0.15 \cdot 20}} = 0.81$$

$$P(A_2) = \frac{e^{-0.15 \cdot 20}}{e^{-0.15 \cdot 10} + e^{-0.15 \cdot 20}} = 0.19$$

We assume that a new mode of transportation joins the route in question. The cost c_3 of a new mode equals 15. We then calculate the market share between three modes of transportation. The probabilities of different modes of transportation being used are:

$$P(A_1) = \frac{e^{-0.15 \cdot 10}}{e^{-0.15 \cdot 10} + e^{-0.15 \cdot 15} + e^{-0.15 \cdot 20}} = 0.58$$

$$P(A_2) = \frac{e^{-0.15 \cdot 20}}{e^{-0.15 \cdot 10} + e^{-0.15 \cdot 15} + e^{-0.15 \cdot 20}} = 0.13$$

$$P(A_3) = \frac{e^{-0.15 \cdot 15}}{e^{-0.15 \cdot 10} + e^{-0.15 \cdot 15} + e^{-0.15 \cdot 20}} = 0.29$$

Table 5.20 Data on the travel cost

Transportation mode	Travel cost in monetary units
A_1	10
A_2	20

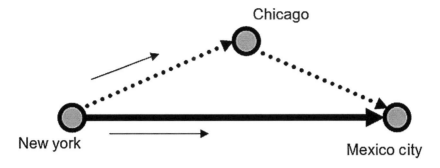

Figure 5.38 Route choice in air transportation

As we can see, by introducing a third mode of transportation on the route in question, the share of the first mode decreases from 81% of all passengers transported to 58%, and the second mode goes from 19% to 13%.

Example 5.20

Two air carriers fly between New York and Mexico City (Figure 5.38).

Carrier 1 flies non-stop between New York and Mexico City. The average ticket price equals $434 (round trip). The flight time equals 5 h 5 min (305 minutes). When flying from New York to Mexico City, carrier 2 makes a stopover in Chicago. The average ticket price equals $374. Flight time (including changing plane in Chicago) equals 8 h 10 min (490 minutes). Air passengers can choose between a non-stop flight and a less expensive flight with one stopover. The following choice function has been calibrated:

$$V_i = -0.05 \cdot c_i - 0.02 \cdot t_i$$

(a) Calculate market share of every air carrier; (b) Carrier 1 wants to increase the average ticket price to $470. Estimate the new market shares.

SOLUTION:

(a) The utilities are:

$$V_1 = -0.05 \cdot 434 - 0.02 \cdot 305 = -27.8$$

$$V_2 = -0.05 \cdot 374 - 0.02 \cdot 490 = -28.5$$

The probabilities that the competitive air carriers will be chosen are:

$$p_1 = \frac{e^{-27.8}}{e^{-27.8} + e^{-28.5}} = 0.668188$$

$$p_2 = \frac{e^{-28.5}}{e^{-27.8}+e^{-28.5}} = 0.33812$$

These probabilities represent carriers' market shares. In other words, carrier 1 can attract approximately 67% of the total market, while carrier 2 captures 33%.

(a) The new market shares are:

$$V_1 = -0.05 \cdot 470 - 0.02 \cdot 305 = -29.6$$

$$V_2 = -0.05 \cdot 374 - 0.02 \cdot 490 = -28.5$$

The new values of probabilities are:

$$p_1 = \frac{e^{-29.6}}{e^{-29.6}+e^{-28.5}} = 0.24974$$

$$p_2 = \frac{e^{-28.5}}{e^{-29.6}+e^{-28.5}} = 0.75026$$

We conclude that, after introducing the new ticket prices, carrier 1 would significantly decrease market share (from approximately 67% of the market to 25% of the total market).

5.5.3 Independence of irrelevant alternatives property

One of the most important characteristics of the Multinomial Logit Model (MNL) is its independence from the irrelevant alternatives (IIA) property. The premise in the MNL is that additional alternatives are irrelevant in deciding between the two alternatives in the pair. Let us clarify this property by using the following example. Passengers on a certain route choose between train, bus, and private cars. The probabilities of different modes of transportation being used are:

$$P(\text{Train}) = \frac{e^{V(\text{Train})}}{e^{V(\text{Train})} + e^{V(\text{Bus})} + e^{V(\text{Car})}}$$

$$P(\text{Bus}) = \frac{e^{V(\text{Bus})}}{e^{V(\text{Train})} + e^{V(\text{Bus})} + e^{V(\text{Car})}}$$

$$P(\text{Car}) = \frac{e^{V(\text{Car})}}{e^{V(\text{Train})} + e^{V(\text{Bus})} + e^{V(\text{Car})}}$$

The ratios of each pair of these probabilities equal:

$$\frac{P(\text{Train})}{P(\text{Bus})} = \frac{\dfrac{e^{V(\text{Train})}}{e^{V(\text{Train})} + e^{V(\text{Bus})} + e^{V(\text{Car})}}}{\dfrac{e^{V(\text{Bus})}}{e^{V(\text{Train})} + e^{V(\text{Bus})} + e^{V(\text{Car})}}} = \frac{e^{V(\text{Train})}}{e^{V(\text{Bus})}} = e^{V(\text{Train})-V(\text{Bus})}$$

$$\frac{P(\text{Train})}{P(\text{car})} = \frac{\dfrac{e^{V(\text{Train})}}{e^{V(\text{Train})} + e^{V(\text{Bus})} + e^{V(\text{Car})}}}{\dfrac{e^{V(\text{Car})}}{e^{V(\text{Train})} + e^{V(\text{Bus})} + e^{V(\text{Car})}}} = \frac{e^{V(\text{Train})}}{e^{V(\text{Car})}} = e^{V(\text{Train})-V(\text{Car})}$$

$$\frac{P(\text{Bus})}{P(\text{Car})} = \frac{\dfrac{e^{V(\text{Bus})}}{e^{V(\text{Train})} + e^{V(\text{Bus})} + e^{V(\text{Car})}}}{\dfrac{e^{V(\text{Car})}}{e^{V(\text{Train})} + e^{V(\text{Bus})} + e^{V(\text{Car})}}} = \frac{e^{V(\text{Bus})}}{e^{V(\text{Car})}} = e^{V(\text{Bus})-V(\text{Car})}$$

As can be seen from the previous relations, the ratio of probabilities for any pair of alternatives depends purely on the attributes of those alternatives. This ratio does not depend on the attributes of the third alternative, fourth alternative, etc. This ratio is unchanged in any case (whether the third, or the fourth transportation mode is available or not). The IIA property permits the addition or elimination of an alternative from the choice set with no influence on the structure or parameters of the model. In contrast, there is a criticism of the MNL, for its IIA property. The unwanted characteristic of the IIA property means that the setting up of a new transportation mode on an observed route will decrease the probability of existing transportation modes proportionally to their probabilities prior to the change.

Some transportation modes could be similar. They could share attributes that are not contained in the utility function. For example, buses and trams in public transport have similar fare structures, similar levels of privacy, etc. The nested logit grouped transportation modes (alternatives) that were more similar to each other. In other words, the NL model forms groups (nests) of similar alternatives. In this way, the unwanted characteristic of the IIA property could be beaten.

5.5.4 The red bus/blue bus paradox

The experience gained in MNL applications showed that, in some situations, the MNL can produce incorrect predictions of choice probabilities. The classical example that illustrates such a situation is the "red bus/blue bus paradox." Let us analyze the case when commuters can choose between the car and the blue bus. The characteristics of the transport system and commuters are such that 2/3 of the users choose the car, and the remaining 1/3 of the users choose a blue bus.

Let us imagine a situation in which a competitive public transport operator introduces a red bus service on the same route. These red buses have the same characteristics as blue buses and the new operator offers the same ticket price, the same schedule, and serves the same bus stops. In other words, competitive blue and red buses differ only in color. It is logical to assume that the commuters who have used the car will continue to use the car, and that the population of commuters who have used the blue bus will split and that one half of them will use the blue bus and the other half the red bus. Logical thinking leads us to the conclusion that the introduction of a red bus cannot change the existing modal split between cars and buses.

Consequently, the probability of choosing specific transportation modes should accordingly be: car, 2/3; blue bus, 1/6 and red bus, 1/6. On the other hand, due to the IIA property, the MNL will preserve the ratio between car probability P_{Car} and blue bus probability $P_{Blue\ bus}$ as:

$$\frac{P_{Car}}{P_{Blue\ bus}} = \frac{\frac{2}{3}}{\frac{1}{3}} = 2$$

i.e.

$$P_{Car} : P_{Blue\ bus} = 2 : 1$$

Assuming that commuters choose a red or blue bus with the same probability, we have the following:

$$P_{Red\ bus} : P_{Blue\ bus} = 1 : 1$$

Consequently, MNL calculates that the share probabilities for the three alternatives under consideration equal:

$$P_{Car} = \frac{1}{2}$$

$$P_{Blue\ bus} = \frac{1}{4}$$

$$P_{Red\ bus} = \frac{1}{4}$$

The MNL generated the unacceptable statement that the probability that commuters will choose car will drop from two-thirds to one half, as a consequence of introducing a new transportation mode (red bus) which is the same as one of the existing transportation modes (blue bus).

5.5.5 Logit model estimation

Let us study travelers that have to choose transportation mode. Let us assign I to the set of individuals, and J to the set of alternatives (transportation modes) to be chosen. We collect the data related to the travelers' choices of transportation modes, as well as the data related to the values of the choice function's variables. We note the i-th traveler, and the j-th alternative (transportation mode). The number of probabilities p_{ij} that are to be calculated in the logit model based on disaggregated data (in the case of disaggregated data, the values of the variables in the choice function are different for all travelers) is very large. This number equals the product of the number of observed travelers and the number of competitive transportation modes. In the case of the logit model based on aggregated data, we assume that the values of the variables in the choice function are equal for all travelers. Consequently, the probability of choosing a specific transportation mode is constant, and does not vary from traveler to traveler. In other words, the following is satisfied in the case of the aggregate logit model:

$$p_{ij} = p_j \qquad i = 1, 2, \ldots, |I|$$

Parameters of the logit model are estimated by the maximum likelihood method. This method discovers parameters that maximize the likelihood that the sample was produced from the model with the chosen parameter values. In other words, the maximum likelihood method finds the values of parameters that are most likely to produce the choices detected in the sample. The likelihood function $L(\alpha)$ for a sample composed of $|I|$ individuals, and $|J|$ alternatives is defined in the following way:

$$L(\alpha) = \prod_{\forall i \in I} \prod_{\forall j \in J} \left(P_{ij}(\alpha) \right)^{\delta_{ij}}$$

where:

$$\delta_{ij} = \begin{cases} 1, & \text{if } i\text{-th individual chooses } j\text{-th alternative} \\ 0, & \text{otherwise} \end{cases}$$

The values of the parameters that maximize the likelihood function are obtained by equating the first derivative of the likelihood function to zero. The log of a likelihood function produces the same maximum as the likelihood function. We introduce into the analysis log of a likelihood function, as a replacement for the likelihood function itself, since it is easier to differentiate log of a likelihood function. Log-likelihood function $LL(\alpha)$ equals:

$$LL(\alpha) = Log\big(L(\alpha)\big)$$

$$LL(\alpha) = \prod_{\forall i \in I} \prod_{\forall j \in J} \delta_{ij} \cdot \ln\big(P_{ij}(\alpha)\big)$$

By equating the first derivative of the likelihood function $LL(\alpha)$ to zero, we obtain the values of the parameters that maximize the likelihood function.

Example 5.21

Drivers taking a trip between two nodes choose between two possible routes on the basis of perceived travel time. Data on the perceived travel time are given in Table 5.21.

A traffic engineer observed the decisions of 1,000 drivers. The total of 700 drivers chose the first route. The remaining 300 drivers chose the second route. Calibrate the binomial aggregate logit model based on these data.

SOLUTION:

Since we have to calibrate the binomial aggregated logit model, we assume that the probability of choosing a specific transportation mode is constant and does not vary from traveler to traveler. The choice function equals:

$$V(i) = a \cdot t_i i = 1, 2$$

where:

a – parameter to be estimated

t_i – perceived travel time of the i-th route

The probabilities of route 1 and route 2 being chosen by any driver are:

$$P_1 = \frac{e^{5 \cdot a}}{e^{5 \cdot a} + e^{10 \cdot a}}$$

$$P_2 = \frac{e^{10 \cdot a}}{e^{5 \cdot a} + e^{10 \cdot a}}$$

Table 5.21 Data on the perceived travel time

Route	Perceived travel time
1	5
2	10

Log-likelihood function $LL(a)$ equals:

$$LL(a) = \prod_{\forall i \in I} \prod_{\forall j \in J} \delta_{ij} \cdot \ln\left(P_{ij}(a)\right)$$

$$LL(a) = 700 \cdot \ln(P_1) + 300 \cdot \ln(P_2)$$

$$LL(a) = 700 \cdot \ln\left(\frac{e^{5 \cdot a}}{e^{5 \cdot a} + e^{10 \cdot a}}\right) + 300 \cdot \ln\left(\frac{e^{10 \cdot a}}{e^{5 \cdot a} + e^{10 \cdot a}}\right)$$

$$LL(a) = 700 \cdot 5a + 300 \cdot 10a - 1{,}000 \cdot \ln\left(e^{5 \cdot a} + e^{10 \cdot a}\right)$$

By equating the first derivative $\dfrac{dLL(a)}{da}$ of the likelihood function $LL(a)$ to zero we get the following equation:

$$1{,}500 \cdot e^{5 \cdot a} - 3{,}500 \cdot e^{10 \cdot a} = 0$$

The solution of this equation equals:

$$a = -0.1694$$

The probabilities of route 1, and route 2 being chosen are respectively equal:

$$P_1 = \frac{e^{5 \cdot (-0.1694)}}{e^{5 \cdot (-0.1694)} + e^{10 \cdot (-0.1694)}} = 0.7$$

$$P_2 = \frac{e^{10 \cdot (-0.1694)}}{e^{5 \cdot (-0.1694)} + e^{10 \cdot (-0.1694)}} = 0.3$$

Example 5.22

Business travelers traveling on a specific route during the winter choose between the airplane and train uniquely on the basis of the schedule reliability. Let there be a possible evaluation from 0 to 10, with 0 denoting a totally unreliable transportation mode, and 10 an exceptionally reliable mode. The transportation planner has to make the travelers' choice prediction between air transportation and train. The following are assumed utility functions of the competitive transportation modes:

$$V_{Air} = a \cdot r_{Air}$$

$$V_{Train} = a \cdot r_{Train}$$

Table 5.22 Estimated schedule reliability and performed choices of seven travelers

Traveler	Estimated airplane schedule reliability	Estimated train schedule reliability	Chosen transportation mode
1	9	7	Aircraft
2	8	6	Aircraft
3	9	8	Aircraft
4	7	10	Train
5	8	9	Train
6	5	7	Train
7	9	10	Train

where:

r_{Air} – perceived air transportation schedule reliability

r_{Train} – perceived train schedule reliability

The transportation planner assumes that the values of the variables in the choice function are different for all travelers. She observed the mode choice of seven travelers. The estimated schedule reliability and performed choices of seven passengers are shown in Table 5.22.

Calculate the value of the parameter a that maximizes the likelihood function.

SOLUTION:

We treat air transportation as the first transportation mode ($j = 1$), and train as the second transportation mode ($j = 2$). For the considered sample of travelers, the log-likelihood reads:

$$LL(a) = \prod_{\forall i \in I} \prod_{\forall j \in J} \delta_{ij} \cdot \ln\left(P_{ij}(a)\right)$$

$$LL(\alpha) = 1 \cdot \ln P_{11} + 0 \cdot \ln P_{12} + 1 \cdot \ln P_{21} + 0 \cdot \ln P_{22} + 1 \cdot \ln P_{31}$$

$$+ 0 \cdot \ln P_{32} + 0 \cdot \ln P_{41} + 1 \cdot \ln P_{42} + 0 \cdot \ln P_{51} + 1 \cdot \ln P_{52}$$

$$+ 0 \cdot \ln P_{61} + 1 \cdot \ln P_{62} + 0 \cdot \ln P_{71} + 1 \cdot \ln P_{72}$$

$$LL(a) = \ln P_{11} + \ln P_{21} + \ln P_{31} + \ln P_{42} + \ln P_{52} + \ln P_{62} + \ln P_{72}$$

The probabilities P_{11}, P_{22}, P_{31}, and P_{42} equal:

$$P_{11} = \frac{e^{9a}}{e^{9a} + e^{7a}}$$

$$P_{21} = \frac{e^{8a}}{e^{8a} + e^{6a}}$$

$$P_{31} = \frac{e^{9a}}{e^{9a} + e^{8a}}$$

$$P_{42} = \frac{e^{10a}}{e^{7a} + e^{10a}}$$

$$P_{52} = \frac{e^{9a}}{e^{8a} + e^{9a}}$$

$$P_{62} = \frac{e^{7a}}{e^{5a} + e^{7a}}$$

$$P_{72} = \frac{e^{10a}}{e^{9a} + e^{10a}}$$

By equating the first derivative $\dfrac{dLL(a)}{da}$ of the likelihood function LL

(a) to zero, we get the following solution:

$$a \approx 10.4$$

5.6 THE EXTRAPOLATION METHODS

There are two major groups of quantitative forecasting techniques: (a) the extrapolation method (time series models); and (b) the explanatory method (regression models). Explanatory models (regression models) are very suitable approaches that allow us to look at various "what-if" scenarios. As we already explained in previous sections, when using the explanatory model, the value of a dependent variable (potential transportation demand, for example) is forecasted on the basis of known and quantifiable factors that induce the transportation demand (socio-economic characteristics and characteristics of the transportation system). By the extrapolation method, future transportation demand is forecasted on the basis of earlier characteristics of the demand over a period of time. The extrapolation method can be, above all, used to forecast the future values of a dependent variable under unchanged conditions.

Generally, transportation demand could have trend components, cyclical components, seasonal peaks, and random variations (Figure 5.39). The trend represents permanent, generally increasing or generally decreasing patterns. Pattern could be initiated by a variety of causes. Seasonal components represent the usual pattern of up-and-down oscillations. For

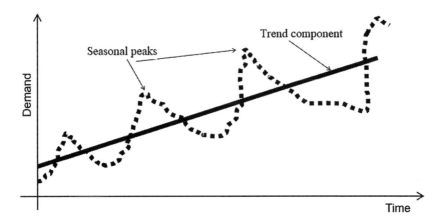

Figure 5.39 Components of transportation demand

example, oscillations take place within a specific year on a lot of touristic routes in air transportation. Cyclical components represent repeated up-and-down oscillations. Cyclical components could be, in addition, caused by a number of reasons. Random variations take place due to unexpected events. These variations are usually short-term changes and they do not repeat.

Forecasting methods for prediction of future events are based on information from the past. Therefore, a forecast is no better than the information contained in the past data. The analyst typically has at his/her disposal sets of equally spaced numerical data. These data are gathered by observing the demand at common time periods (days, months, years).

Knowledge acquired in forecasting different phenomena has shown that all forecasts are always inaccurate. The important question is: how big is the forecast error? Any forecast is more accurate over shorter than longer time periods.

5.6.1 Time series models

Time series forecasting is based completely on data from the past. The basic hypothesis in time series forecasting is that the principal factors from the past will continue their influence in the future.

The moving average model uses previous t periods in order to predict demand in the time period $t+1$. Analysts use simple moving average and weighted moving average methods. The forecast value in the simple moving average method equals:

$$F_{t+1} = \frac{A_t + A_{t-1} + \cdots + A_{t-n}}{n}$$

where:

    t is the present time period

    F_{t+1} is the forecast for the subsequent period

    n is the forecasting horizon

    A_i is the actual demand in the i-th time period

The forecast, within the moving average method, represents linear combinations of the demand from the past time periods.

Example 5.23

The data on the number of passengers on U.S. Airlines are given in Table 5.23.

    Predict the number of passengers in the year 2016 by using the 3-year moving average.

SOLUTION:

$$F_{2016} = \frac{A_{2015} + A_{2014} + A_{2013}}{3}$$

$$F_{2016} = \frac{798 + 763 + 743}{3}$$

$$F_{2016} = 768 \text{ million passengers}$$

The weighted moving average method permits the analyst to give more importance to some of the historical data. The forecast value in the weighted moving average method equals:

Table 5.23 The number of passengers on U.S. airlines, from 2009–2018

Year	Number of passengers (in million)
2010	720
2011	731
2012	737
2013	743
2014	763
2015	798
2016	824
2017	849
2018	889
2019	926

$$F_{t+1} = w_t \cdot A_t + w_{t-1} \cdot A_{t-1} + \cdots + w_{t-n} \cdot A_{t-n}$$

$$w_t + w_{t-1} + \cdots + w_{t-n} = 1$$

where
   t is the present time period
   F_{t+1} is the forecast for the subsequent period
   n is the forecasting horizon
   A_i is the actual demand in the i-th time period
   w_i is the importance (weight) the analyst gives to the i-th time period

Example 5.24

Predict the number of passengers in 2019 by using the weighted moving average method. Use the 3-year moving average, and the following set of weights:

$$w_{2015} = 0.6, \quad w_{2014} = 0.3, \quad w_{2013} = 0.1$$

SOLUTION:

$$F_{2016} = w_{2015} \cdot A_{2015} + w_{2014} \cdot A_{2014} + w_{2013} \cdot A_{2013}$$

$$F_{2016} = 0.6 \cdot 798 + 0.3 \cdot 763 + 0.1 \cdot 743$$

$$F_{2016} = 782 \text{ million passengers}$$

The actual number of passengers in 2016 was 768 million. The simple moving average method predicted 768 million passengers, whereas the weighted moving average method forecasted 782 million passengers. We decreased the forecast error by giving more importance to the events that happened recently.

The weighted moving average method is more practical for the analysis of the transportation demand because of the capability to modify the time periods' weights. An engineer can decide on the set of weights based on the importance that she/he thinks that the data from the past have.

Exponential smoothing is based on the idea of giving greater weights to the latest observations than to the observations from the distant past. In other words, in the case of exponential smoothing, the weights decrease exponentially with the past. In this way, the lowest weights are given to the most distant observations. The forecast value in the exponential smoothing equals:

$$F_{t+1} = \alpha \cdot A_t + (1 - \alpha) \cdot F_t$$

i.e.:

$$F_{t+1} = F_t + \alpha \cdot \left(A_t - F_t \right)$$

$$F_{t+1} = F_t + \alpha \cdot e_t$$

where
   t is the present time period
   F_t is the forecast for the period t
   A_i is the actual demand in the i-th time period
   e_t is the forecast error for the i-th time period
   α is the smoothing constant

Exponential smoothing is based on the idea that the forecast should depend on the latest observation, as well as on the forecast error of the latest forecast. As we can see, from the last equation, the new forecast is equal to the old forecast, plus the correction for the error that happened in the last forecast. The smoothing constant α articulates the analyst's reaction to the forecast error. The lower the α value, the lower the reaction to forecast error. The higher the α value, the higher the analyst's reaction to the difference between the actual and the forecasted values.

Example 5.25

The data on the number of passengers on U.S. airlines are given in Table 5.24.
   Predict the number of passengers on U.S. airlines, by using the exponential smoothing method. The actual number of passengers is given in Table 5.24. Use the smoothing constant $\alpha = 0.8$. Assume that $F_1 = A_1$.

Table 5.24 The number of passengers on U.S. airlines, from 2009 to 2019

Year	Number of passengers (in million)
2010	720
2011	731
2012	737
2013	743
2014	763
2015	798
2016	824
2017	849
2018	889
2019	926

SOLUTION:

Since $F_1 = A_1$, we get:

$$F_1 = A_1 = 720$$

The first forecast error e_1 equals:

$$e_1 = A_1 - F_1 = 720 - 720 = 0$$

The second forecast value F_2 equals:

$$F_2 = F_1 + \alpha \cdot e_1 = 720 + 0.8 \cdot 0 = 720$$

The second forecast error e_2 equals:

$$e_2 = A_2 - F_2 = 731 - 720 = 11$$

The third forecast value F_3 equals:

$$F_3 = F_2 + \alpha \cdot e_2 = 720 + 0.8 \cdot 11 = 728.8$$

The third forecast error e_3 equals:

$$e_3 = A_3 - F_3 = 737 - 728.8 = 8.2$$

The fourth forecast value F_4 equals:

$$F_4 = F_3 + \alpha \cdot e_3 = 728.8 + 0.8 \cdot 8.2 = 735.36$$

All forecast values and all forecast errors are given in Table 5.25.

Table 5.25 Forecasted values of the number of passengers on U.S. airlines

Year T	Actual number of passengers A_t	Forecasted number of passengers F_t	Forecast error $e_t = A_t - F_t$
2010	720	720	0
2011	731	720	11
2012	737	728.8	8.2
2013	743	735.36	7.64
2014	763	741.472	21.528
2015	798	758.6944	39.3056
2016	824	790.1389	33.86112
2017	849	817.2278	31.77222
2018	889	842.6456	46.35444
2019	926	879.7291	46.27089

5.6.2 Trend projection

Trends should be understood as data tendencies to increase or decrease gradually over time. A long-term trend is most frequently modeled as a linear, quadratic, or exponential function. An extrapolation is simply made of trends noted by the number of passengers transported, or the number of passenger-kilometers effected, and a forecast on this basis is made of the number of passengers, or passenger-kilometers in the next period. This is known as the independent estimation method, since factors that affect the number of passengers are not taken into consideration when estimating the number of users.

Increase in the number of users is an essential property of many transportation systems. We denote time by t, and the number of trips that changes over time by $D(t)$. We also denote by $F(t)$ the forecast of the transportation demand $D(t)$ at some future time t.

The most frequently used trend models of the transportation demand in a period t are given in Table 5.26.

In the case of a linear trend, it is assumed that the growth in demand for transportation is constant within a unit of time (Figure 5.40). Model calibration, i.e. estimation of the parameters a and b, is made using the least squares method.

A nonlinear trend model is shown in Figure 5.41. Growth phenomena in transportation can be successfully described by the Gompertz curve, as well as by the logistic curve (Figure 5.42). Many growth phenomena in transportation show an "S"-shaped pattern. Initially the growth is slow. In the second stage, growth speeds up, and, finally, in the third stage, growth slows down and approaches a limit. As t increases, the population approaches k. A logistic curve is typically S-shaped, or sigmoidal. A typical logistic curve is shown in Figure 5.42.

Table 5.26 The most frequently used trend models of the transportation demand

Model name	Mathematical description
Linear trend	$F(t) = a + b \cdot t$
Exponential curve	$F(t) = a \cdot b^t$
Modified exponential curve	$F(t) = k + a \cdot b^t$
Gompertz curve	$F(t) = k \cdot a^{b^t}$
Logistic function	$F(t) = \dfrac{k}{1 + b \cdot e^{-a \cdot t}}$

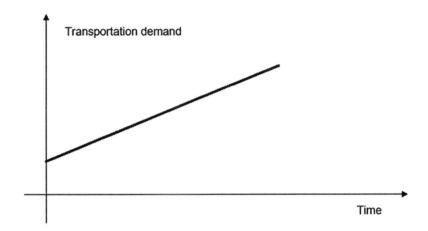

Figure 5.40 Linear trend model

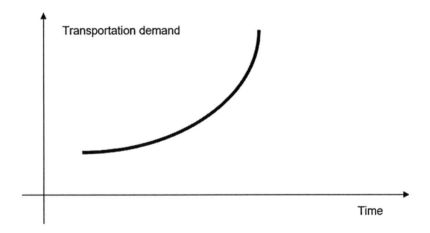

Figure 5.41 Nonlinear trend model

The total passenger traffic at the Belgrade Airport, Serbia, has been described as a function of time using a logistic curve (Grilihes, 1978). Data on passenger traffic from 1962–1978 are given in Table 5.27.

The following logistic equation was obtained:

$$F(t) = \frac{9,023,394}{1 + 39.88 \cdot e^{-0.176 \cdot t}}$$

A logistic curve could be a very convenient tool to describe number of tourists, as well as the demand of the number of air passengers on tourist routes. Lundtorp (www2.dst.dk/internet/4thforum/docs/c2-5.doc) developed the tourism destination life cycle model based on a logistic curve.

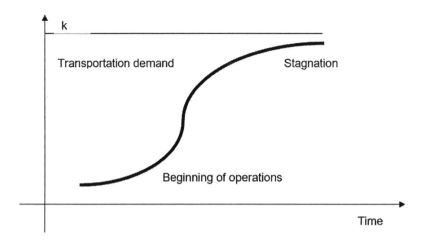

Figure 5.42 Logistic curve

Table 5.27 Data on passenger traffic at the Belgrade
Airport, Serbia, from 1962–1978

Year	Number of passengers
1962	220,726
1963	282,873
1964	329,619
1965	405,191
1966	335,999
1967	399,066
1968	462,919
1969	602,257
1970	838,156
1971	1,036,311
1972	1,155,166
1973	1,434,454
1974	1,688,247
1975	2,020,291
1976	2,047,016
1977	2,280,972

Using statistical data related to Danish package tours to Portugal in the period 1976–1994, Lundtorp showed that the following logistic curve explained these touristic trips:

$$AP = 13,000 + \frac{52,000}{1 + e^{-0.54 \cdot (t-1986)}}$$

where AP represents the number of Danish air passengers to Portugal.

Prediction of airport growth could successfully be achieved using a logistic curve. It is reasonable to assume that the population of passengers will not grow without end. As with any other market or environment, it will always eventually reach airport "saturation" level. As many natural populations compete for scarce food resources or living space, and as many new technologies compete in a limited market, in the same way, various destinations compete for travelers. Even if airline tickets become very cheap, nobody will travel to Miami for vacation seven times per month. In other words, there is always a limit (saturation level) to how many times a passenger wants to travel to a specific destination.

The Mean Forecast Error and the Mean Absolute Deviation are common measures for measuring the accuracy of the forecast. The Mean Forecast Error (MFE) represents the average error in the observations. The MFE equals:

$$\text{MFE} = \frac{\sum_{i=1}^{n}(A_i - F_i)}{n}$$

The Mean Absolute Deviation (MAD) represents the average absolute error in the observations. The MAD equals:

$$\text{MAD} = \frac{\sum_{i=1}^{n}|A_i - F_i|}{n}$$

The smaller the MFE and MAD values, the better the forecast.

PROBLEMS

5.1 The speeds of 45 vehicles were measured. The following are the values of the measured speeds (in mph), ranked in increasing order:
64, 65, 65, 65, 65, 66, 66, 67, 67, 67, 67, 68, 68, 68, 68, 68, 69, 69, 69, 69, 69, 69, 69, 69, 69, 69, 70, 70, 70, 70, 70, 70, 71, 71, 71, 71, 72, 72, 72, 73, 73, 74, 74.
Present the recorded speeds and their corresponding number of appearance (frequencies) in a frequency table. Create the frequency polygon.

5.2 The one hundred recorded vehicle speed values (veh/h) are given in Table 5.28.
Put individual speed measurements into 6 class intervals. Use the left-end inclusion convention. Rearrange the observed data and show them in the frequency table. Create a histogram.

5.3 Table 5.29 is a frequency table for the data set that contains 45 measured vehicle speeds (mph).

5.4 The sample is composed of the following 10 observations of the number of left-turning cars, waiting at the red light:
6, 7, 9, 8, 10, 9, 9, 8, 7, 11
Calculate the sample mean.

Table 5.28 One hundred recorded traffic flow values (veh/h)

63	55	46	70	55
48	56	58	57	50
55	66	57	51	64
70	45	66	45	57
69	58	65	47	61
56	47	63	47	62
59	59	52	70	62
46	53	66	57	66
58	45	47	60	55
61	45	56	55	66
49	60	47	47	62
64	60	68	63	62
51	55	63	48	61
45	52	58	62	63
66	68	49	68	66
49	58	70	66	53
49	68	59	69	51
50	67	68	50	56
52	56	60	58	63
65	65	53	62	57

Table 5.29 A frequency table for the data set that contains 45 measured vehicle speeds

Vehicle speed (mph)	Frequency
54	2
55	3
56	3
57	4
58	5
59	11
60	5
61	4
62	4
63	2
64	2

Table 5.30 A frequency table for the data set that of
50 waiting passengers at the bus stop

Number of the waiting passengers at the bus stop	Frequency
10	8
15	8
18	12
20	16
22	6

5.5   Table 5.30 is a frequency table for the data set that contains 50 waiting passengers at the bus stop.
   Determine the mean of the number of the waiting passengers at the
bus stop.

5.6   The distance, d, between two points is divided into three sections
d_1, d_2, d_3 $(d = d_1 + d_2 + d_3)$. Car velocities on these sections are v_1, v_2, v_3,
respectively. Calculate the average car velocity v that will enable the
car to travel the distance d in the same time as in the case of three difference velocities.

5.7   The distance between point A and the point B equals d. When driving
from A to B, the driver had an average speed 60 km/h. On the way
back, the average speed was 80 km/h. Determine the average speed of
the whole journey.

5.8   The geometric mean G of the numbers $x_1, x_2, ..., x_n$ is defined as:

$$G = \sqrt[n]{x_1 \cdot x_2 \cdot .... x_n}$$

   Calculate the geometric mean of the numbers 5, 25, 125.

5.9   Gross weights of seaborne freight handled in a new port (in tons
per inhabitant) in the past three years are 5, 20, 45, respectively. By
using the geometric mean, determine the average growth of seaborne
freight.

5.10 The sample is composed of 10 recorded numbers of passengers at a
bus stop, waiting for a bus which should show up at 8:30 a.m. The
numbers of passengers equal:
15, 10, 8, 9, 12, 9, 11, 14, 8, 14
   Determine the average number and the standard deviation of the
number of passengers that wait for the bus.

5.11 The sample contains recorder speeds of 100 vehicles. The sample
mean and the variance are:

$\bar{x} = 64$ mph

$\sigma = 4$ mph

Find the 95% confidence interval and the 99% confidence interval estimator of the population mean.

5.12 The survey of 100 residents of the city area showed that 35% of the residents ride a metro on a daily basis. Estimate, with the 95% confidence interval, the percentage of potential metro passengers from the observed city area.

5.13 Two independent isolated intersections are analyzed. There are 100 measurements ($n_1 = 100$) from the first intersection and 100 measurements ($n_2 = 100$) from the second intersection. The average delays per vehicle and standard deviations are as follows:

$\bar{x}_1 = 18$ seconds

$\bar{x}_2 = 15$ seconds

$\sigma_1 = 2$ seconds

$\sigma_2 = 2$ seconds

Compare average waiting time per vehicle on two intersections, using a 95% confidence interval.

5.14 The survey of 100 randomly chosen citizens in city area #1 indicated 45 car owners. The same survey of 100 randomly chosen citizens in city area #2 indicated 20 car owners. Estimate, with the 95% confidence interval, the difference between the proportions of car owners in city area #1 and city area #2.

5.15 Test the hypothesis that the population mean of the number of passengers on the bus, running between the stop A and the stop B, is equal to 10, i.e. that $H_0(\mu = 10)$. The sample contains $n = 100$ passengers with the mean value $\bar{x} = 12$ and the standard deviation $s = 2$. The alternative hypothesis is $H_1(\mu \neq 10)$. The probability of incorrectly rejecting the null hypothesis is $\alpha = 0.05$.

5.16 Previous transportation surveys and studies showed that 20% of students ride a bicycle when going to university. The latest sample, consisting of $n = 100$ students, showed that $m = 16$ of them use a bicycle. Let p be the proportion of bicycle riders in the student population. Test the null hypothesis $H_0(p = 0.20)$, against the alternative hypothesis $H_1(p < 0.20)$.

5.17 Two independent samples are taken at two airports. Each sample contained $n_1 = n_2 = 100$ transfer passengers. The walking distances between arrivals and departure gate were measured. The average walking distance in the case of the airport was equal to $\bar{x}_1 = 380$ m,

Table 5.31 The values of the independent variable x and
          dependent variable y

x_i	-2	0	1	2	4
y_i	0.5	1	1.5	2	3

Table 5.32 The values of the independent variable
           x and dependent variable y to be used
           to calculate the correlation coefficient

x	y
1	1
3	2
4	4
6	4
8	5
9	7
11	8
14	9

Table 5.33 The values of the independent variable x and dependent
           variable y used to calculate the correlation coefficient

x_i	35	13	12	30	8	22	20
y_i	82	42	22	62	26	58	38

with a standard deviation of $s_1 = 20$ m. The average walking distance in the case of the second bus station was equal to $x_2 = 350$ m,
with a standard deviation of $s_1 = 10$ m. Compare two population means,
i.e. $H_0 (\mu_1 - \mu_2 = 0)$, against the alternative hypothesis $H_1 (\mu_1 - \mu_2 \neq 0)$.

5.18 Table 5.31. shows the values of the independent variable x and depen-
      dent variable y. Assume that there is a linear trend. Determine the
      coefficients a and b.

5.19 The values of independent variable x and dependent variable y are
      shown in Table 5.32. Determine the correlation coefficient between
      these variables.

5.20 The values of independent variable x and dependent variable y are
      shown in Table 5.33. Determine the correlation coefficient between
      these variables.

BIBLIOGRAPHY

Barnes, J. Wesley, *Statistical Analysis for Engineers and Scientists: A Computer
    Based Approach*, McGraw-Hill, New York, 1994.

Ben-Akiva, M.E. and Lerman, S.R., *Discrete Choice Analysis: Theory and Application to Travel Demand*, MIT Press, Cambridge, MA, 1985.

Brockwell, P. J. and Davis, R. A., *Introduction to Time Series and Forecasting*, Springer Verlag, New York, 1996.

Bureauof Transportation Statistics, www.bts.gov/figure-1-january-december-pa ssengers-us-airlines-2003-2019.

Burr, Irving W., *Applied Statistical Methods*, Academic Press, New York, 1974.

Crow, E.L., Davis, F.A. and Maxfield, M.W., *Statistics Manual*, Dover Publications, New York, 1960.

FHWA, www.ops.fhwa.dot.gov/congestion_report/executive_summary.htm.

de Ortuzar, J. D. and Willumsen, L. G., *Modelling Transport*, John Willey & Sons, New York, 1990.

Edwards, R.V., *Processing Random Data*, World Scientific, Hackensack, NJ, 2006.

Haldar, A. Mahadevan, *Probability, Reliability, and Statistical Methods in Engineering Design*, John Wiley & Sons, New York, 2000.

Kanafani, A., *Transportation Demand Analysis*, Mc Graw Hill, New York, 1983.

Kennedy, J. B. and Neville, A. N., *Basic Statistical Methods for Engineers and Scientists*, 3rd ed., Harper & Row, New York, 1986.

Kokoska, S., *Introductory Statistics: A Problem-Solving Approach*, W.H. Freeman and Company, New York, 2011.

Lundtorp, S., Testing a Model for Tourism Development: The Case of the Island of Bornholm, Working paper, Research Centre of Bornholm, Denmark, ww2.d st.dk/internet/4thforum/docs/c2-5.doc.

Ogunnaike, B.A., *Random Phenomena*, CRC Press, Boca Raton, FL, 2009.

Ross, S., *Introduction to Probability and Statistics for Engineers and Scientists*, 2nd ed., Academic Press, 2000.

Ross, S., *A First Course in Probability*, 8th ed., Pearson Prentice Hall, 2010.

Rumsey, D., *Statistics for Dummies*, Wiley Publishing, Hoboken, NJ, 2003.

Rumsey, D., *Probability for Dummies*, Wiley Publishing, Hoboken, NJ, 2006.

Sheffi, Y., *Urban Transportation Networks*, Prentice Hall, Englewood Cliffs, NJ, 1985.

Taha, H., *Operations Research*, MacMillan Publishing Co., Inc., New York, 1982.

Teodorović, D. and Janić, M., *Transportation Engineering: Theory, Practice and Modeling*, Elsevier, New York, 2016.

Vardeman, S. B., *Statistics for Engineering Problem Solving*, PWS Publishing, Boston, 1994.

Vukadinović, S. and Popović, J., *Mathematical Statistics*, University of Belgrade, 2008 (in Serbian).

Walpole, R. E. and Myers, R.H., *Probability and Statistics for Engineers and Scientists*, 7th ed., Prentice Hall, Upper Saddle River, NJ, 2002.

Washington, S.P., Karlaftis, M.G., and Mannering, F.L., *Statistical and Econometric Methods for Transportation Data Analysis*, Chapman & Hall/ CRC, Boca Raton, FL, 2003.

Ziemer, Rodger E., *Elements of Engineering Probability & Statistics*, Prentice Hall, Upper Saddle River, NJ, 1997.

Chapter 6

Simulation

Simulation models imitate the behavior of real-world systems, most frequently *via* computers. These models study interactions among elements of the system, as well as the system's operating characteristics. Based on the results obtained from the simulation, engineers make various conclusions and take appropriate actions.

Simulation should be considered as a statistical experiment. Simulation experiments could be performed entirely on a computer. The results generated by a simulation experiment represent observations. There is a logical question as to why we are developing simulation models, carrying out computer experiments, and collecting "observations" from the simulations.

Often, in reality, the system we want to study still does not exist. In order to more closely observe the behavior and characteristics of the future system, we simulate its behavior. So, for example, if we want to determine the potential congestion points at a future airport, we develop a simulation model, and try to describe, as closely as possible, the future traffic operations at the airport.

In some other situations, the system we want to study does exist, but studying this system directly is often disruptive and/or very costly. We do not develop simulation models in cases where there is an analytic solution for the considered problem. Simulation research power is best seen when studying a complex system for which there are no analytic approaches.

The simulation can be static or dynamic. In a simulation model, we can assume that changes in an observed system occur either continuously, or only at discrete points in time.

Simulation models have been used to study the choice of routes by drivers, the choice of transport mode, the design of parking lots, harbors, and airports, and to investigate distribution system design, bus scheduling, taxi and truck dispatching, etc.

The Monte Carlo simulation method, which we will describe in the next section, is based on the idea of using sampling in order to estimate performances of an observed system.

6.1 THE MONTE CARLO SIMULATION METHOD

The *Monte Carlo* simulation method has been used in engineering applications since the late 1950s. In this method, *random numbers* are used to obtain samples from probability distributions. Sampling from the relevant probability distribution is based on the utilization of the [0, 1] random numbers.

The [0, 1] random numbers are uniformly distributed in the interval [0, 1]. In other words, every one of the values in the interval [0, 1] has the same chance to happen. The [0, 1] random numbers are generated in an entirely random manner.

We illustrate the Monte Carlo method with the following example. There are two paths between node A and node B (Figure 6.1).

We assume that there are equal chances of choosing the left path (L) or the right path (R) by the driver who travels from A to B. In other words, we assume that any driver chooses a path between A and B with the following probabilities:

$$p(L) = 0.5 \qquad p(R) = 0.5$$

We denote by r the generated random number from the interval [0, 1]. Given that the [0, 1] random numbers are uniformly distributed in the interval [0, 1], we formulate the following rules for determining the driver's choice:

If $0 \le r \le 0.5$ the driver chooses L
If $0.5 < r \le 1$ the driver chooses R

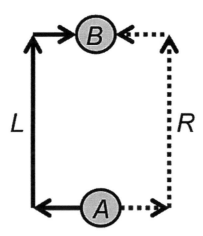

Figure 6.1 Two paths between the node A and the node B

Let us study the route choice of the first 10 drivers. The choices of the first 10 drivers are equivalent to generating 10 random numbers from the interval [0, 1]. Let us assume that we generated the following random numbers:

0.051455, 0.627205, 0.084273, 0.82207, 0.298202, 0.203535, 0.535325, 0.359749, 0.701533, 0.116597

The drivers' choices will be: L, R, L, R, L, L, R, L, R, L

Let us consider the case when drivers can choose one of five routes. The routes are denoted, respectively, as 1, 2, 3, 4, and 5. We assume that there are equal chances of the driver who travels from A to B choosing any path. In other words, we assume that any driver chooses the path between A and B with the following probabilities:

$$p(1) = 1/5, \quad p(2) = 1/5, \quad p(3) = 1/5, \quad p(4) = 1/5, \quad p(5) = 1/5$$

We denote by x the outcome (driver's choice of the route). We also designate $p(x)$ and $F(x)$ as the probability density function and the cumulative density function, respectively. The possible outcomes and the corresponding values of $p(x)$ and $F(x)$ are shown in Table 6.1, as well as in Figure 6.2.

The following rules determine the driver's choice:

If $0 \leq F(x) \leq 0.2$, the driver chooses route 1

If $0.2 < F(x) \leq 0.4$, the driver chooses route 2

If $0.4 < F(x) \leq 0.6$, the driver chooses route 3

If $0.6 < F(x) \leq 0.8$, the driver chooses route 4

If $0.8 < F(x) \leq 1$, the driver chooses route 5

Figure 6.2 also shows the experiment in which we generated the random $r = 0.7$. We assign this number to $F(x)$. We obtain the outcome (driver's route choice) by *inverting* $F(x)$. In our case, $r = 0.7$ and $F(x) = 0.7$, and consequently, $x = 4$. This method is called the *method of inversion*. The method of inversion is used for all probability distributions. We illustrate the use of this method in the case where we have to perform sampling of the exponential distribution.

Table 6.1 The possible outcomes and corresponding values of $p(x)$ and $F(x)$

X	1	2	3	4	5
$p(x)$	0.2	0.2	0.2	0.2	0.2
$F(x)$	0.2	0.4	0.6	0.8	1

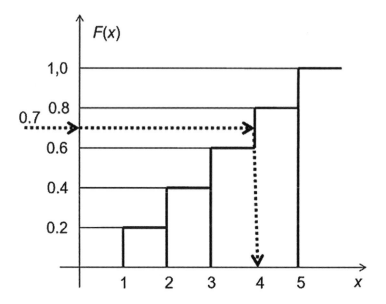

Figure 6.2 Outcomes x and F(x)

6.2 SAMPLING THE BERNOULLI DISTRIBUTION

In the case of the Bernoulli distribution, the probability of experimental success is denoted by p. The probability of failure is equal to q $(p+q=1)$. The Bernoulli random variable can take on two values, 1 or 0. The probability mass function reads:

$$X = \begin{Bmatrix} 0 & 1 \\ q & p \end{Bmatrix}$$

The mean and variance of the Bernoulli random variable are, respectively, equal to:

$$E(X) = 0 \cdot q + 1 \cdot p = p$$

$$\text{var}(X) = E(X^2) - (E(X))^2 = 1^2 \cdot p - p^2 = p - p^2 = p \cdot (1-p) = p \cdot q$$

We generate random number r from the interval [0, 1]. We simulate the values of the random variable X by using the following rules:

The case when $r < p$ corresponds to the case when $x = 1$
The case when $r \geq p$ corresponds to the case when $x = 0$

Table 6.2 The random numbers generated, tests performed, random variable values, and vehicle types that appeared at the specific point on the road

Random number r	Test r<p or r≥p	Random variable value x	Vehicle that appeared
0.6108823742	0.6108823742>0.6	0	Car
0.5538116148	0.5538116148<0.6	1	Heavy vehicle
0.2311375077	0.2311375077<0.6	1	Heavy vehicle
0.6005780048	0.6005780048>0.6	0	Car
0.3950340415	0.3950340415<0.6	1	Heavy vehicle
0.8047315060	0.8047315060>0.6	0	Car
0.4622154186	0.4622154186<0.6	1	Heavy vehicle
0.6982231190	0.6982231190>0.6	0	Car
0.6099311085	0.6099311085>0.6	0	Car
0.5786548156	0.5786548156<0.6	1	Heavy vehicle

Example 6.1

On a road that connects a factory in the suburbs with the city, 60% of the traffic is heavy vehicles. The Bernoulli random variable can take on two values: 1 (a heavy vehicle appeared at the specific point of a road) and 0 (a car appeared). The probability mass function reads:

$$X = \begin{Bmatrix} 0 & 1 \\ 0.4 & 0.6 \end{Bmatrix}$$

We generated the following random numbers (r):

0.6108823742	0.8047315060
0.5538116148	0.4622154186
0.2311375077	0.6982231190
0.6005780048	0.6099311085
0.3950340415	0.5786548156

The random numbers generated, the tests performed, the random variable values, and the vehicle types that appeared at the specific point on the road are listed in Table 6.2.

6.3 SAMPLING THE BINOMIAL DISTRIBUTION

The probability of events that success happens x times in n trials equals:

$$P(X = x) = \frac{n!}{x!(n-x)!} \cdot p^x \cdot q^{n-x}$$

The mean and variance of the binomial random variable are, respectively, equal to:

$$E(X) = n \cdot p$$

$$\text{var}(X) = n \cdot p(1 - p)$$

When n is small, or moderate, we use the following routine to obtain the values of the random variable X that has a binomial distribution:

Set $x = 0$

For $i = 1$ to n

    Generate the random number r from the interval $[0,1]$

    If $r < p$

    Then $x = x + 1$

Next i

Example 6.2

The probability that the driver-beginner (with less than one year of driving experience) appears at the specific intersection equals 0.1. This value is estimated by simulation of the number of drivers who are beginners among 100 drivers that appeared at the intersection during the observed period of time.

SOLUTION:

We generated the following random numbers (r):

0.2007977801	**0.0440369832**	0.6552641925	0.1081171832
0.2174485891	0.1874824390	0.7086092743	0.6981552641
0.6007497664	0.2562565365	0.5132961926	0.2496055127
0.9082743975	0.2320218042	0.5292657864	0.6699024176
0.2811631632	0.7627785609	0.9062099403	0.7710130332
0.7898902881	0.1062038671	0.4164879024	0.1196892602
0.7922939152	0.9148924721	0.4417625976	0.3334589012
0.4046797813	0.5804740987	0.5888232928	0.6759322678
0.9364791893	0.9856731905	0.7157582646	0.8186925170
0.5685087970	0.4927006137	0.5959306820	0.9489581104
0.3324681362	0.4348109535	0.5635451599	0.1575000182
0.4321374558	0.3370955947	0.3034116818	0.1060804421
0.6919007539	**0.0533517805**	0.3008104603	0.4096711811
0.8387828006	0.2997462297	0.2628753095	0.1350703580
0.6982247773	0.2748033455	**0.0850214077**	0.5561237521
0.8500631656	0.6972546777	0.5797603264	0.5269232188
0.8448251958	0.1345519841	0.5469592911	0.7220551149
0.6785302222	0.2136322294	0.7456457228	0.8627066052
0.7696227279	0.3007715879	0.6127772151	0.2260457093
0.2151306435	0.3702933576	0.9644499540	0.6318283182
0.3625131580	0.8211274062	0.6251212142	0.8524950736
0.3516149582	0.9364815947	0.8009532416	0.5900977705
0.6243198138	0.7304327270	0.2686181754	0.2750747857
0.0695044627	0.5162895718	0.7598245655	0.3453761404
0.9826599837	0.8903259102	0.9569706883	0.3049653362

Four of the random numbers (denoted in bold with a larger font size) are less than 0.1. We conclude that four driver-beginners appeared at the intersection during the observed period of time.

Example 6.3

Let us designate, by X, the binomial random variable that represents the number of aircraft delayed on arrival at one hub. The probability that an aircraft arrives late at the airport is equal to 0.35. Perform one simulation run and simulate the arrivals of five aircraft. Determine how many of them will arrive late.

SOLUTION:

All five aircraft that we analyze belong to one of the two categories, namely (a) delayed aircraft, or (b) aircraft that arrive on time. We generate five random numbers and we formulate the following rules for determining the category of each aircraft:

if $0 < r_i \leq 0.35$ aircraft i is delayed,
if $0.35 < r_i \leq 1$ aircraft i is not delayed

We generated the following random numbers: 0.6679, 0.1040, 0.1778, 0.0385, and 0.3300. Now we have:

$0.35 < r_1 = 0.6679 \leq 1$ the first aircraft is not delayed,

$0 < r_2 = 0.1040 \leq 0.35$ the second aircraft is delayed,

$0 < r_3 = 0.1778 \leq 0.35$ the third aircraft is delayed,

$0 < r_4 = 0.0385 \leq 0.35$ the fourth aircraft is delayed,

$0 < r_4 = 0.3300 \leq 0.35$ the fifth aircraft is delayed

In total, four aircraft are delayed, and one was not delayed ($x = 4$).

We note that we have to generate one random number for each aircraft. This can be time consuming if we have to perform the simulation many times. The simulation could also be performed in the following way.

We calculate the following probabilities: $p_0 = P(X = 0)$, $p_1 = P(X = 1)$, $p_2 = P(X = 2)$, $p_3 = P(X = 3)$, $p_4 = P(X = 4)$ and $p_5 = P(X = 5)$. Now, we generate one random number (r) and determine x in the following way:

if $0 < r \leq p_0$ then $x = 0$

if $p_0 < r \leq p_0 + p_1$ then $x = 1$

if $p_0 + p_1 < r \leq p_0 + p_1 + p_2$ then $x = 2$

if $p_0 + p_1 + p_2 < r \leq p_0 + p_1 + p_2 + p_3$ then $x = 3$

if $p_0 + p_1 + p_2 + p_3 < r \leq p_0 + p_1 + p_2 + p_3 + p_4$ then $x = 4$

if $p_0 + p_1 + p_2 + p_3 + p_4 < r \leq p_0 + p_1 + p_2 + p_3 + p_4 + p_5$ then $x = 5$

Let us make ten simulation runs for the previous example. The probabilities $P(X = k)$, $k = 0,...,5$ are, respectively, equal to:

$$p_0 = P(X = 0) = \frac{5!}{0! \cdot (5-0)!} \cdot 0.35^0 \cdot 0.65^{5-0} = 1 \cdot 0.35^0 \cdot 1 = 0.1160$$

$$p_1 = P(X = 1) = \frac{5!}{1! \cdot (5-1)!} \cdot 0.35^1 \cdot 0.65^{(5-1)} = 5 \cdot 0.35 \cdot 0.1785 = 0.3124$$

$$p_2 = P(X = 2) = \frac{5!}{2! \cdot (5-2)!} \cdot 0.35^2 \cdot 0.65^{(5-2)} = 10 \cdot 0.1225 \cdot 0.2746 = 0.3364$$

$$p_3 = P(X = 3) = \frac{5!}{3! \cdot (5-3)!} \cdot 0.35^3 \cdot 0.65^{(5-3)} = 10 \cdot 0.0429 \cdot 0.4225 = 0.1811$$

$$p_4 = P(X = 4) = \frac{5!}{4! \cdot (5-4)!} \cdot 0.35^4 \cdot 0.65^{(5-4)} = 5 \cdot 0.015 \cdot 0.65 = 0.0488$$

$$p_5 = P(X = 5) = \frac{5!}{5! \cdot (5-5)!} \cdot 0.35^5 \cdot 0.65^{(5-5)} = 1 \cdot 0.0525 \cdot 1 = 0.0525$$

Now,

if $0 < r \le 0.1160$	then $x = 0$
if $0.1160 < r \le 0.4284$	then $x = 1$
if $0.4284 < r \le 0.7648$	then $x = 2$
if $0.7648 < r \le 0.9459$	then $x = 3$
if $0.9459 < r \le 0.9947$	then $x = 4$
if $0.9947 < r \le 1$	then $x = 5$

We generated 10 random numbers and we obtained the following simulation outcome:

$r_1 = 0.0705,$	$x_1 = 0$
$r_2 = 0.5369,$	$x_2 = 2$
$r_3 = 0.4835,$	$x_3 = 2$
$r_4 = 0.1550,$	$x_4 = 1$
$r_5 = 0.4552,$	$x_5 = 2$
$r_6 = 0.2884,$	$x_6 = 1$
$r_7 = 0.2567,$	$x_7 = 1$
$r_8 = 0.8189,$	$x_8 = 3$
$r_9 = 0.5936,$	$x_9 = 2$
$r_{10} = 0.6324,$	$x_{10} = 2$

6.4 SAMPLING THE EXPONENTIAL DISTRIBUTION

The time between vehicle arrivals is the random variable T that has an exponential distribution with parameter λ:

$$f(t) = \lambda e^{-\lambda t}$$

Let us assume that $\lambda = 0.25 \left[\dfrac{\text{veh}}{\text{s}} \right]$. The cumulative density function $F(x)$ equals:

$$F(x) = \int_0^t \lambda \cdot e^{-\lambda t} dt = 1 - e^{-\lambda t}$$

We generate random number r from the interval $[0, 1]$. We get:

$$r = F(x)$$

$$r = 1 - e^{-\lambda t}$$

$$t = -\frac{1}{\lambda} \cdot \ln(1 - r)$$

Since r is a random number, $R = 1 - r$ is also a random number, so we can write:

$$t = -\frac{1}{\lambda} \cdot \ln R$$

We generated the following random numbers (R):

0.312230, 0.28341, 0.297506, 0.510998, and 0.220226

The time intervals between vehicle arrivals are equal to:

$$t_1 = -\frac{1}{0.25} \cdot \ln(0.312230) = 4.66\,\text{s}$$

$$t_2 = -\frac{1}{0.25} \cdot \ln(0.283410) = 5.04\,\text{s}$$

$$t_3 = -\frac{1}{0.25} \cdot \ln(0.297506) = 4.85\,\text{s}$$

$$t_4 = -\frac{1}{0.25} \cdot \ln(0.510998) = 2.29\,\text{s}$$

$$t_5 = -\frac{1}{0.25} \cdot \ln(0.220226) = 6.05\,\text{s}$$

Example 6.4

Aircraft arrive at the airport. Aircraft interarrival time is a random variable that has an exponential distribution with parameter $\lambda = 10$ aircraft/hour. Suppose that the last aircraft arrived at time point $T_o = 10$ min. Simulate the arrival times of the next 5 aircraft.

SOLUTION:

$$t_i = -\frac{1}{\lambda} \ln R_i$$

We generate random numbers: 0.1983, 0.6568, 0.0635, 0.7270, and 0.4823

The interarrival times are, respectively:

$$t_1 = -\frac{1}{10} \cdot \ln 0.1983 = -0.1 \cdot (-1.618) = 0.1618\,\text{h} = 9.708\,\text{min}$$

$$t_2 = -\frac{1}{10} \cdot \ln 0.6568 = -0.1 \cdot (-0.4204) = 0.04204\,\text{h} = 2.522\,\text{min}$$

$$t_3 = -\frac{1}{10} \cdot \ln 0.0635 = -0.1 \cdot (-2.757) = 0.2757\,\text{h} = 16,54\,\text{min}$$

$$t_4 = -\frac{1}{10} \cdot \ln 0.7270 = -0.1 \cdot (-0.3188) = 0.03188\,\text{h} = 1.91\,\text{min}$$

$$t_5 = -\frac{1}{10} \cdot \ln 0.4823 = -0.1 \cdot (-0.7292) = 0.07292\,\text{h} = 4.375\,\text{min}$$

Arrival time of the first aircraft is:

$$T_1 = T_0 + t_1 = 10 + 9.708 = 19.708\,\text{min}$$

Arrival time of the second aircraft is:

$$T_2 = T_1 + t_2 = 19.708 + 2.522 = 22.23\,\text{min}$$

Arrival time of the third aircraft is:

$$T_3 = T_2 + t_3 = 22.23 + 16.54 = 38.77\,\text{min}$$

Arrival time of the fourth aircraft is:

$$T_4 = T_3 + t_4 = 38.77 + 1.91 = 40.68 \, \text{min}$$

Finally, arrival time of the fifth aircraft is:

$$T_5 = T_4 + t_5 = 40.68 + 4.375 = 45.055 \, \text{min}$$

6.5 SAMPLING THE POISSON DISTRIBUTION

The Poisson random variable represents counts of the number of times that the observed event (for example, vehicle arrival) happens within the given time interval t. The probability that the event happened k times during the time interval t equals:

$$P(X = k) = \frac{(\lambda t)^k}{k!} \cdot e^{-\lambda t}$$

In the case of a Poisson distribution of the number of events, the intervals between two events are exponentially distributed, i.e.:

$$f(t) = \lambda \cdot e^{-\lambda \cdot t}$$

In order to sample the Poisson distribution, we have to sample the exponential distribution. We will sample the exponential distribution as many times as needed (Figure 6.3).

We will stop sampling when the sum of the exponential random variables exceeds t for the first time. The sampled Poisson value k will be equal to the number of times we sampled the exponential distribution minus one, i.e.:

$$\sum_{i=1}^{k} t_i \leq t < \sum_{i=1}^{k+1} t_i$$

When the exponential distribution was sampled, we found the following:

$$t = -\frac{1}{\lambda} \cdot \ln R$$

After substitution, we get:

$$\sum_{i=1}^{k} -\frac{1}{\lambda} \cdot \ln R_i \leq t < \sum_{i=1}^{k+1} -\frac{1}{\lambda} \cdot \ln R_i$$

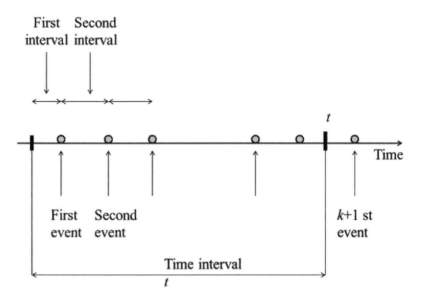

Figure 6.3 Sampling the Poisson distribution

$$\sum_{i=1}^{k} -\ln R_i \le \lambda \cdot t < \sum_{i=1}^{k+1} -\ln R_i$$

$$-\sum_{i=1}^{k} \ln R_i \le \lambda \cdot t < -\sum_{i=1}^{k+1} \ln R_i$$

i.e.:

$$\sum_{i=1}^{k+1} \ln R_i < -\lambda \cdot t \le \sum_{i=1}^{k} \ln R_i$$

$$\ln \prod_{i=1}^{k+1} R_i < -\lambda \cdot t \le \ln \prod_{i=1}^{k} R_i$$

$$\prod_{i=1}^{k+1} R_i < e^{-\lambda \cdot t} \le \prod_{i=1}^{k} R_i$$

Example 6.5

Sample a Poisson distribution. The mean rate is $\lambda = 5$ events per hour. The observation period is $t = 1.2$ hours.

SOLUTION:

We use the following relation:

$$\prod_{i=1}^{k+1} R_i < e^{-\lambda \cdot t} \leq \prod_{i=1}^{k} R_i$$

The $\lambda \cdot t$ equals:

$$\lambda \cdot t = 5 \text{ events per hour} \cdot 1.2 \text{ hours} = 6$$

i.e.:

$$e^{-\lambda \cdot t} = e^{-6} = 0.00247875217$$

We estimated that we should generate 10 random numbers; if 10 random numbers were insufficient, we would generate more random numbers. We generated the following 10 random numbers:

0.3358262765
0.2843463959
0.0589969646
0.8010143465
0.6187307869
0.3652426042
0.9432070586
0.9248110238
0.6248122041
0.8207757980

The random numbers and the calculated values of $\prod_{i=1}^{k} R_i$ and $\prod_{i=1}^{k+1} R_i$ are given in Table 6.3.

From Table 6.3, we see that the relation $\prod_{i=1}^{k+1} R_i < e^{-\lambda \cdot t} \leq \prod_{i=1}^{k} R_i$ is satisfied in the 8th row of the Table, i.e.:

$$\prod_{i=1}^{k+1} R_i = 0.0016559472 < e^{-\lambda \cdot t} = 0.00247875217 \leq \prod_{i=1}^{k} R_i = 0.0026488714$$

We can conclude that the sampled Poisson value equals 8.

6.6 SAMPLING THE NORMAL DISTRIBUTION

It has been shown, in the previous chapters, that the distribution of the sum of n identically distributed random variables can be approximated by a normal distribution when the number of elements in the sample n approaches infinity.

Table 6.3 The random numbers and the calculated values of $\prod_{i=1}^{k} R_i$ and $\prod_{i=1}^{k+1} R_i$

k	Random numbers	$\prod_{i=1}^{k} R_i$	$\prod_{i=1}^{k+1} R_i$
1	0.3358262765	0.3358262765	0.2843463959
2	0.2843463959	0.2843463959	0.0167755743
3	0.0589969646	0.0167755743	0.0134374757
4	0.8010143465	0.0134374757	0.0083141799
5	0.6187307869	0.0083141799	0.0030366927
6	0.3652426042	0.0030366927	0.0028642299
7	0.9432070586	0.0028642299	0.0026488714
8	0.9248110238	**0.0026488714**	**0.0016550472**
9	0.6248122041		
10	0.8207757980		

The relation $\prod_{i=1}^{k+1} R_i < e^{-\lambda \cdot t} \leq \prod_{i=1}^{k} R_i$ is satisfied in the 8th row of the Table 6.3

Let us generate the sequence of uniform random numbers $R_1, R_2, ..., R_n$ from the interval [0,1]. We denote by R the sum of these uniform random numbers, i.e.:

$$R = R_1 + R_2 + \cdots + R_n$$

It has been shown (based on the Central Limit Theorem) that the random variable R is normally distributed with mean $n/2$ and variance $n/12$. In other words, $R \sim N\left(\dfrac{n}{2}, \sqrt{\dfrac{n}{12}}\right)$. The corresponding standard normal variable is:

$$Z = \frac{R - \dfrac{n}{2}}{\sqrt{\dfrac{n}{12}}}$$

The standard random variable of any normal distribution with mean μ and standard deviation σ is:

$$Z = \frac{X - \mu}{\sigma}$$

We can write:

$$z = \frac{x - \mu}{\sigma} = \frac{R - \dfrac{n}{2}}{\sqrt{\dfrac{n}{12}}}$$

We obtain the following:

$$x = \mu + \frac{\sigma}{\sqrt{\dfrac{n}{12}}} \cdot \left(R - \frac{n}{2} \right)$$

In order to simplify the last formula, the recommendation is that n should be equal to 12. We obtain the following:

$$x = \mu + \sigma \cdot \left(\sum_{i=1}^{12} R_i - 6 \right)$$

Example 6.6

Sample a normal distribution. The mean is $\mu = 25$. The standard deviation is $\sigma = 5$.

SOLUTION:

We generated the following 12 random numbers:

0.74, 0.03, 0.87, 0.17, 0.63, 0.05, 0.95, 0.12, 0.55, 0.46, 0.80, and 0.58

Since:

$$x = \mu + \sigma \cdot \left(\sum_{i=1}^{12} R_i - 6 \right)$$

we have:

$$x = 25 + 5 \cdot \left(0.74 + 0.003 + 0.87 + \ldots + 0.45 + 0.80 + 0.58 - 6 \right)$$

$$x = 25 + 5 \cdot \left(5.95 - 6 \right)$$

We can conclude that the sampled normal distribution value equals:

$$x = 24.75$$

Example 6.7

The number of trucks that arrive at the container terminal exhibits a Poisson distribution. The average arrival rate of trucks is $\lambda = 5$ trucks/hour. Truck service time follows anormal distribution, with the following parameters, $\mu = 10$ min, $\sigma = 1.5$ min. Once a truck has been serviced, it leaves the terminal. Trucks are serviced in the order in which they arrive at the terminal. Servicing the truck starts immediately after the service of previous trucks is completed. Simulate the arrivals and service of trucks that enter the terminal during the time interval of $t = 0.5$ hour. Suppose that the first truck arrives at the moment $T_1 = 0$ min.

SOLUTION:

We can simulate the moment when truck i arrives in the following way:

$$T_i = T_{i-1} + t_i$$

where

T_{i-1} – moment when truck $i-1$ arrives

t_i – interarrival time between truck $i-1$ and i

Since the total number of trucks that arrive during some period of time has a Poisson distribution, the truck interarrival times follow an exponential distribution. The interarrival time, t_i, can be simulated in the following way:

$$t_i = -\frac{1}{\lambda} \cdot \ln r$$

$$\lambda = 5 \frac{\text{truck}}{\text{hour}} = \frac{1}{12} \frac{\text{truck}}{\text{min}}$$

$$t_i = -12 \cdot \ln r$$

We simulate the service time of truck i in this way:

$$t_i^s = \mu + \sigma \cdot \left(\sum_{j=1}^{12} R_j - 6 \right)$$

Also, the beginning and end of the service of truck i are denoted by times T_i^{sb} and T_i^{se}, respectively. We calculate these times in the following way:

$$T_i^{sb} = \max\{T_i, T_{i-1}^{se}\}$$

$$T_i^{se} = T_i^{sb} + t_i^s$$

For the first truck, we know that it arrives at the moment $T_1 = 0$. We simulate only its service time. The service of this truck can start at time 0 ($T_i^{sb} = 0$). We generated the following 12 random numbers: 0.8424, 0.8163, 0.7629, 0.2743, 0.4872, 0.0366, 0.0284, 0.3887, 0.9561, 0.0862, 0.4979, and 0.1030. The sum of these random numbers equals 5.28. The duration of the service time of the first truck is:

$$t_1^s = \mu + \sigma \cdot \left(\sum_{j=1}^{12} R_j - 6 \right) = 10 + 1.5 \cdot (5.28 - 6) = 10 - 1.08 = 8.92 \, \text{min}$$

Consequently:

$$T_1^{se} = T_1^{sb} + t_1^s = 0 + 8.92 = 8.92 \, \text{min}$$

The arrival time of the second truck is:

$$T_2 = T_1 + t_2$$

To simulate t_2, we generate one random number: 0.8188. Now, t_2 is equal to:

$$t_2 = -12 \cdot \ln 0.8188 = -12 \cdot (-0.19992) = 2.4 \, \text{min}$$

Truck 2 arrives at the moment:

$$T_2 = T_1 + t_2 = 0 + 2.4 = 2.4 \, \text{min}$$

Service of truck 2 starts at the moment:

$$T_2^{sb} = \max \{ T_2, T_1^{se} \} = \max \{ 8.92, 2.4 \} = 8.92 \, \text{min}$$

In order to simulate the duration of the truck 2 service, we generated the following 12 random numbers: 0.4633, 0.2853, 0.8572, 0.9380, 0.4511, 0.1755, 0.2452, 0.4719, 0.0569, and 0.3274. The sum of these numbers is 4.2718.

The duration of the service time of the second truck is:

$$t_2^s = \mu + \sigma \cdot \left(\sum_{j=1}^{12} R_j - 6 \right) = 10 + 1.5 \cdot (4.2718 - 6) = 10 - 2.59 = 7.41 \, \text{min}$$

The second truck will finish service at:

$$T_2^{se} = T_2^{sb} + t_2^s = 8.92 + 7.41 = 16.33 \, \text{min}$$

To simulate t_3 we generated one random number: 0.7037. Now, t_3 is equal to:

$$t_3 = -12 \cdot \ln 0.7037 = -12 \cdot (-0.3514) = 4.22 \, \text{min}$$

Truck 3 arrives at the moment:

$$T_3 = T_2 + t_3 = 2.4 + 4.22 = 6.62 \, \text{min}$$

Service of truck 2 starts at the moment:

$$T_3^{sb} = \max\left\{T_3, T_2^{se}\right\} = \max\left\{6.62, 16.33\right\} = 16.33\,\text{min}$$

To simulate the duration of the truck 3 service, we generated the following 12 random numbers: 0.8228, 0.6183, 0.8636, 0.3561, 0.1488, 0.9649, 0.6069, 0.0026, 0.2141, and 0.2942. The sum of these numbers equals 4.8923.

The duration of the service time of the third truck is:

$$t_3^s = \mu + \sigma \cdot \left(\sum_{j=1}^{12} R_j - 6\right) = 10 + 1.5 \cdot \left(4.8923 - 6\right) = 10 - 1.66 = 8.34\,\text{min}$$

Truck 3 will finish service at:

$$T_3^{se} = T_3^{sb} + t_3^s = 16.33 + 8.34 = 24.67\,\text{min}$$

In the same manner, we continue for truck 4.

$$r = 0.4899$$

$$t_4 = -12 \cdot \ln 0.4899 = -12 \cdot \left(-0.7136\right) = 8.56\,\text{min}$$

$$T_4 = T_3 + t_4 = 6.62 + 8.56 = 15.18\,\text{min}$$

$$T_4^{sb} = \max\left\{T_4, T_3^{se}\right\} = \max\left\{15.18, 24.67\right\} = 24.67\,\text{min}$$

The random numbers R_j are: 0.9358, 0.2108, 0.1702, 0.6221, 0.3607, 0.0419, 0.4848, 0.8007, 0.1576, and 0.8116. The sum of these numbers is 4.5962.

$$t_4^s = \mu + \sigma \cdot \left(\sum_{j=1}^{12} R_j - 6\right) = 10 + 1.5 \cdot \left(4.5962 - 6\right) = 10 - 2.11 = 7.89\,\text{min}$$

$$T_4^{se} = T_4^{sb} + t_4^s = 24.67 + 7.89 = 32.56\,\text{min}$$

We continue simulation for truck 5.

$$r = 0.1395$$

$$t_5 = -12 \cdot \ln 0.1395 = -12 \cdot \left(-1.9697\right) = 23.64\,\text{min}$$

$$T_5 = T_4 + t_5 = 15.18 + 23.64 = 38.82\,\text{min}$$

We note that truck 5 arrives at the terminal after $t = 0.5$ hour $= 30$ min and we should stop with the simulation.

6.7 SAMPLING THE UNIFORM DISTRIBUTION

The probability density function of the uniform distribution is:

$$f(x) = \begin{cases} 0 & \text{for } x < a \\ \dfrac{1}{b-a} & \text{for } a < x < b \\ 0 & \text{for } x > b \end{cases}$$

The cumulative density function $F(x)$ equals:

$$F(x) = \begin{cases} 0, & x < a \\ \dfrac{x-a}{b-a}, & a < x < b \\ 1, & x > b \end{cases}$$

In order to simulate a random variable X, that has a uniform distribution $(a < x < b)$, we use the cumulative density function:

$$F(x) = \frac{x-a}{b-a}$$

By using the method of inversion, we obtain:

$$F(x) = r$$

$$\frac{x-a}{b-a} = r$$

$$x - a = (b-a) \cdot r$$

Finally, we have:

$$x = a + (b-a) \cdot r$$

Example 6.8

Passenger waiting time has a uniform distribution, with parameters $a = 0$ and $b = 10$ minutes $(0 < T_w < 10)$. Simulate the waiting times of 5 passengers.

SOLUTION:

We generated the following five random numbers: 0.7900, 0.8948, 0.2266, 0.1186, and 0.9619. The waiting times are, respectively:

$$t_{wi} = a + (b - a) \cdot r_i$$

$$t_{w1} = a + (b - a) \cdot r_1 = 0 + 10 \cdot 0.7900 = 7.9 \, \text{minutes}$$

$$t_{w2} = a + (b - a) \cdot r_2 = 0 + 10 \cdot 0.8948 = 8.948 \, \text{minutes}$$

$$t_{w3} = a + (b - a) \cdot r_3 = 0 + 10 \cdot 0.2266 = 2.266 \, \text{minutes}$$

$$t_{w4} = a + (b - a) \cdot r_4 = 0 + 10 \cdot 0.1186 = 1.186 \, \text{minutes}$$

$$t_{w5} = a + (b - a) \cdot r_5 = 0 + 10 \cdot 0.9619 = 9.619 \, \text{minutes}$$

6.8 SAMPLING THE ERLANG DISTRIBUTION

Let us denote by $T_i \, (i = 1, 2, \dots, k)$ the random variables that follow an exponential distribution with the parameter λ. The sum of these variables is the new random variable $(T = T_1 + T_2 + \dots + T_k)$ that has an Erlang distribution, with the parameters k and λ. The fact that T is the sum of T_i can be utilized to simulate T.

To simulate T_i, we use the following relation:

$$T_i = -\frac{1}{\lambda} \cdot \ln r_i$$

Now, T equals:

$$T = T_1 + T_2 + \dots + T_k = -\frac{1}{\lambda} \cdot \ln r_1 - \frac{1}{\lambda} \cdot \ln r_2 - \dots - \frac{1}{\lambda} \cdot \ln r_k$$

$$= -\frac{1}{\lambda} \cdot (\ln r_1 + \ln r_2 + \dots + \ln r_k) = -\frac{1}{\lambda} \cdot \ln (r_1 \cdot r_2 \cdot \dots \cdot r_k)$$

From the last expression, we see that the random variable T, that has an Erlang distribution with the parameters k and λ, can be simulated by using the following expression:

$$T = -\frac{1}{\lambda} \cdot \ln \left(\prod_{i=1}^{k} r_i \right)$$

Example 6.9

Ships arrive at the port for loading/unloading operations. The total time that the ship spends at the port is a random variable that has an Erlang distribution, with parameters $k = 2$ and $\lambda = 0.2$ h^{-1}. Simulate the duration of the service times of five ships.

SOLUTION:

To simulate the five random variables that have Erlang distribution, with parameter $k = 2$, we need 10 random numbers. We generated the following random numbers: 0.1278, 0.9317, 0.3209, 0.6660, 0.1255, 0.4675, 0.4113, 0.1578, 0.7702, and 0.0746. The service times of five ships are equal, respectively, to:

$$T_1 = -\frac{1}{\lambda} \cdot \ln(r_1 \cdot r_2) = -\frac{1}{0.2} \cdot \ln(0.1278 \cdot 0.9317) = -5 \cdot \ln(0.1191) = 10.64 \text{ h}$$

$$T_2 = -\frac{1}{\lambda} \cdot \ln(r_3 \cdot r_4) = -\frac{1}{0.2} \cdot \ln(0.3209 \cdot 0.6660) = -5 \cdot \ln(0.2137) = 7.72 \text{ h}$$

$$T_3 = -\frac{1}{\lambda} \cdot \ln(r_5 \cdot r_6) = -\frac{1}{0.2} \cdot \ln(0.1255 \cdot 0.4675) = -5 \cdot \ln(0.0587) = 14.18 \text{ h}$$

$$T_4 = -\frac{1}{\lambda} \cdot \ln(r_7 \cdot r_8) = -\frac{1}{0.2} \cdot \ln(0.4113 \cdot 0.1578) = -5 \cdot \ln(0.0649) = 13.67 \text{ h}$$

$$T_5 = -\frac{1}{\lambda} \cdot \ln(r_9 \cdot r_{10}) = -\frac{1}{0.2} \cdot \ln(0.7702 \cdot 0.0746) = -5 \cdot \ln(0.0574) = 14.28 \text{ h}$$

PROBLEMS

1. The truck interarrival times have an exponential distribution, with the parameter $\lambda = 10$ trucks per hour. Simulate 5 interarrival times if the random numbers are 0.125, 0.367, 0.259, 0.957, and 0.688.
2. The random variable that represents the number of ships that arrive at the port has a Poisson distribution. The mean rate is $\lambda = 5$ ships per hour. How many ships will arrive at the port if the observation time period is 1.5 hours, and we have the following generated random numbers: 0.6195, 0.6352, 0.6508, 0.0389, 0.1108, 0.9248, 0.7368, 0.1653, 0.4926, and 0.1925?
3. Passenger waiting time at the bus station has a uniform distribution with the parameters $a = 5$ minutes and $b = 15$ minutes. Make a simulation of 5 waiting times of passengers where the generated random numbers are: 0.8944, 0.6309, 0.3354, 0.8992, and 0.7613.

Table 6.4 Random numbers that should be used for the simulation of ship service time

Random numbers	Simulation				
	1	*2*	*3*	*4*	*5*
R_1	0.38624	0.91016	0.37973	0.08881	0.68012
R_2	0.69801	0.47067	0.7936	0.00742	0.8365
R_3	0.34328	0.55121	0.02846	0.72549	0.69942
R_4	0.00261	0.49561	0.73105	0.39976	0.26811
R_5	0.30015	0.69664	0.32978	0.07454	0.94687
R_6	0.57432	0.88385	0.72829	0.09427	0.04115
R_7	0.31126	0.74259	0.04416	0.58018	0.27833
R_8	0.32397	0.49186	0.01427	0.85625	0.21941
R_9	0.57072	0.02211	0.21832	0.60649	0.32697
R_{10}	0.70703	0.08918	0.9028	0.42886	0.07374
R_{11}	0.05995	0.87431	0.03486	0.14048	0.94939
R_{12}	0.39901	0.26976	0.83455	0.82883	0.79889

4. Ship service time has a normal distribution, with the parameters $\mu = 10$ hours and $\sigma = 1.5$ hours. Simulate 5 service times using random numbers given in Table 6.4.

BIBLIOGRAPHY

Benton, D.J., *Monte Carlo Simulation: The Art of Random Process Characterization*, Kindle Edition, Knoxville, TN, 2018.

Brandimarte, P., *Handbook in Monte Carlo Simulation: Applications in Financial Engineering, Risk Management, and Economics*, Wiley, Hoboken, NJ, 2014.

Hillier, F. and Lieberman, G., *Introduction to Operations Research*, McGraw–Hill Publishing Company, New York, 1990.

Hubbard, Douglas and Samuelson, Douglas A., Modeling Without Measurements, *OR/MS Today: 28–33*, October 2009.

Kroese, D.P., Taimre, T., and Botev, Z.I., *Handbook of Monte Carlo Methods*, Wiley Series in Probability and Statistics, John Wiley and Sons, New York, 2011.

Larson, R. and Odoni, A., *Urban Operations Research*, Prentice Hall, Englewood Cliffs, 1981.

Mazhdrakov, M., Benov, D. and Valkanov, N., *The Monte Carlo Method, Engineering Applications*, ACMO Academic Press, Sofia, Bulgaria 2018.

Newell, G. F., *Applications of Queueing Theory*, 2nd ed., Chapman and Hall, 1982.

Taha, H., *Operations Research*, MacMillan Publishing Co., Inc., New York, 1982.

Vukadinović, S., *Introduction to Probability Theory*, Privredni pregled, Belgrade (in Serbian).

Winston, W., *Operations Research*, Duxbury Press, Belmont, CA, 1994.

Chapter 7

Queueing theory

There are numerous situations where aircraft, vessels, trains, buses, cars, drivers, passengers, or pedestrians are forced to wait. Drivers wait because of accidents and incidents, as well as traffic signals. Cars also wait in a queue for toll booths and at ramp meters. Pedestrians frequently wait prior to crossing the street until traffic has stopped from both directions or the street is clear. In situations where the airport is busy, the aircraft is accommodated into a holding pattern with other aircraft that are waiting to land. Passengers at the airports wait in front of ticketing counters, security checkpoints, and baggage systems. Waiting time for security passenger search has become very long at many airports. Trains sometimes wait in order to provide minimum safe separation distances. Vessel bunching causes significant port congestion. Locks and dams also cause significant waiting times for many vessels. Trucks wait in ports to pick up loaded containers, or to drop off empty containers. Queueing in various transportation systems happens on a daily basis. Obviously, clients' total waiting time depends on the number of other requests for service, as well as on the number of servers.

Why do we wait? The answer is straightforward: Queues arise in situations where demands for service during certain time periods are greater than the available capacity of the servers. Various clients (pedestrians, drivers, cars, aircraft, vessels) request a range of services on a daily basis (landing on a runway, crossing the street, passing through the intersection, unloading at the dock, etc.). The basic characteristic of service demand is the customer arrival rate. Arrival rates in transportation systems fluctuate considerably, depending on the time of day, the day of the week, or the month of a year. Service rate represents the number of clients that can be served in a given time unit. For example, the average service time for check-in is about 2–3 minutes on many domestic flights. In the case of a group arrival (such as a whole family), this service time is significantly longer. There are variations in arrival rates and service times in many queueing systems. These variations generate queues and reduce the level of service offered to the clients.

Queueing theory is the mathematical analysis of queues. Queueing models help us to predict queue lengths, as well as waiting times in queues. The beginnings of the queueing theory are associated with the Danish engineer and applied mathematician Agner Krarup Erlang, who analyzed telephone traffic problems and published, in 1909,the first paper devoted to queueing theory. Queueing theory has been used in models related to crowd dynamics, public transit, urban and road traffic control, elevator traffic control, airport operations, air traffic control, emergency egress analysis, railway, telephone, and internet traffic. This theory has been also widely used to analyze and plan operations in a variety of areas like bank services, supermarkets, health services, manufacturing, amusement parks, or mining.

Queueing theory helps us to assess levels of service and operational performance of the various transportation systems. The usual metrics for the level of service are average waiting time a client spends in a queue, and the average number of clients in a queue. Utilization of the service facility has often been used as a metric for the system operational performance. Planners and engineers bring into play queueing theory techniques in different design stages of the future service facility (calculation of the number of lanes at an intersection, the estimation of the length of left-turning bays, calculation of the size of the check-in area at an airport, calculation of the required number of parking spaces, etc.). In essence, the analysts use queueing theory techniques to answer the following questions: what is the operational efficiency of the observed queueing system? what is the level of service offered to the clients? should transportation capacity be enlarged in response to expected demand?

7.1 BASIC ELEMENTS OF QUEUEING SYSTEMS

There are numerous queueing systems. The following are the major characteristics of every queueing system:

(a) arrival process type;
(b) service process;
(c) number of servers;
(d) queue discipline;
(e) queue capacity

Arrivals in a queueing system could be deterministic or random. The arrival rate is constant, during specific time periods, in the case of deterministic arrivals. Bottlenecks at highways occur due to traffic accidents or work zones. Bottleneck analysis in transportation systems is based on deterministic queueing. The service process could be characterized by deterministic service time, or by service time that represents random variable. Stochastic queueing is related to the cases when client arrival and service times are described by probability distributions known as arrival and service time distributions. In the case of stochastic queueing, queues are described by probabilistic metrics, such as the expected number of entities in the system, expected waiting time per client, percentage of time when the server is busy, etc.

Queueing systems could have one or more servers. The number of servers equals one, for example, in the case of an airport with only one runway. The number of servers equals, for example, four when four toll booths are open on a highway.

In the greater part of queueing systems, queue discipline is *First In, First Out (FIFO)*. The queue discipline *FIFO* is also frequently called *First Come, First Served (FCFS)*. There are also other queue disciplines, like *Last In, First Out (LIFO)*, and *Service In Random Order (SIRO)*.

In a number of queueing systems, there are no specific constraints related to the permitted queue capacity. In such queueing systems, queue capacity is treated as equal to infinity. On the other hand, there are many other queueing systems where the queue has limited capacity. For example, shock waves occur when left-turning vehicles are needed to slow down in the through lanes. In such situations, as a consequence, through traffic also slows down. Left-turn bays could notably shrink the negative shock wave effect. The lengths of the left-turn bay (queue capacity) must be suitable to meet the storage requirements of left-turners.

The subsequent traditional format has been utilized to denote the major queueing system characteristics (arrival process type, service process type, the number of servers, etc.):

$$(A / B / C):(D / E / F)$$

where:
A – arrival distribution
B – service time distribution
C – number of servers
D – service discipline
E –maximum number of clients allowed to be in the queueing system at any one time
F – size of calling source

Usually, we use notation A/B/C to explain arrival distribution, service time distribution, and the number of servers, respectively. The uniform, deterministic distribution of arrivals or departures is denoted by D, while the exponential distribution is denoted by M. A "general distribution" is denoted by G. For example, the notation M/M/1 indicates a queueing system that is characterized by exponentially distributed interarrival times, exponentially distributed service times, and the existence of one server, whereas notation D/D/1 describes a queueing system with deterministic arrivals, deterministic departures, and one server. Every one of

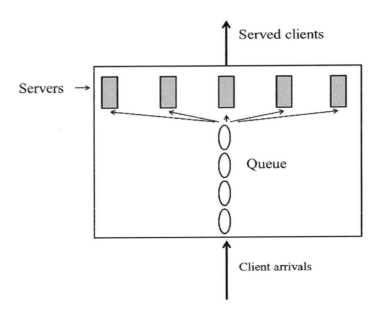

Figure 7.1 Queueing system

the queueing systems can be graphically represented in the way shown in Figure 7.1.

Example

The simple, signalized intersection (Figure 7.2) has two approaches. There is only one lane in both approaches.

We treat approach 1 as a queueing system. The vehicles coming from approach 1 are clients. In our queueing system, there is only one server. The server is the traffic light. The server is busy (red light for approach 1) in cases when vehicles from approach 2 are passing through the intersection. The service passes through the intersection. We could treat service time (time needed to pass through the intersection) as a deterministic quantity. In many cases, vehicle arrivals are random. In other words, vehicles appear from approach 1 at random time points. A vehicle that requires service (passes through the intersection) is instantly accepted if there is a green light given to approach 1 and if there is no queue of vehicles from approach 1 already exists. If there is a queue, the latest-appearing vehicle will enter the queue. The queue discipline is the FIFO discipline.

7.2 D/D/1 QUEUEING

In the case of deterministic arrivals there is no randomness. This means that the time points of appearance of the first, second, third, ...customer are precisely known (Figure 7.3).

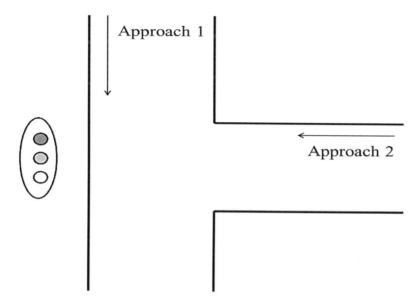

Figure 7.2 Signalized intersection

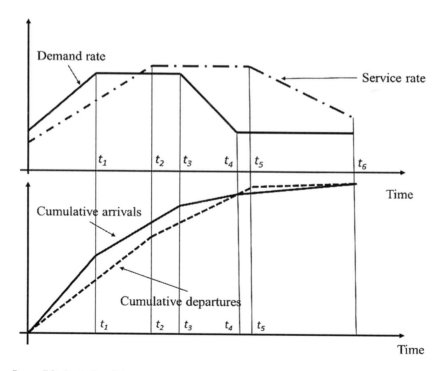

Figure 7.3 Arrival and departure rates

The total number of customers that will appear in a certain time period is known in the case of constant arrival rate, as well as in the case of a deterministic arrival rate that changes over time. Deterministic queueing is a technique that has been utilized, among other techniques, for bottleneck analysis in transportation systems. In the case of the D/D/1 queueing systems, service process is also characterized by deterministic service time. In the case shown in Figure 7.3, the demand rate [clients/hour] is known.

This rate increases from the beginning of the observation until t_1. The rate is constant between t_1 and t_3. The demand rates decline between t_3 and t_4. Finally, after t_4, the rate is constant. The service rate [clients/hour] could be portrayed in a similar way. It is straightforward to produce cumulative number of arrivals and cumulative number of departures for known arrivals and departure times, respectively. The cumulatives provide information about the total number of arrived customers and the total number of departed customers up to a certain time point. The statement that the cumulative number of vehicles arrived at the toll booth at 8:00 a.m. equals 180 means that by 8:00 a.m., a total of 180 vehicles will have appeared.

The basics of queueing theory can be easily understood when he simple D/D/1 queueing system is studied. Deterministic queueing is similar to a continuous flow of entities passing over a point over time. We use continuous lines to represent departure and arrival cumulatives in the D/D/1 queueing systems. It has been shown that these lines are a very good approximation for the cumulative stepped lines that are the real lines that represent the number of customers arriving in or departing from the queueing system.

Example 7.1

Work zones generate very high non-recurring delays. Each year, thousands of work zones produce substantial losses in terms of capacity, poor traffic flow, and much dissatisfaction from drivers. One should also take into account the delay cost when calculating the total cost of future work zones. The cumulative arrivals and the cumulative departures are shown in Figure 7.4.

The queue length equals the ordinate distance between the cumulative arrivals and cumulative departures. The longest queue happens 45 minutes after the beginning of observation. The customers entering the system 45 minutes after it begins to suffer the longest delays. The waiting time (delay) represents the horizontal distance between the two cumulative curves.

Example 7.2

Vehicles travel through the work zone. After analyzing the statistical data, it has been concluded that both arrival and service rates are deterministic. Service rate is constant. Arrival rate changes over time. These rates [vehicle/min] are given by the following functions:

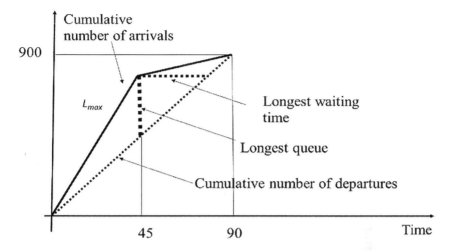

Figure 7.4 Client arrivals modeling: deterministic arrivals

$$\lambda(t) = -\frac{1}{6}t + 15$$

$$\mu = 9$$

where t is the time (in minutes) after the start of the observation of the queueing system. Cumulative numbers of arrivals $A(t)$ and cumulative number of departures $D(t)$ are equal:

$$A(t) = \int_0^t \lambda(t)\,dt = \int_0^t \left(-\frac{1}{6}t + 15\right)dt = -\frac{1}{6} \cdot \frac{t^2}{2} + 15 \cdot t = -\frac{1}{12}t^2 + 15 \cdot t$$

$$D(t) = \int_0^t \mu(t)\,dt = \int_0^t 9\,dt = 9 \cdot t$$

Cumulative numbers of arrivals and departures are shown in Figure 7.5.

The queue will dissipate at the time point when the total number of cumulative arrivals equals the total number of cumulative departures, i.e.:

$$A(t) = D(t)$$

After substitution we get the following equation:

$$-\frac{1}{12}t^2 + 15 \cdot t = 9 \cdot t$$

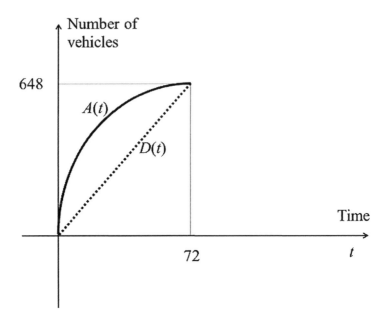

Figure 7.5 Cumulative arrivals and cumulative departures

$$-\frac{1}{12}t^2 + 6 \cdot t = 0$$

The solutions of the equation are $t_1 = 0$ and $t_2 = 72$. The cumulative number of arrivals $A(t)$ is equal to the cumulative number of departures $D(t)$ for the first time at the beginning of our observation $(t = 0)$. These cumulatives are equal for a second time when the queue dissipates. We conclude that the queue will dissipate after 72 minutes.

The total number of arrivals during the 72-minute period equals:

$$A(72) = D(72) = 648$$

The total delay D of all vehicles is represented by the area between the cumulative number of arrivals and the cumulative number of departures. In other words:

$$D = \int_0^{72} A(t) \cdot dt - \int_0^{72} D(t) \cdot dt$$

$$D = \int_0^{72} \left[-\frac{1}{12}t^2 + 15t \right] dt - \int_0^{100} 9t \cdot dt$$

$$D = -\frac{1}{36} \cdot t^3 + 15 \cdot \frac{t^2}{2} - 9 \cdot \frac{t^2}{2} \bigg|_0^{72}$$

$$D = 5,184$$

The average delay d per vehicle equals:

$$d = \frac{5,184}{648} = 8$$

Because of the work zone, the average delay per vehicle equals 8 minutes. Queue length $L(t)$ at any moment t represents the difference between the cumulative number of arrivals, $A(t)$, at moment t and the cumulative number of departures, $D(t)$, by moment t:

$$L(t) = A(t) - D(t)$$

$$L(t) = -\frac{1}{12} t^2 + 15 \cdot t - 9 \cdot t = -\frac{1}{12} t^2 + 6 \cdot t$$

We determine the maximum queue length as follows:

$$\frac{d[L(t)]}{dt} = 0$$

$$\frac{d\left[-\frac{1}{12} t^2 + 6 \cdot t\right]}{dt} = 0$$

$$-\frac{1}{6} \cdot t + 6 = 0$$

$$t = 36$$

We conclude that the maximum queue length happens 36 minutes after the beginning of observations. The maximal queue length equals:

$$L_{max} = L(36) = -\frac{1}{12} \cdot 36^2 + 6 \cdot 36$$

$$L_{max} = 108 \text{ vehicles}$$

7.3 M/M/1 QUEUEING

Many of the queueing systems in transportation are M/M/1 queueing systems. The examples could be one open toll booth at the highway, one bridge toll booth, one open check-in counter at the airport, a vehicle inspection facility with one repairman working, etc. The M/M/1queueing system has the following characteristics: (a) Poisson distribution arrivals (exponential interarrival times); (b) exponential service times; (c) one server; (d) FIFO queue discipline.

Let us introduce the following notation:

λ – mean arrival rate
μ – mean service rate

The following relations describe M/M/1 queueing:

Probability of having no customers in the queueing system equals:

$$P_0 = 1 - \frac{\lambda}{\mu}$$

Probability of having n customers in the queueing system equals:

$$P_n = \left(\frac{\lambda}{\mu}\right)^n \cdot P_0$$

The average number of customers in the queue:

$$L_q = \frac{\lambda^2}{\mu \cdot (\mu - \lambda)}$$

The average number of customers in the queueing system:

$$L_s = L_q + \frac{\lambda}{\mu} = \frac{\lambda^2}{\mu \cdot (\mu - \lambda)} + \frac{\lambda}{\mu} = \frac{\lambda}{(\mu - \lambda)}$$

The average waiting time a customer spends in the queue:

$$W_q = \frac{L_q}{\lambda} = \frac{\frac{\lambda^2}{\mu \cdot (\mu - \lambda)}}{\lambda} = \frac{\lambda}{\mu \cdot (\mu - \lambda)}$$

The average waiting time a customer spends in the queueing system:

$$W_s = W_q + \frac{1}{\mu} = \frac{\lambda}{\mu \cdot (\mu - \lambda)} + \frac{1}{\mu} = \frac{1}{\mu - \lambda}$$

Example 7.3

Most US states collect taxes from trucks and commercial vehicles based on the weight of the goods transported. The truck weight is checked

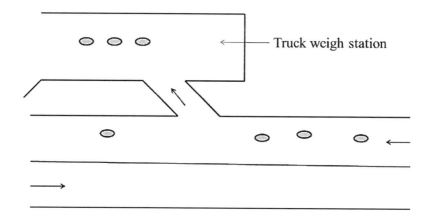

Figure 7.6 Truck weigh station

to make certain that it falls inside the safety guidelines. Some of the trucks from the traffic flow must enter truck weigh stations for additional safety inspections. Consider the truck weigh station in Figure 7.6. The drivers can expect one or two truck measurements, and additional safety inspections. Service time varies, but on average is equal to 15 minutes. On average, 3 drivers per hour are requested to submit to the additional safety inspections.

Treat the truck weigh station as a M/M/1 queueing system and calculate the following:

The average number of trucks in the queue
The average number of trucks in the queueing system
The average waiting time a truck spends in the queue
The average waiting time a truck spends in the queueing system

SOLUTION:

The average arrival rate λ equals $3\left(\dfrac{\text{veh}}{\text{h}}\right) = \dfrac{1}{20}\left(\dfrac{\text{veh}}{\text{min}}\right)$. The average service rate equals $\mu = \dfrac{1}{15}\left(\dfrac{\text{veh}}{\text{min}}\right)$.

The average number of clients in the queue equals:

$$L_q = \frac{\lambda^2}{\mu \cdot (\mu - \lambda)} = \frac{\left(\dfrac{1}{20}\right)^2}{\dfrac{1}{15}\cdot\left(\dfrac{1}{15} - \dfrac{1}{20}\right)} = 2.25 \text{ vehicles}$$

The average number of clients in the queueing system equals:

$$L_s = \frac{\lambda}{(\mu - \lambda)} = \frac{\dfrac{1}{20}}{\left(\dfrac{1}{15} - \dfrac{1}{20}\right)} = 3 \text{ vehicles}$$

The average waiting time a client spends in the queue equals:

$$W_q = \frac{\lambda}{\mu \cdot (\mu - \lambda)} = \frac{\dfrac{1}{20}}{\dfrac{1}{15} \cdot \left(\dfrac{1}{15} - \dfrac{1}{20}\right)} = 45 \text{ min}$$

The average waiting time a client spends in the queueing system equals:

$$W_s = \frac{1}{\mu - \lambda} = \frac{1}{\dfrac{1}{15} - \dfrac{1}{20}} = 60 \text{ min}$$

Example 7.4

There are various unannounced roadside inspections of commercial vehicles. These inspections could include a check of the documents, a check of brakes, emissions, and the vehicle's overall conditions. Vehicles arrive for roadside inspections according to a Poisson distribution, with a mean of 2 per hour. The time for inspections is a random variable that has exponential distribution, with a mean of 20 minutes per commercial vehicle. Treat a roadside inspection facility as a M/M/1 queueing system. Calculate the number of parking spaces needed, so that incoming commercial vehicle will be able to park in a parking space at least 75% of the time.

SOLUTION:

The values of λ and μ are equal:

$$\lambda = 2 \text{ Veh/h}$$

$$\mu = \frac{60}{20} = 3 \text{ Veh/h}$$

The average number of clients in the queue equals:

$$L_q = \frac{\lambda^2}{\mu \cdot (\mu - \lambda)} = \frac{2^2}{3 \cdot (3 - 2)} = 1.33 \text{ cars.}$$

The number of commercial vehicles in the queue could be fewer or more than 1.33 cars. In other words, the number of vehicles in the queue could be 0, 1, 2,... vehicles. We denote by m the number of parking spaces needed for commercial vehicles. Taking into account the requirement that incoming commercial vehicles should be able to park at parking space at least 75% of the time, we write the following:

$$P_0 + P_1 + P_2 + \cdots + P_m \geq 0.75$$

We got:

$$P_0 + \left(\frac{\lambda}{\mu}\right)^1 \cdot P_0 + \left(\frac{\lambda}{\mu}\right)^2 \cdot P_0 + \cdots + \left(\frac{\lambda}{\mu}\right)^m \cdot P_0 \geq 0.75$$

$$P_0 \cdot \left[1 + \left(\frac{\lambda}{\mu}\right)^1 + \left(\frac{\lambda}{\mu}\right)^2 + \cdots + \left(\frac{\lambda}{\mu}\right)^m\right] \geq 0.75$$

$$\left(1 - \frac{2}{3}\right) \cdot \left[1 + \left(\frac{2}{3}\right)^1 + \left(\frac{2}{3}\right)^2 + \cdots + \left(\frac{2}{3}\right)^m\right] \geq 0.75$$

$$\left(1 - \frac{2}{3}\right) \cdot \frac{1 - \left(\frac{2}{3}\right)^{m+1}}{1 - \frac{2}{3}} \geq 0.75$$

$$\left(\frac{2}{3}\right)^{m+1} \leq 0.25$$

$$(m+1) \cdot \log\left(\frac{2}{3}\right) \leq \log(0.25)$$

$$(m+1) \cdot (-0.174) \leq -0.602$$

$$(m+1) \geq 3.46$$

$$m \geq 2.46$$

We conclude that $m \geq 3$. In words, we must provide at least 3 parking spaces, so that incoming commercial vehicles will be able to park at a parking space at least 75% of the time.

7.4 M/M/S QUEUEING

In the M/M/s queueing system, arrivals happen according to a Poisson distribution, whereas service times are exponentially distributed. The total number of servers is equal to s. This means that a maximum of s customers can be served at once. Clearly, the system's service rate is much higher than in the case of one server. Examples of the M/M/s queueing system are parking lots where each parking place represents one server, airport operations in the case of multiple runways, or a small number of toll booths on the highway. The utilization factor ρ of the facilities is defined in the following way:

$$\rho = \frac{\lambda}{s \cdot \mu}$$

The following relations describe M/M/s queueing:

$$p_0 = \frac{1}{\displaystyle\sum_{k=0}^{s-1} \frac{\left(\dfrac{\lambda}{\mu}\right)^k}{k!} + \frac{\left(\dfrac{\lambda}{\mu}\right)^s}{s!} \cdot \frac{1}{1 - \dfrac{\lambda}{s \cdot \mu}}}$$

$$p_k = \frac{\left(\dfrac{\lambda}{\mu}\right)^k}{k!} \cdot p_0 \qquad \text{for } 0 \le k \le s$$

$$p_k = \frac{\left(\dfrac{\lambda}{\mu}\right)^k}{s! \cdot s^{k-s}} \cdot p_0 \qquad \text{for } k \ge s$$

$$L_q = \frac{(\rho)^{s+1}}{(s-1)! \cdot (s-\rho)^2} \cdot p_0$$

$$W_q = \frac{L_q}{\lambda}$$

$$L_s = \lambda \cdot W_s = \lambda \cdot \left(W_q + \frac{1}{\mu}\right) = L_q + \frac{\lambda}{\mu}$$

Example 7.5

The security check point at a small airport has two X-ray machines. On average, a passenger needs 40 s to go through the check system (exponential distribution service time).The arrival process could be described by a Poisson distribution, with a mean arrival rate of one passenger every 30 s.

What is the current utilization of the existing X-ray machines?

SOLUTION

The queueing system that we analyze has multiple servers and an infinite source.

The mean arrival rate equals one passenger every 30 s, i.e.

$$\lambda = \frac{1}{\dfrac{30}{3,600}} \left[\frac{\text{pass}}{\text{h}} \right] = 120 \left[\text{pass/h} \right]$$

On average, a passenger takes 40 s to pass through the system. The service rate equals:

$$\mu = \frac{1}{\dfrac{40}{3,600}} \left[\frac{\text{pass}}{\text{h}} \right] = 90 \left[\text{pass/h} \right]$$

The utilization of the facility equals:

$$\rho = \frac{\lambda}{s \cdot \mu}$$

$$\rho = \frac{120}{2 \cdot 90} = 0.67$$

The following are values of the queueing parameters for a multi-server queueing system with an infinite population:

Idle probability:

$$p_0 = \cfrac{1}{\displaystyle\sum_{k=0}^{s-1} \cfrac{\left(\dfrac{\lambda}{\mu}\right)^k}{k!} + \cfrac{\left(\dfrac{\lambda}{\mu}\right)^s}{s!} \cdot \cfrac{1}{1 - \dfrac{\lambda}{s \cdot \mu}}}$$

$$p_0 = \cfrac{1}{\displaystyle\sum_{k=0}^{2-1} \cfrac{\left(\dfrac{120}{90}\right)^k}{k!} + \cfrac{\left(\dfrac{120}{90}\right)^2}{2!} \cdot \cfrac{1}{1 - \dfrac{120}{2 \cdot 90}}}$$

$$p_0 = 0.2$$

Expected number of clients in the queue:

$$L_q = \frac{(\rho)^{s+1}}{(s-1)! \cdot (s-\rho)^2} \cdot p_0$$

$$L_q = \frac{(0.67)^{2+1}}{(2-1)! \cdot (2-0.67)^2} \cdot 0.2$$

$$L_q = 1.13$$

Expected number of customers in system:

$$L_s = L_q + \frac{\lambda}{\mu}$$

$$L_s = 1.13 + \frac{120}{90}$$

$$L_s = 2.46$$

Average waiting time in queue:

$$W_q = \frac{L_q}{\lambda}$$

$$W_q = \frac{2.46}{120} = 0.0205[\text{h}] = 69.84 \text{ s} \approx 74 \text{ s}$$

Average waiting time in system:

$$W_s = W_q + \frac{1}{\mu}$$

$$W_s = 74 + 45 = 119 \text{ s}$$

Example 7.6

Departure flight schedule during an observed time interval contains information related to the number of flights, the number of passengers, and the start time of check-in for each flight. The passengers' arrival pattern is recorded at the airport's arrival section. Requests for check-in happen according to a Poisson distribution with a mean of 360 requests per hour. Each passenger check-in officer can handle an average of 30 requests per hour. The cost of adding a new passenger check-in officer is estimated at 10 monetary units per hour. The cost of waiting time (monetary units per hour) per passenger is estimated at 15 monetary units per hour. Determine the optimal number of check-in officers.

SOLUTION:

There are operator's cost and passengers' cost of waiting time. Both depend on the number of operational check-in officers. The operator's cost $O(s)$ equals:

$$O(s) = 10 \cdot s$$

The passengers' cost $P(s)$ equals:

$$P(s) = 15 \cdot L_s(s)$$

where $L_s(s)$ represents the expected number of passengers in the system in the case when s check-in officers work

The total costs $T(s)$ equal:

$$T(s) = O(s) + P(s) = 10 \cdot s + 15 \cdot L_s(s)$$

The optimal number of operational check-in officers must satisfy the following two relations:

$$T(s) \le T(s-1)$$

$$T(s) \le T(s+1)$$

Taking into account that:

$$\rho = \frac{\lambda}{s \cdot \mu} < 1 \quad \Rightarrow \quad s > \frac{\lambda}{\mu} = \frac{360}{30} = 12$$

we conclude that s should be greater than or equal to 13. The determination of the s optimal value is accomplished by performing the computations as shown in Table 7.1. We see from Table 7.1 that the optimal s value equals 15. In the case of 15 operational check-in officers, the total cost equals 349.15 monetary units.

7.5 LITTLE'S LAW

Little's law is the most significant result of the queueing theory. This law is valid for *any* queueing system that is in a stable condition (stable conditions do not assume, for example, the beginning of the operations in the system). The relationship described by *Little*'s law requires no assumptions about the probability distributions of the interarrival and service times. Several

Table 7.1 The check-in system parameters for various s values

s	p_0	L_q	$L_s(s)$	$O(s)$	$P(s)$	$T(s)$
13	3.1535E-06	8.453	20.453	130	306.790	436.790
14	4.67255E-06	2.890	14.890	140	223.353	363.353
15	5.41827E-06	1.277	13.277	150	199.151	349.151
16	5.78775E-06	0.614	12.614	160	189.206	349.206
17	5.97098E-06	0.304	12.304	170	184.559	354.559
18	6.06128E-06	0.151	12.151	180	182.268	362.268

queueing systems have been created for the queueing subsystems. *Little*'s law is valid for queueing subsystems, as well as for an entire system.

An interpretation of *Little*'s Law follows. Suppose that N customers arrive in the queueing system during the time interval $(0,T)$. Figure 7.7 shows the cumulative arrivals and departures.

The queue length, over whichever time point, corresponds to the maximum ordinate distance between the cumulative departure and cumulative arrival curves. It is very easy to "read" from the figure by visual inspection of the waiting time for every customer. The shaded area shown in Figure 7.7 represents the total waiting time (the total delay) of all customers. The average delay time, W, is the quotient of the total delay and the number of clients, i.e.:

$$W = \frac{\text{Area}}{N}$$

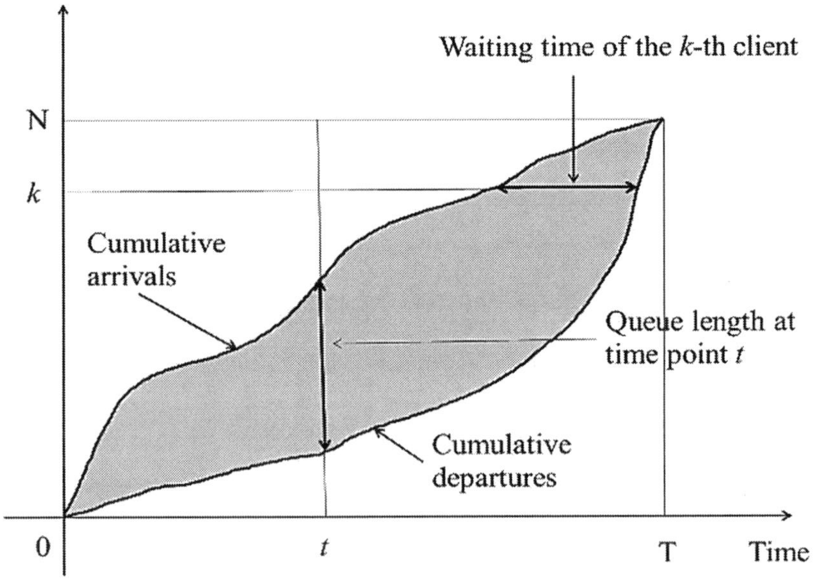

Figure 7.7 Graphical illustration of the queue length and waiting time

In addition, the shaded area also signifies the total length of all queues over all time points. The average queue length L equals:

$$L = \frac{\text{Area}}{T}$$

We conclude the following:

$$W \cdot N = L \cdot T$$

i.e.:

$$L = \frac{N}{T} \cdot W$$

The ratio N/T actually represents the arrival rate λ.
  We can write:

$$L = \lambda \cdot W$$

This last relationship has become known as *Little*'s law. This law offers the following conclusion: the average number of clients in a queueing system (over a particular time interval) is equal to their average arrival rate multiplied by their average time in the system. Correspondingly, we can also write the following:

$$L_q = \lambda \cdot W_q$$

$$W = W_q + \frac{1}{\mu}$$

7.6 QUEUEING THEORY AND CONGESTION PRICING

The number of trips by private cars has increased considerably in recent decades in many cities, and on many highways. This is a consequence of the growth of metropolitan areas, increases in population, and increases in the number of licensed drivers. Simultaneously, road network capacities have not kept up with this increase in travel demand. Urban road networks in many countries are heavily congested, resulting in increased travel times, increased number of stops, unexpected delays, greater travel costs, inconvenience to drivers and passengers, increased air pollution and noise level, and increased number of traffic accidents. A similar situation has occurred in air transportation. There has been an increase in the number of cancelled flights, delayed flights, greater airline and airport operating costs, a decrease in the quality of air traffic service, and a potential decrease in airspace safety.

Increasing the size of traffic network capacities by building additional roads and additional runways is very costly, as well as being environmentally destructive. Researchers, engineers, and policy makers have begun to consider and implement strategies that concentrate on demand management. One of the most significant demand management strategies is congestion pricing. Congestion pricing initiates various fees for road usage. The fees or tolls vary with the location in the network, the time of day and/or the level of traffic congestion. Users pay for using a certain road, corridor, or bridge, or for entering a specific area for the duration of the high-traffic-volume time period.

The major task of any congestion pricing scheme is to shift some users to alternative routes, alternative departure times, alternative airports, or alternative time slots. The basic goal of congestion pricing (to distribute transportation demand more evenly over time and space) is graphically illustrated in Figure 7.8.

Drivers and passengers usually recognize just their own travel time costs, whereas bus operators and airlines correspondingly recognize primarily their own operating costs. They do not consider the supplementary delay their trips impose on other passengers. Noting these facts, Vickrey (1969) claimed that "Charges should reflect as closely as possible the marginal social cost of each trip in terms of the impacts on others. There is no excuse for charges below marginal social cost". It appears entirely logical that users should pay a price equal to the delay cost they impose on others.

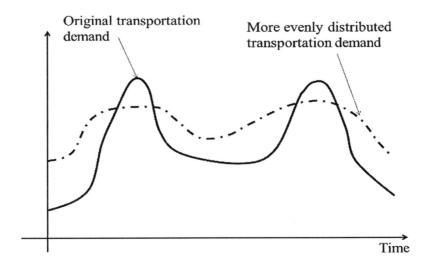

Figure 7.8 More evenly distributed passenger demand

Let us denote by c the delay cost per unit time per client. The total delay cost per unit time, C, equals:

$$C = c \cdot L_q$$

where L_q is the average number of clients in the queue.

We conclude, based on *Little*'s law, that the total cost of delay per unit time equals:

$$C = c \cdot \lambda \cdot W_q$$

where λ is the average arrival rate and W_q the average waiting time a client spends in the queue.

The more vehicles on a highway, the higher the congestion level, and the higher are vehicle delay costs. Marginal cost imposes an additional increase in queueing cost with every extra customer. The marginal delay cost, MC, imposed by an additional client equals:

$$MC = \frac{dC}{d\lambda}$$

i.e.:

$$MC = c \cdot W_q + c \cdot \lambda \cdot \frac{dW_q}{d\lambda}$$

We see from the last equation that the marginal delay cost MC has two terms. The term $c \cdot W_q$ represents the internal cost experienced by the additional customer. The second term in the last relationship, $c \cdot \lambda \cdot \frac{dW_q}{d\lambda}$, represents the "*external*" cost he/she imposes. The majority of clients in queueing systems in transportation do not understand the existence of these "external" costs. When we enter the congested highway, we are exposed to significant delays and increased travel costs. At the same time, we must understand that, by entering the congested highway, we additionally increase delay costs for all other users.

The marginal delay cost in the case of the M/M/1 queueing system equals:

$$MC = c \cdot W_q + c \cdot \lambda \cdot \frac{dW_q}{d\lambda}$$

In the case of the M/M/1 queueing system, the average waiting time, W_q, equals:

$$W_q = \frac{\lambda}{\mu \cdot (\mu - \lambda)}$$

The marginal delay cost MC equals:

$$MC = c \cdot \frac{\lambda}{\mu \cdot (\mu - \lambda)} + c \cdot \lambda \cdot \frac{d\left[\dfrac{\lambda}{\mu \cdot (\mu - \lambda)}\right]}{d\lambda}$$

$$MC = \frac{c \cdot \lambda \cdot (2 \cdot \mu - \lambda)}{\mu (\mu - \lambda)^2}$$

Example 7.7

There are two competitive routes between point A and point B. A tunnel is located on the shorter route (Figure 7.9). This route is characterized by a high congestion level.

By analyzing traffic conditions, traffic engineers concluded that the shorter route with the tunnel could be treated as a M/M/1 queueing system. The average arrival rate equals 180[veh/h]. The average service rate equals $\mu = 200 \left[\dfrac{\text{veh}}{\text{h}}\right]$. For safety reasons, the traffic authorities

and local government intend to introduce congestion pricing in order to decrease the congestion level along the shorter route with the tunnel. Calculate the congestion price (marginal delay cost) that should be charged to drivers who use the congested route.

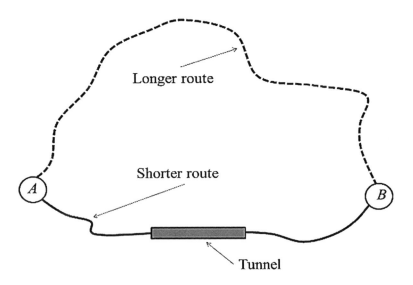

Figure 7.9 Shorter and longer route

Table 7.2 Congestion price MC
(marginal delay cost, $) as a
function of the delay cost per
unit time per driver, c

c [$/h]	MC ≈ 0.5·c [$]
18	9
15	7.5
12	6
9	4.5
6	3

SOLUTION:

Marginal delay cost MC imposed by an additional vehicle equals:

$$MC = \frac{c \cdot \lambda \cdot (2 \cdot \mu - \lambda)}{\mu \cdot (\mu - \lambda)^2}$$

$$\frac{c \cdot \lambda \cdot (2 \cdot \mu - \lambda)}{\mu \cdot (\mu - \lambda)^2} = \frac{180 \cdot (2 \cdot 200 - 180)}{200 \cdot (200 - 180)^2} \cdot c$$

$$MC = 0.495 \cdot c \approx 0.5 \cdot c$$

Table 7.2 shows congestion price MC [$] (marginal delay cost) as a function of the delay cost per unit time per driver c.

7.7 QUEUEING THEORY AND CAPITAL INVESTMENTS IN TRANSPORTATION

Queueing theory enables us to measure the operational efficiency of the observed queueing system, as well as the level of service offered to the clients. By varying the number of servers in the system, we perform sensitivity analysis and study changes in the service facility idle time, queue length, and the average waiting time, caused by the changes in the number of servers. In many situations, we can considerably improve queueing system operations by adding more servers (providing more throughlanes on a highway, a new runway at an airport, or expanding the dock in a harbor).

Let us suppose that we are at the stage of increasing the number of toll booths on a highway. We need to make a decision about the number of toll booths (Figure 7.10).

It is clear that the queue length through the rush hour will decrease with the increase in the number of toll booths. How many tool booths do we

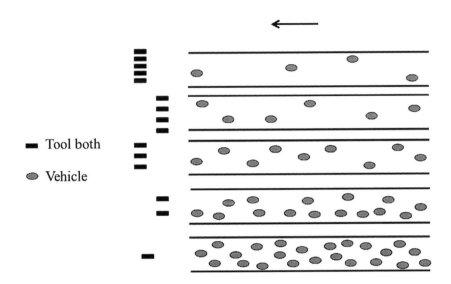

Figure 7.10 Queue length as a function of the number of toll booths

need? By using queueing theory techniques, we can calculate average queue length during the rush hour, average waiting time per client, percentage of server idle time, etc. The expected user waiting cost, and the total toll booth construction and maintenance costs are shown in Figure 7.11.

The larger the number of toll booths, the smaller the drivers' waiting cost, but the higher the toll booths construction and maintenance cost. This is also applicable for every other transportation facility (highway, airport, railway station, port). The curves shown in Figure 7.11 describe operations for all transportation facilities. High level-of-service (short waiting times, short queue lengths) is costly, but, as a result, users' waiting costs are very low, and *vice versa*. The "optimal" number of servers represents the compromise between queue lengths, user waiting times, and construction and maintenance costs. Queueing theory helps us in performing a complete analysis of the queueing phenomenon, and to examine the trade-off between the service costs and the costs of waiting for the service.

7.8 SIMULATION OF QUEUEING SYSTEMS

Simulation is also used in the queueing system analysis. It has been shown that, using simulation techniques, we can analyze very complex, real-world queueing systems. A simulation approach can easily handle any distribution of interarrival and service times. By simulation, which we perform "off-line", we can analyze a longer period of activities of the queueing system over a short period of time.

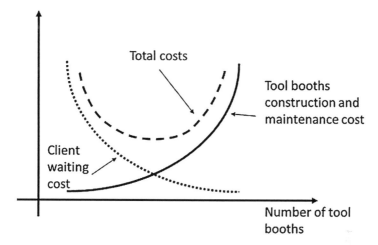

Figure 7.11 User waiting cost, construction cost and total cost in the queueing system

We illustrate the basic steps in the simulation of queueing systems by using the following simple example.

The ship to be served arrives at the part of the port that handles containerized cargo. The part for the containerized cargo has two berths and places for two ships that wait for a service. Ship interarrival time has uniform distribution with the parameters $a = 8$ hand $b = 12$ h. Service time has an Erlang distribution, with the parameters $k = 2$ and $\mu = 1/6 \ h^{-1}$. Simulate arrival and service of the first 10 ships that arrive at the port. Consider that first ship arrives at time point $T = 0$, when the port is empty.

Since the interarrival time has uniform distribution, the interarrival time t_i between arrival of ship $i - 1$ and ship i can be simulated in the following way:

$$t_i = a + (b - a) \cdot r = 8 + 4 \cdot r$$

The arrival moment T_i of the i-th ship equals:

$$T_i = T_{i-1} + t_i$$

We simulate service time $t_{ser,i}$ of the i-th ship by using the following relationship:

$$t_{ser,i} = -\frac{1}{\lambda} \cdot \ln(r_1 \cdot r_2)$$

The flow chart of the simulation process is given in Figure 7.12.

In the first part of the simulation process, we determine the values of t_i, T_i, and t_{ser}. This part of the simulation is given in Table 7.3. Table 7.3

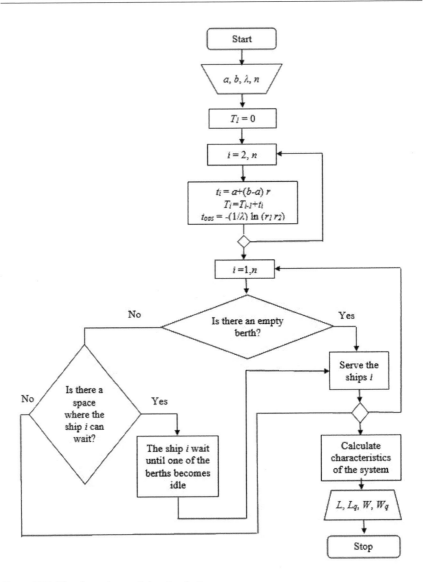

Figure 7.12 The flow chart of the simulation process

contains the ordinary ship numbers, generated random numbers, simulated moments of arrivals, and simulated duration of service.

In the second part of the simulation process, we analyze how ships should be served, taking into account the occupancies of the berths. Based on the data contained in Table 7.3, it is possible to record the movement of every arrived ship. In some cases, the arrived ship was immediately accepted for service, whereas, in other cases, the arrived ship was sent to join the queue

Table 7.3 Ordinary ship numbers, generated random numbers, simulated moments of arrivals, and simulated duration of service

i	$r\grave{}$	t_i	T_i	r_1	r_2	t_{ser}
1	–	–	0	0.37442	0.20002	15.5505
2	0.6858	10.7432	10.7432	0.8052	0.8755	2.09744
3	0.7293	10.9174	21.6606	0.6633	0.5218	6.36616
4	0.7806	11.1223	32.7829	0.0264	0.3240	28.5796
5	0.9947	11.9789	44.7617	0.7706	0.2906	8.97751
6	0.4249	9.6996	54.4613	0.0008	0.5075	46.7384
7	0.8668	11.4674	65.9287	0.8979	0.9977	0.66004
8	0.3210	9.2840	75.2127	0.0304	0.3086	28.0054
9	0.1388	8.5553	83.7680	0.8966	0.9149	1.18799
10	0.4370	9.7480	93.5160	0.9275	0.5910	3.60716
11	0.67586	10.7034	104.2194	0.13336	0.52754	15.9254
12	0.03736	8.1494	112.3689	0.74012	0.14074	13.5705

for service. The allocation of arrived ships to the berths, or to the queue, and the timing of the start and end of the service and the start and end of the waiting time in the queue, are given in Table 7.4.

The changes in the number of ships in the port during the observed period of time are given in Figure 7.13.

Finally, we have to calculate the characteristics of the system. We calculate the following characteristics: (a) the probability that the ship that

Table 7.4 The allocation of arrived ships to the berths or to the queue, the start and end time of the service, and the start and end of the waiting time in the queue

i	First berth (server)		Second berth (server)		First place in the queue		Second place in the queue	
	start	end	start	end	start	end	start	end
1	0	15.5505						
2			10.7432	12.8407				
3	21.6606	28.0267						
4	32.7829	61.3625						
5			44.7617	53.7393				
6			54.4613	101.2				
7	65.9287	66.5887						
8	75.2127	103.218						
9			101.2	102.388	83.768	101.2		
10			102.388	105.995	101.2	102.388	93.516	101.2
11	104.219	120.145						
12			112.369	125.939				

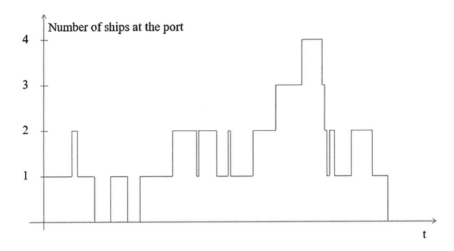

Figure 7.13 The changes in the number the ships in the port during the observed period of time

arrives at the port will be served; (b)the average number of ships waiting for the service (L_q); (c) the average number of ships at the port (L); (d) the average waiting time that ships spend waiting for the service (W_q); and (e) the average time that ships spend at the port (W).

From Table 7.4, we conclude that all the arrived ships are served. The probability that an arrived ship will be served is:

$$P = \frac{\text{Number of served ships}}{\text{Number of all ships}} = \frac{12}{12} = 1$$

The probability that the port is empty (i.e. that there are no ships in the port) equals:

$$p_0 = \frac{\text{Sum of periods when there were zero ships in the port}}{\text{Total duration of simulation process}}$$

$$= \frac{(21.6606 - 15.5505) + (32.7829 - 28.0267)}{125.9394} = 0.0863$$

In a similar way we calculate the probabilities p_1, p_2, p_3, and p_4.

$$p_1 = \frac{58.88}{125.939} = 0.4675$$

$$p_2 = \frac{37.74}{125.939} = 0.2997$$

$$p_3 = \frac{10.936}{125.939} = 0.087$$

$$p_4 = \frac{7.6837}{125.939} = 0.061$$

The average queue length equals:

$$L_q = 1 \cdot p_3 + 2 \cdot p_4 = 0.209$$

The average number of ships in the port equals:

$$L = 1 \cdot p_1 + 2 \cdot p_2 + 3 \cdot p_3 + 4 \cdot p_4 = 1.57$$

The average time that a ship spends in the port equals:

$$W = \frac{\text{Total time ships spend at the port}}{\text{Number of ships}} = \frac{197.57}{12} = 16.46 \text{ h}$$

PROBLEMS

7.1. Describe the basic elements of the queueing system (arrival process type, service process, number of servers, and queue discipline) in the following situations:
pay toll at gates on a highway
washing facility
aircraft arriving at an airport
vessels entering a port
calls for police assistance in a city

7.2. A 100-foot long car wash tunnel can handle up to one hundred vehicles per hour. On average, 90 drivers per hour request a car wash.
   Treat the car wash tunnel as an M/M/1 queueing system and calculate the following:
The average number of cars in the queue
The average number of cars in the queueing system
The average waiting time a car spends in the queue
The average waiting time a car spends in the queueing system

7.3. Clients arrive in the system according to a Poisson process. Every 12 minutes, one client arrives. Service time is exponentially distributed. Average service time equals 8 minutes per client. Determine the basic characteristics of the queueing system.

7.4. The queueing system has a single server. Clients arrive in the system with a Poisson arrival distribution. The mean arrival rate is $\lambda = 2.5$ clients per minute. The single server has an exponential service time distribution. The mean service rate is $\mu = 3$ clients per minute. Decide which option is better: (a) a single server works two times faster; (b) the system has two servers and each server works at the original rate of $\mu = 3$ clients per minute.

BIBLIOGRAPHY

Andreatta, G. and Odoni, A. R., Analysis of Market-Based Demand Management Strategies for Airports and En Route Airspace. In T. Ciriani (Ed.), *Operations Research in Space and Air*, Kluwer Academic Publishers, Boston, 2003.

Hillier, F. and Lieberman, G., *Introduction to Operations Research*, McGraw–Hill Publishing Company, New York, 1990.

Larson, R. and Odoni, A., *Urban Operations Research*, Prentice Hall, Englewood Cliffs, 1981.

Newell, G. F., *Applications of Queueing Theory*, 2nd ed., Chapman and Hall, London, 1982.

Ortuzar, J.D. de and Willumsen, L. G., *Modelling Transport*, John Wiley and Sons, Toronto, 1990.

Taha, H., *Operations Research*, MacMillan Publishing Co., Inc., New York, 1982.

Teodorović, D. and Janić, M., *Transportation Engineering: Theory, Practice and Modeling*, Elsevier, New York, 2016.

Vickrey, W., A Proposal for Revising New York's Subway Fare Structure, *Journal of the Operations Research Society of America*, 3, 38–68, 1955.

VickreyW., Pricing in Urban and Suburban Transport, *The American Economic Review* (Papers and Proceedings), 53, 452–65, 1963.

Vickrey, W., Congestion Theory and Transport Investment, *American Economic Review* (Papers and Proceedings), 59, 251–261, 1969.

VickreyW., Statement to the Joint Committee on Washington, DC, Metropolitan Problems. *Journal of Urban Economics*, 36, 42–65, 1994.

Winston, W., *Operations Research*, Duxbury Press, Belmont, CA, 1994.

Chapter 8

Heuristic and metaheuristic algorithms

A large number of real-world transportation problems have been formulated and solved by integer programming, dynamic programming, and graph theory techniques over the past five decades. It is essential to note, however, that the majority of the real-world problems solved by some of the optimization techniques were low dimensional in nature. A lot of engineering and control problems are combinatorial by their very nature. The majority of the combinatorial optimization problems are hard to solve, either because of their high dimensionality or because it is not easy to break them down into smaller sub-problems. Representatives of this type of problem are vehicle fleet planning, vehicle routing and scheduling, crew scheduling, transportation network design, location problems, and optimizing alignments for highways and public transportation routes through complex geographic spaces, etc.

In many cases, optimal solution of the problem cannot be achieved within an acceptable central processing unit (CPU) time. Often, there is a combinatorial explosion of the promising combinations of the decision variables that could be optimal; for instance, in a problem, if we have considered 1,000 binary variables that can take values 0 or 1, the total number of all possible solutions is equal to 2^{1000}. In some other situations, it is hard to estimate defined objective function values. In order to conquer these difficulties, various *heuristic* (`ευρισκω, Greek for "I find") *algorithms* were proposed over the past five decades. Many of the heuristic algorithms developed have been able to generate sufficiently good solution(s) in acceptable amounts of CPU time. On the other hand, heuristic algorithms that depend primarily on the experience and/or judgment of the analyst frequently do not produce the optimal solution.

8.1 COMPLEXITY OF ALGORITHMS

Practically, all algorithms could be classified as either exact or *heuristic*. Heuristic algorithms, that are able to produce high-quality solution(s) in an acceptable amount of CPU time, could be portrayed as a blend of scientific methods, invention, experience, and intuition for problem solving. The real-life engineering, management, and control problems are commonly solved by a range of heuristic algorithms. In order to solve a specific problem, the analyst could develop various heuristic algorithms. The question remains, which one of these heuristic algorithms is the best. The quality of an algorithm is primarily described by its complexity. The complexity of any algorithm is regularly measured by the number of *elementary operations* (addition, subtraction, multiplication, division, comparison between two numbers, and execution of a branching instruction) that have to be executed by the algorithm to arrive at the solution under the *worst-case conditions*.

Let us assume that the number of nodes in an analyzed transportation network equals n. The number of network nodes could represent the dimensions of the problem that we analyze in the transportation network. Let us also suppose that the total number of elementary operations E, to be performed in order to execute the proposed algorithm, is equal to:

$$E = 3n^3 + 6n^2 + 4n + 9$$

Obviously, the value of E is primarily determined by n^3 as n increases. It is usual to say that the complexity of the proposed algorithm is proportional to n^3. More frequently, analysts say that the algorithm requires $O(n^3)$ time (under the assumption that each elementary operation requires one unit of time). The complexity of *polynomial algorithms* is proportional to, or bounded by, a polynomial function of the dimension of the input. For instance, the algorithm that needs $O(n^3)$ time is a polynomial algorithm.

Nondeterministic polynomial (exponential) algorithms violate all polynomial limits (in the case of the large sizes of the input). For instance, the algorithm that demands $O(3^n)$ time is exponential. It is common in computer science to think of polynomial algorithms as good algorithms. Simultaneously, the exponential algorithms are recognized as bad algorithms. Two of the most- important criteria for evaluation of a specific algorithm are the quality of the generated solution and the CPU time. There is always a trade-off between the quality of the solutions generated by the algorithm and the search time for the optimal or near-optimal solution (Figure 8.1).

Certainly, getting a better solution requires a longer CPU time. In the case where, for example, we perform real-time traffic control of an isolated intersection, we are ready to accept a poorer solution, if it can be generated quickly. On the other hand, generating the best network of metro lines, for example, implies that we are ready to accept longer CPU times in order to generate a high-quality solution.

When evaluating specific heuristic algorithms, it is also important to analyze the simplicity of the algorithm and the complexity of the algorithm implementation. The closer the objective function value produced by the algorithm is to the optimal value, the better the proposed algorithm.

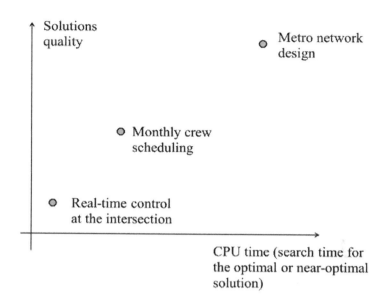

Figure 8.1 Trade-off between quality of the solutions generated by the algorithm and the search time for the optimal or near-optimal solution

Within the *worst-case analysis*, we perform the analysis of such numerical examples that show the worst-possible results that can be obtained by the analyzed algorithm. Such numerical examples are usually atypical within the problem being considered. For instance, we can more easily evaluate the proposed algorithm if we know that, in the worst case, the algorithm generates a solution that has an objective function value 5% higher (in a case of minimization) than the optimal solution value.

Within the *average-case analysis*, analysts typically test large numbers of problem instances that can happen in real life and carry out statistical analysis on the algorithm performances. It is always vital to test the developed heuristic algorithm on real-life examples.

8.2 GREEDY HEURISTICS

Engineers frequently use greedy heuristic algorithms to solve combinatorial optimization problems. The application of these algorithms is justified when, due to the dimensions of the problem under study, exact methods result in a very long computer time. Greedy heuristic algorithms are fast and generate solutions that are, generally, not optimal. The basic characteristic of the greedy algorithm is that, when trying to discover a global optimum, a greedy algorithm always takes the locally optimal choice at each stage. At each stage, a greedy algorithm looks for what is the best thing to do at that stage, without considering future stages. For example, the greedy heuristic for the Traveling Salesman Problem reads: always go to the nearest city that has not been visited.

In the case when the solution obtained by the greedy algorithm is not high quality, it can be improved by applying other algorithms (mainly local search algorithms). Greedy heuristic algorithms are also frequently used to generate initial solutions when using metaheuristic algorithms (simulated annealing, tabu search, bee colony optimization, etc.).

Let us illustrate the greedy algorithm by considering the traveling salesman problem (TSP) and the transit network design problem (TNDP).

The TSP route should start at one node, go through all the other nodes, and end in the node from which it began. The TSP route should be generated in such a way that its length is as short as possible.

In the case of the transit network design problem, a set of bus routes must be designed in such a way that the total travel time of passengers is as short as possible. The origin-destination matrix represents the basic input data in this problem. When traveling from their origin to their destination, passengers make transfers when it is necessary.

Example 8.1

Determine the travelling salesmen route by the greedy heuristic. The route should start from node 0. Distances between the nodes are given in Table 8.1. Figure 8.2 shows the node locations.

SOLUTION:

The route should start at node 0. Let us discover the node nearest to node 0.

Table 8.1 Distances between the nodes

	0	1	2	3	4	5	6	7	8	9	10	11	12
0	0	19	22	52	31	40	21	21	10	15	21	17	43
1	19	0	21	62	48	59	18	22	15	33	39	5	62
2	22	21	0	42	36	56	3	38	27	36	40	16	58
3	52	62	42	0	30	58	45	73	62	57	57	57	56
4	31	48	36	30	0	29	38	50	41	29	28	44	27
5	40	59	56	58	29	0	57	50	46	27	21	57	5
6	21	18	3	45	38	57	0	35	25	35	40	13	59
7	21	22	38	73	50	50	35	0	11	24	29	24	54
8	10	15	27	62	41	46	25	11	0	19	25	15	49
9	15	33	36	57	29	27	35	24	19	0	6	31	30
10	21	39	40	57	28	21	40	29	25	6	0	37	25
11	17	5	16	57	44	57	13	24	15	31	37	0	59
12	43	62	58	56	27	5	59	54	49	30	25	59	0

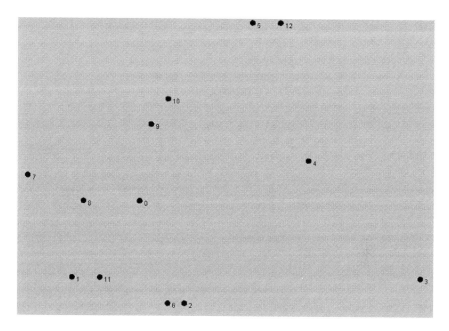

Figure 8.2 Node locations

Since:

$$\min\{19, 22, 52, 31, 40, 21, 21, 10, 15, 21, 17, 43\} = 10$$

we conclude that the nearest node is node 8.

The traveling salesman goes to node 8. Now, we should find the nearest node to node 8, but it cannot be node 0, i.e.:

$$\min\{15, 27, 62, 41, 46, 25, 11, 19, 25, 15, 49\} = 11$$

we conclude that the nearest node is node 7.

We continue the process. Now, we should find the node nearest to node 7, but it cannot be nodes 0, 8 or 7, i.e.:

$$\min\{22, 38, 73, 50, 50, 35, 24, 29, 24, 54\} = 22$$

The next node to be visited is node 1.

We continue this process:

$$\min\{21, 62, 48, 59, 18, 33, 39, 5, 62\} = 5$$

Next node is 11.

$$\min\{16, 57, 44, 57, 13, 31, 37, 59\} = 13$$

Next node is 6.

$$\min\{3, 45, 38, 57, 35, 40, 59\} = 3$$

Next node is 2.

$$\min\{42, 36, 56, 36, 40, 58\} = 36$$

Next node is 4.

$$\min\{30, 29, 29, 28, 27\} = 27$$

Next node is 12.

$$\min\{56, 5, 30, 25\} = 5$$

Next node is 5.

$$\min\{58, 27, 21\} = 21$$

Next node is 10.

$$\min\{57, 6\} = 6$$

Next node is 9.

$$\min\{57\} = 57$$

Next node is 3.

We do not have unvisited nodes, and, consequently, we finish the route by returning to node 0. Finally, the obtained route reads (as illustrated in Figure 8.3):

(0, 8, 7, 1, 11, 6, 2, 4, 12, 5, 10, 9, 3, 0)

The total length of this route equals 268 distance units.

We use greedy heuristics very often to find an initial solution, and, after that, we try to improve the initial solution by using some other algorithm. For example, we can use greedy heuristics to find TSP route, and, after that, we apply 2-opt local search algorithms in order to improve the obtained solution. The basic idea of this algorithm is to try to find a better, feasible solution by replacing two edges from the route by two new edges. If the new solution is better than the previous one, it should be kept, otherwise the previous solution should be retained. The process continues until no improvement is made.

By applying a 2-opt algorithm on the TSP route: (0, 8, 7, 1, 11, 6, 2, 4, 12, 5, 10, 9, 3, 0), we obtain the following solution: (0, 8, 7, 1, 11, 6, 2, 3, 4, 12, 5, 10, 9, 0). This new route is shown in Figure 8.4.

Example 8.2

The transportation network shown in Figure 8.5 consists of 15 nodes and 21 links. Travel time values are shown on each link. The origin-destination matrix is given in Table 8.2. By applying the greedy heuristics given in the paper of Nikolić and Teodorović (2013), design 5 bus routes. Each route cannot have less than 3 or more than 7 nodes. The description of the Nikolić and Teodorović (2013) algorithm is given below.

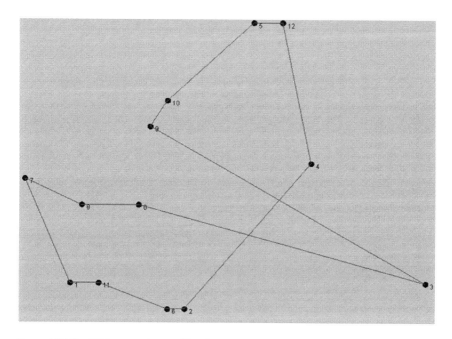

Figure 8.3 The TSP route obtained by the greedy heuristic

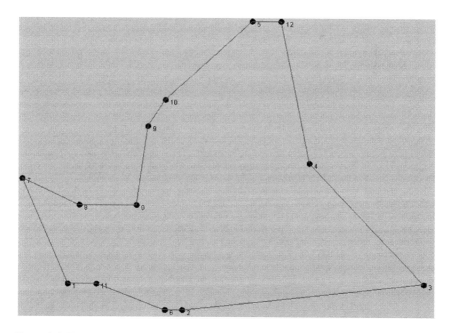

Figure 8.4 Travelling salesman route obtained by 2-OPT algorithm

SOLUTION:

The greedy algorithm proposed in the paper of Nikolić and Teodorović (2013) consists of the following steps:

Step 1: Prescribe the total number of bus routes NBL in the network. Denote the set of bus routes by Y. Set $Y = \phi$. Let $m = 1$.

Step 2: Find the pair of nodes that has the highest ds_{ij} value. Let this pair be the pair of nodes (a, b). The nodes a and b are the terminals of the new bus route. Find the shortest path between these two nodes. The nodes that belong to the shortest path represent stations on the bus routes. Add route l to set Y.

Step 3: Update the matrix DS, without taking into account passenger travel demands that are already satisfied.

Step 4: If $m = NBL$, stop; otherwise, set $m = m + 1$ and return to Step 2.

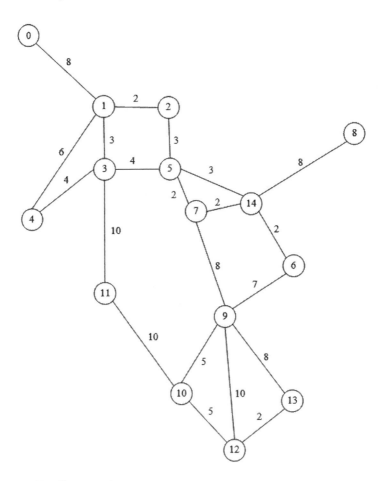

Figure 8.5 Mandl's network

Table 8.2 Origin-destination matrix

	0	1	2	3	4	5	6	7	8	9	10	11	12	13	14
0	0	400	200	60	80	150	75	75	30	160	30	25	35	0	0
1	400	0	50	120	20	180	90	90	15	130	20	10	10	5	0
2	200	50	0	40	60	180	90	90	15	45	20	10	10	5	0
3	60	120	40	0	50	100	50	50	15	240	40	25	10	5	0
4	80	20	60	50	0	50	25	25	10	120	20	15	5	0	0
5	150	180	180	100	50	0	100	100	30	880	60	15	15	10	0
6	75	90	90	50	25	100	0	50	15	440	35	10	10	5	0
7	75	90	90	50	25	100	50	0	15	440	35	10	10	5	0
8	30	15	15	15	10	30	15	15	0	140	20	5	0	0	0
9	160	130	45	240	120	880	440	440	140	0	600	250	500	200	0
10	30	20	20	40	20	60	35	35	20	600	0	75	95	15	0
11	25	10	10	25	15	15	10	10	5	250	75	0	70	0	0
12	35	10	10	10	5	15	10	10	0	500	95	70	0	45	0
13	0	5	5	5	0	10	5	5	0	200	15	0	45	0	0
14	0	0	0	0	0	0	0	0	0	0	0	0	0	0	0

Value ds_{ij}, represents the total number of passengers who can travel without transfer, using route r, where i and j are terminals of route r, and ds_{ij}, can be calculated in the following way:

$$ds_{ij} = \sum_{m \in N_r} \sum_{n \in N_r} d_{mn}$$

Let us demonstrate how we calculate values of ds_{ij} on examples, for the nodes $i = 0$ and $j = 5$. Taking into consideration that we have to find the shortest paths between all pairs of nodes, we apply Floyd's algorithm. The final matrix of shortest distances between the nodes is given in Table 8.3, and the matrix of predecessors is given in Table 8.4.

Let us discover the shortest path between node 0 and node 5. Let us observe the row of node 0 in Table 8.4. The path from node 0 should be finished at node 5. The predecessor of node 5 is node 2. The predecessor of node 2 is node 1. The predecessor of node 1 is node 0. The shortest path from node 0 to node 5 reads: $(0, 1, 2, 5)$. We calculate the value of ds_{05} in the following way:

$$ds_{05} = d_{01} + d_{02} + d_{05} + d_{10} + d_{12} + d_{15} + d_{20} + d_{21} + d_{25} + d_{50} + d_{51} + d_{52}$$

$$= 400 + 200 + 150 + 400 + 50 + 180 + 200 + 50 + 180 + 150 + 180 + 180$$

$$= 2320$$

In the same way, we calculate ds_{ij} values for each pair of nodes i and j. Table 8.5 shows the values of ds_{ij} for each pair of i and j.

We notice that the route between nodes 0 and 5 has 4 nodes. This is important since we have a constraint that the route cannot have less

Table 8.3 The shortest distances between the nodes (node 0 is the warehouse, the shops are denoted as nodes 1, 2, ..., 5)

	0	1	2	3	4	5	6	7	8	9	10	11	12	13
0	8	10	11	14	13	18	15	24	23	28	21	33	31	16
1	0	2	3	6	5	10	7	16	15	20	13	25	23	8
2	2	0	5	8	3	8	5	14	13	18	15	23	21	6
3	3	5	0	4	4	9	6	15	14	19	10	24	22	7
4	6	8	4	0	8	13	10	19	18	23	14	28	26	11
5	5	3	4	8	0	5	2	11	10	15	14	20	18	3
6	10	8	9	13	5	0	4	10	7	12	19	17	15	2
7	7	5	6	10	2	4	0	10	8	13	16	18	16	2
8	16	14	15	19	11	10	10	0	17	22	25	27	25	8
9	15	13	14	18	10	7	8	17	0	5	15	10	8	9
10	20	18	19	23	15	12	13	22	5	0	10	5	7	14
11	13	15	10	14	14	19	16	25	15	10	0	15	17	17
12	25	23	24	28	20	17	18	27	10	5	15	0	2	19
13	23	21	22	26	18	15	16	25	8	7	17	2	0	17
14	8	6	7	11	3	2	2	8	9	14	17	19	17	0

Table 8.4 The matrix of predecessors

	0	1	2	3	4	5	6	7	8	9	10	11	12	13	14
0	0	0	1	1	1	2	14	5	14	7	9	3	9	9	5
1	1	1	1	1	1	2	14	5	14	7	9	3	9	9	5
2	1	2	2	1	1	2	14	5	14	7	9	3	9	9	5
3	1	3	1	3	3	3	14	5	14	7	9	3	9	9	5
4	1	4	1	4	4	3	14	5	14	7	9	3	9	9	5
5	1	2	5	5	3	5	14	5	14	7	9	3	9	9	5
6	1	2	5	5	3	14	6	14	14	6	9	3	9	9	6
7	1	2	5	5	3	7	14	7	14	7	9	3	9	9	7
8	1	2	5	5	3	14	14	14	8	6	9	3	9	9	8
9	1	2	5	5	3	7	9	9	14	9	9	10	9	9	6
10	1	2	5	5	3	7	9	9	14	10	10	10	10	12	6
11	1	3	1	11	3	3	14	5	14	10	11	11	10	12	5
12	1	2	5	5	3	7	9	9	14	12	12	10	12	12	6
13	1	2	5	5	3	7	9	9	14	13	12	10	13	13	6
14	1	2	5	5	3	14	14	14	14	6	9	3	9	9	14

than 3 or more than 7 nodes. Table 8.6 lists the number of nodes in the route for each pair of nodes.

Now, we are ready to design the first route. For that purpose, we have to find, among the pairs of nodes that have the values in Table 8.5 between 3 and 7, the pair of nodes (i, j) that have the largest values of ds_{ij}, i.e.:

$$\max \{1300, 1160, 1000, ..., 1129\} = 7870$$

Table 8.5 Values of ds_{ij} for all pairs of i and j

	0	1	2	3	4	5	6	7	8	9	10	11	12	13	14
0	0	800	1,300	1,160	1,000	2,320	3,030	3,030	2,500	6,340	7,870	1,280	7,500	6,790	2,320
1	800	0	100	240	40	820	1,380	1,380	940	4,370	5,840	310	5,460	4,820	820
2	1,300	100	0	420	260	360	740	740	450	3,470	4,900	510	4,540	3,910	360
3	1,160	240	420	0	100	200	500	500	290	3,620	5,090	50	4,690	4,060	200
4	1,000	40	260	100	0	400	750	750	510	4,110	5,620	180	5,190	4,550	400
5	2,320	820	360	200	400	0	200	200	60	2,840	4,230	280	3,890	3,270	0
6	3,030	1,380	740	500	750	200	0	100	30	880	2,150	600	1,900	1,290	0
7	3,030	1,380	740	500	750	200	100	0	30	880	2,150	600	1,900	1,290	0
8	2,500	940	450	290	510	60	30	30	0	1,190	2,500	380	2,210	1,600	0
9	6,340	4,370	3,470	3,620	4,110	2,840	880	880	1,190	0	1,200	1,850	1,000	400	880
10	7,870	5,840	4,900	5,090	5,620	4,230	2,150	2,150	2,500	1,200	0	150	190	310	2,150
11	1,280	310	510	50	180	280	600	600	380	1,850	150	0	480	600	280
12	7,500	5,460	4,540	4,690	5,190	3,890	1,900	1,900	2,210	1,000	190	480	0	90	1,900
13	6,790	4,820	3,910	4,060	4,550	3,270	1,290	1,290	1,600	400	310	600	90	0	1,290
14	2,320	820	360	200	400	0	0	0	0	880	2,150	280	1,900	1,290	0

Table 8.6 The number of nodes in the routes which can be generated for each pair of nodes as terminals

	0	1	2	3	4	5	6	7	8	9	10	11	12	13	14
0	1	2	3	3	3	4	6	5	6	6	7	4	7	7	5
1	2	1	2	2	2	3	5	4	5	5	6	3	6	6	4
2	3	2	1	3	3	2	4	3	4	4	5	4	5	5	3
3	3	2	3	1	2	2	4	3	4	4	5	2	5	5	3
4	3	2	3	2	1	3	5	4	5	5	6	3	6	6	4
5	4	3	2	2	3	1	3	2	3	3	4	3	4	4	2
6	6	5	4	4	5	3	1	3	3	2	3	5	3	3	2
7	5	4	3	3	4	2	3	1	3	2	3	4	3	3	2
8	6	5	4	4	5	3	3	3	1	4	5	5	5	5	2
9	6	5	4	4	5	3	2	2	4	1	2	3	2	2	3
10	7	6	5	5	6	4	3	3	5	2	1	2	2	3	4
11	4	3	4	2	3	3	5	4	5	3	2	1	3	4	4
12	7	6	5	5	6	4	3	3	5	2	2	3	1	2	4
13	7	6	5	5	6	4	3	3	5	2	3	4	2	1	4
14	5	4	3	3	4	2	2	2	2	3	4	4	4	4	1

The value of 7870 corresponds to the pair of nodes (0,10). The route for this pair of nodes is: (0, 1, 2, 5, 7, 9, 10). We note that this route has 7 nodes, which means that the constraint of minimal and maximal number of nodes is not broken.

Now, we set that $d_{ij}=0$ where $i,j \in \{0,1,2,5,7,9,10\}$, i.e.: $d_{01} = d_{02} = d_{05} = d_{07} = d_{07} = d_{09} = d_{0,10} = d_{10} = d_{12} = d_{15} = d_{17} = d_{19} = d_{1,10} = \cdots$ $\cdots = d_{10,7} = d_{10,9} = 0$. We again calculate the ds_{ij} values, and these values are given in Table 8.7.

We design the next route by discovering the maximal value in Table 8.7, and by taking into account the number of nodes given in Table 8.6. The new best value is 2350. This value corresponds to the pair of nodes (4, 12). The new route reads: (4, 3, 5, 7, 9, 12). This route contains 6 nodes. We again set that $d_{ij}=0$, where $i,j \in \{4,3,5,7,9,12\}$. The updated ds_{ij} values are given in Table 8.8.

The pair of nodes (8, 13) has the highest value of ds_{ij}. This value is 1600. The new route is: (8,14,6,9,13). This route includes 5 nodes and it can be accepted. We again set that $d_{ij}=0$ where $i,j \in \{8,14,6,9,13\}$. The new values of ds_{ij} are given in Table 8.9.

Taking the same procedure, we can generate the fourth route. The highest ds_{ij} value involves the pair of nodes (0,6), and, because of that, the new route is: (0,1,2,5,14,6). We again calculate ds_{ij} values. These values are given in Table 8.10.

The last, fifth route can be generated finding the largest ds_{ij} value among the elements given in Table 8.10. We note that this is the value

Table 8.7 Updated values of ds_{ij} for all pairs of i and j after the first route is generated

	0	1	2	3	4	5	6	7	8	9	10	11	12	13	14
0	0	0	0	360	200	0	710	0	180	0	0	480	1,160	450	0
1	0	0	0	240	40	0	560	0	120	0	0	310	1,090	450	0
2	0	0	0	320	160	0	380	0	90	0	0	410	1,070	440	0
3	360	240	320	0	100	200	500	300	290	780	860	50	1,850	1,220	200
4	200	40	160	100	0	400	750	550	510	1,270	1,390	180	2,350	1,710	400
5	0	0	0	200	400	0	200	0	60	0	0	280	1,050	430	0
6	710	560	380	500	750	200	0	100	30	880	950	600	1,900	1,290	0
7	0	0	0	300	550	0	100	0	30	0	0	400	1,020	410	0
8	180	120	90	290	510	60	30	30	0	1,190	1,300	380	2,210	1,600	0
9	0	0	0	780	1,270	0	880	0	1,190	0	0	650	1,000	400	880
10	0	0	0	860	1,390	0	950	0	1,300	0	0	150	190	310	950
11	480	310	410	50	180	280	600	400	380	650	150	0	480	600	280
12	1,160	1,090	1,070	1,850	2,350	1,050	1,900	1,020	2,210	1,000	190	480	0	90	1,900
13	450	450	440	1,220	1,710	430	1,290	410	1,600	400	310	600	90	0	1290
14	0	0	0	200	400	0	0	0	0	880	950	280	1,900	1,290	0

Table 8.8 Values of ds_{ij} for all pairs of i and j after two generated routes

	0	1	2	3	4	5	6	7	8	9	10	11	12	13	14
0	0	0	0	360	200	0	710	0	180	0	0	480	110	450	0
1	0	0	0	240	40	0	560	0	120	0	0	310	40	450	0
2	0	0	0	320	160	0	380	0	90	0	0	410	20	440	0
3	360	240	320	0	0	0	300	0	90	0	80	50	0	440	0
4	200	40	160	0	0	0	350	0	110	0	120	80	0	440	0
5	0	0	0	0	0	0	200	0	60	0	150	80	0	430	0
6	710	560	380	300	350	200	0	100	30	880	0	400	900	1,290	0
7	0	0	0	0	0	0	100	0	30	0	0	100	0	410	0
8	180	120	90	90	110	60	30	30	0	1,190	1,300	180	1,210	1,600	0
9	0	0	0	0	0	0	880	0	1,190	0	0	650	0	400	880
10	0	0	0	80	120	150	0	0	1,300	0	0	150	190	310	950
11	480	310	410	50	80	80	400	100	180	650	150	0	480	600	80
12	110	40	20	0	0	0	900	0	1,210	0	190	480	0	90	900
13	450	450	440	440	440	430	1,290	410	1,600	400	310	600	90	0	1,290
14	0	0	0	0	0	0	0	0	0	880	950	80	900	1,290	0

Table 8.9 Values of ds_{ij} for all pairs of i and j after three generated routes

	0	1	2	3	4	5	6	7	8	9	10	11	12	13	14
0	0	0	0	360	200	0	710	0	180	0	0	480	110	50	0
1	0	0	0	240	40	0	560	0	120	0	0	310	40	50	0
2	0	0	0	320	160	0	380	0	90	0	0	410	20	40	0
3	360	240	320	0	0	0	300	0	90	0	80	50	0	40	0
4	200	40	160	0	0	0	350	0	110	0	120	80	0	40	0
5	0	0	0	0	0	0	200	0	60	0	0	80	0	30	0
6	710	560	380	300	350	200	0	100	0	0	70	400	20	0	0
7	0	0	0	0	0	0	100	0	30	0	0	100	0	10	0
8	180	120	90	90	110	60	0	30	0	0	110	180	20	0	0
9	0	0	0	0	0	0	0	0	0	0	0	650	0	0	0
10	0	0	0	80	120	0	70	0	110	0	0	150	190	310	70
11	480	310	410	50	80	80	400	100	180	650	150	0	480	600	80
12	110	40	20	0	0	0	20	0	20	0	190	480	0	90	20
13	50	50	40	40	40	30	0	10	0	0	310	600	90	0	0
14	0	0	0	0	0	0	0	0	0	0	70	80	20	0	0

Table 8.10 Values of ds_{ij} for all pairs of i and j after four routes are generated

	0	1	2	3	4	5	6	7	8	9	10	11	12	13	14
0	0	0	0	360	200	0	0	0	180	0	0	480	110	50	0
1	0	0	0	240	40	0	0	0	120	0	0	310	40	50	0
2	0	0	0	320	160	0	0	0	90	0	0	410	20	40	0
3	360	240	320	0	0	0	100	0	90	0	80	50	0	40	0
4	200	40	160	0	0	0	150	0	110	0	120	80	0	40	0
5	0	0	0	0	0	0	0	0	60	0	0	80	0	30	0
6	0	0	0	100	150	0	0	100	0	0	70	200	20	0	0
7	0	0	0	0	0	0	100	0	30	0	0	100	0	10	0
8	180	120	90	90	110	60	0	30	0	0	110	180	20	0	0
9	0	0	0	0	0	0	0	0	0	0	0	650	0	0	0
10	0	0	0	80	120	0	70	0	110	0	0	150	190	310	70
11	480	310	410	50	80	80	200	100	180	650	150	0	480	600	80
12	110	40	20	0	0	0	20	0	20	0	190	480	0	600	80
13	50	50	40	40	40	30	0	10	0	0	310	600	600	0	20
14	0	0	0	0	0	0	0	0	0	0	70	80	80	20	0

of 650, corresponding to the pair of nodes (9, 11). The fifth route reads: (9, 10, 11).

The final set of routes reads:

- bus route 1: 0, 1, 2, 5, 7, 9, 10
- bus route 2: 4, 3, 5, 7, 9, 12
- bus route 3: 8, 14, 6, 9, 13
- bus route 4: 0, 1, 2, 5, 14, 6
- bus route 5: 9, 10, 11.

8.3 METAHEURISTIC ALGORITHMS

When we plan, design, or control transportation systems, we try to minimize travel times, waiting times, or transportation costs. Operators are trying to maximize the number of passengers transported, market share, or profit achieved. For example, there are many possible ways to design a public transport network, to design an airline schedule, or to design a set of vehicle routes in a distribution system. Problems of this type are complex, and they are high dimensional. It is usually not possible to find an optimal solution to these problems within a reasonable computer time. Various approximate algorithms are used to solve various problems in traffic and transportation. The approximate algorithms contain two main groups of algorithms: (a) heuristic algorithms and (b) metaheuristic algorithms.

Many specific heuristic techniques have been developed over the past few decades. The specific heuristics are problem dependent. In other words, they are able to solve only one particular problem.

On the other hand, metaheuristics are general-purpose techniques. Metaheuristics efficiently explore the search space. They are successful at discovering (near)-optimal solutions, of difficult combinatorial optimization problems, in a reasonable computer time.

When searching for the best solution, analysts frequently discover a local optimum. A local optimum represents a solution that is optimal inside a neighboring set of feasible solutions. In contrast, a global optimum represents the optimal solution among all possible feasible solutions (Figure 8.6).

Metaheuristic techniques possess various mechanisms that avoid being trapped in a local optimum.

During the past three decades, researchers in the fields of operations research, mathematical programming, computational intelligence, and soft computing proposed and successfully implemented various metaheuristics techniques, such as simulated annealing (SA), genetic algorithms (GA), tabu search (TS), variable neighborhood search (VNS), ant colony optimization (ACO), particle swarm optimization (PSO), and bee colony optimization (BCO).

Typical representatives of traffic and transportation problems solved by metaheuristics are vehicle fleet planning, and the routing and scheduling of vehicles and crews for airlines, railroads, truck operations, and public

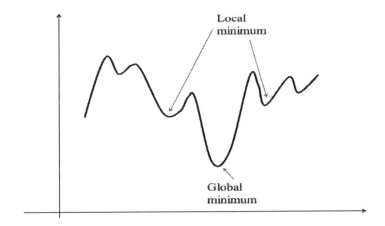

Figure 8.6 Local versus global optimum

transportation services. The set of problems solved by metaheuristic techniques also includes designing transportation networks, as well as optimizing alignments for highways and public transportation routes, different location problems, etc. In many cases, optimal solution cannot be discovered in a reasonable CPU time.

Metaheuristic algorithms could be single-solution-based, or population-based solutions. The single-solution-based techniques start from the initial solution and continuously modify a single solution during the search for the optimal solution. Single-solution-based metaheuristics are characterized by an investigation of the solution space in the neighborhood of the current solution. Each step in these metaheuristics represents a move from the current solution to another potentially good solution in the neighborhood of the current solution. Widely accepted single-solution-based techniques include simulated annealing (SA), tabu search (TS), and variable neighborhood search (VNS).

Contrary to these techniques, population-based metaheuristics use initial populations of solutions that evolve during the search process. In other words, in the case of population-based metaheuristic algorithms, the search is run in parallel from a population of solutions. In this way, these algorithms diversify the search within the search space. Widely used population-based metaheuristics are genetic algorithms (GA), ant colony optimization (ACO), particle swarm optimization (PSO), and bee colony optimization (BCO).

Several metaheuristic techniques are inpired by various phenomena in nature (genetic algorithms, ant colony optimization, bee colony optimization, and artificial immune systems, etc.).

Some of the metaheuristic techniques are deterministic, while others are stochastic. In the case of deterministic techniques, the decisions to be made during the search process are deterministic. In the case of stochastic techniques, various random rules could be implemented during the search for the optimal solution. It is important to note that, in the case of any deterministic technique, the same initial solution will always generate the same final solution. This is not true in the case of stochastic techniques.

Many factors impact the use of a specific metaheuristic algorithm (the frequency of making decisions, the time available for generating problem solutions, the number of decision variables, etc.).

8.4 SIMULATED ANNEALING

Simulated annealing (SA) is a probabilistic technique that has been intensively used in recent decades to solve complex combinatorial problems (Kirkpatrick et al., 1983; Černy, 1985). The SA method is based on the similarity with certain problems in the field of statistical mechanics (Metropolis et al., 1953). The name of the simulated annealing comes from its analogy with annealing in metallurgy, which involves heating and controlled cooling of a material. The process of annealing consists of decreasing the temperature of a material, which, in the beginning of the process, is in the molten state, until the lowest state of energy is reached. At certain points during the process, the so-called thermal equilibrium is reached. In the case of physical systems, we seek to establish the order of particles that has the lowest state of energy. For this process the temperatures, at which the material remains for some time, should be specified in advance.

The fundamental idea of simulated annealing consists of performing small perturbations (small changes in the positions of particles) in a random manner and computing the energy changes between the new and the old configurations of particles, ΔE. In the case where $\Delta E < 0$, it can be concluded that the new configuration of particles has lower energy than the old configuration. The new configuration then becomes a new initial configuration for performing small perturbations. The case when $\Delta E > 0$ means that the new configuration has higher energy. In spite of this, the new configuration is not to be routinely eliminated from the opportunity to become a new initial configuration. "Jumps", from lower to higher energy levels, are possible in physical systems. The higher the temperatures are, the higher the probability that the system will "jump" to a higher energy state. As the temperature declines, the probability that such a "jump" will happen becomes smaller. Metropolis et al. (1953) used Boltzmann's distribution of probabilities to compute the probability of such a "jump". Probability P that, at temperature T, the energy will increase by ΔE, equals:

$$P = e^{-\frac{\Delta E}{T}}$$

The decision whether a new configuration of particles, for which $\Delta E > 0$, should be recognized as a new initial configuration is made upon the generation of a random number r from the interval $[0, 1]$. If $r < P$, the new configuration is accepted as a new initial configuration. In the reverse case, the generated configuration of particles is eliminated from further consideration. As we can see, the probability of accepting a worse state depends on both the temperature T and the change in the energy, ΔE. We also see that, with the decrease in the temperature, the probability of accepting worse moves decreases.

In this way, a successful simulation of reaching thermal equilibrium at a particular temperature is completed. Thermal equilibrium is considered to be reached when, after a number of random perturbations, a considerable decrease in energy is not possible. When thermal equilibrium has been reached, the temperature is decreased, and the explained process is replicated at a new temperature.

The explained procedure can also be used in solving combinatorial optimization problems. A specific configuration of particles can be interpreted as one feasible solution. Similarly, the energy of a physical system can be interpreted as the value of an objective function, whereas temperature plays the role of a control parameter. If $r < P$, the new generated solution is accepted as a new initial solution. In the reverse case, the generated solution is eliminated from further consideration.

Accepting worse solutions, from time to time, is a crucial characteristic of SA, as it permits an extensive search for the global optimal solution.

We start our search for the optimal solution from a high temperature and then repeatedly lower the temperature. The initial temperature is set in a way that a high percentage of bad moves is accepted. With the decrease in temperature, the probability of accepting worse moves decreases. In the case when $T = 0$, no worse moves are accepted. There are various cooling schedules. The most frequently used are exponential schedules and linear schedules. The exponential cooling schedule reads:

$$T(t) = T_0 \alpha^t$$

The linear cooling schedule reads:

$$T(t) = T_0 - \alpha \cdot t$$

where:

T_0 – initial temperature
$0 < \alpha < 1$ – parameter whose value is chosen by the analyst
t – time (step count)

8.5 TABU SEARCH

The tabu search technique (Glover, 1986) belongs to the class of single-solution-based techniques. The basic idea of the tabu search technique is to prohibit or penalize moves that take the solution, in the next iteration, to points in the solution space that have been visited earlier.

We present the basic principles of the tabu search technique by the subsequent illustrative example.

The coordinates of the seven nodes to be visited by the traveling salesman, that starts and returns to node 1, are given in Table 8.11.

The nodes are shown in Figure 8.7.

The traveling salesman problem (TSP) consists of finding the shortest route the salesman should take when serving the nodes. The Euclidean distances between specific pairs of nodes are given in Table 8.12.

The TSP is a classic permutation problem. Discovering the shortest traveling salesman route assumes finding a permutation that corresponds to the route of the shortest length. Since the traveling salesman starts and finishes his tour at depot 1 and is to visit seven nodes, the total number of permutations (various traveling salesman routes) equals $7! = 5040$. When there is a large enough number of nodes to be visited, the number of possible permutations can be astronomical.

We generate the initial solution to our problem by using the nearest-neighbor greedy algorithm. The initial solution reads:

$$(1, 2, 8, 7, 3, 5, 6, 4, 1)$$

The length of the route is 37.838 distance units. The initial solution is shown in Figure 8.8.

Glover and Laguna (1993) note: "Tabu search methods operate under the assumption that a neighborhood can be constructed to identify "adjacent solutions" that can be reached from any current solution." When permutation problems are solved, for the intention of creating a neighborhood, the

Table 8.11 The coordinates of node 1 and seven subsequent nodes to be visited by the traveling salesman

Node	x-coordinate	y-coordinate
1	5	4
2	4	3
3	4	7
4	10	9
5	5	10
6	1	9
7	1	2
8	6	2

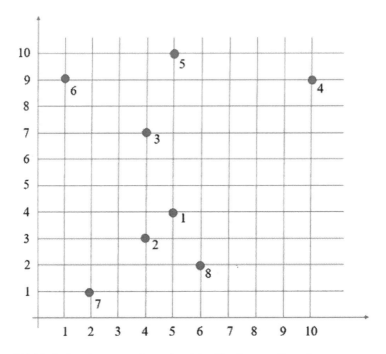

Figure 8.7 Node 1 and seven nodes to be visited by the traveling salesman

Table 8.12 Distances between the nodes

	1	2	3	4	5	6	7	8
1	0	1.414	3.162	7.071	6	6.403	4.472	2.236
2	1.414	0	4	8.485	7.071	6.708	3.162	2.236
3	3.162	4	0	6.325	3.162	3.606	5.831	5.385
4	7.071	8.485	6.325	0	5.099	9	11.402	8.062
5	6	7.071	3.162	5.099	0	4.123	8.944	8.062
6	6.403	6.708	3.606	9	4.123	0	7	8.602
7	4.472	3.162	5.831	11.402	8.944	7	0	5
8	2.236	2.236	5.385	8.062	8.062	8.602	5	0

so-called swaps, XE "Swaps" representing pairwise exchanges, are most frequently employed. In the case of the TSP, a swap exchanges the positions of two nodes in the traveling salesman route. When a swap has been made, a move has essentially been made going ahead from one solution to another one. If, for example, nodes 8 and 6 swap positions, from the initial solution that read

$$(1, 2, 8, 7, 3, 5, 6, 4, 1)$$

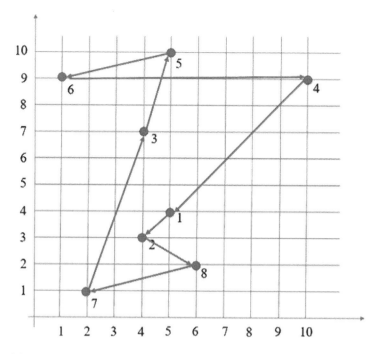

Figure 8.8 The initial solution of the TSP to be solved by tabu search technique

a move has been made that leads us toward the following solution (1, 2, 6, 7, 3, 5, 8, 4, 1).

After the swap, the generated solution has a smaller, equal or larger objective function value, as compared to the prior solution. In our case, after performing a swap of nodes 8 and 6, the objective function takes the following larger value: $1.414 + 6.708 + 7 + 5.831 + 3.162 + 8.062 + 8.062 + 7.0$ $71 = 47.31$.

Given that there are seven nodes to be visited, with two nodes taking part in a swap, the total number of various solutions obtainable by swaps equals 21. In other words, an entire neighborhood of the existing solution comprises twenty-one different solutions.

The tabu search XE "Tabu search" technique uses the concept of the so-called flexible memory. The essential idea is to declare the subset of the moves in a neighborhood to be forbidden. Which moves are to be forbidden (tabu) is decided, according to the recency or frequency that specific moves have contributed to creating the earlier solutions.

By introducing the tabu moves, we make an effort to keep away from the swaps made in the recent past. For instance, it can be specified that nodes 2 and 4, that have already swapped positions in the traveling salesman route, cannot swap positions in the next two, three, or five iterations. Consequently, the swap of nodes 2 and 4 is considered to be tabu in the

Table 8.13 Tabu list at the beginning of the algorithm

	3	4	5	6	7	8
2	0	0	0	0	0	0
	3	0	0	0	0	0
		4	0	0	0	0
			5	0	0	0
				6	0	0
					7	0

next two, three, or five iterations. Glover and Laguna (1993) proposed the following tabu data structure to represent the remaining tabu tenure XE "Tabu tenure" for node pairs (Table 8.13).

> Every cell in the structure represents the number of iterations to be carried out until the corresponding nodes are permitted to swap places another time. If, for instance, the value in cell (3, 5) equals 2, it indicates that nodes 3 and 5 cannot swap positions during the next two iterations. In the same way, if the value in cell (4, 7) equals zero, nodes 4 and 7 are currently permitted to swamp positions, and so on. Tabu restrictions XE "Tabu restrictions" are not relevant in all situations. If, for instance, a tabu move generates a solution with the shortest route so far discovered, the tabu classification will be ignored and the move will be permitted.

We demonstrate the key points of the tabu search technique by the following few iterations. The initial solution reads (1, 2, 8, 7, 3, 5, 6, 4, 1). The total number of different solutions that can be obtained by swaps equals twenty-one.

Iteration I:

The table contains twenty possible swaps, corresponding new routes, the lengths of the new routes, and the gain accomplished when a new route substitutes for an existing one (the gain represents the difference in length between the existing route and the new route). The candidates for swap moves are shown in Table 8.14 in a descending order of gain.

Twenty-one possible routes represent neighboring solutions to the current solution (1, 2, 8, 7, 3, 5, 6, 4, 1). We see from Table 8.14 that the biggest gain is attained if we swap the positions of nodes 5 and 6, so let us swap the positions of nodes 5 and 6. This means that the new traveling salesman route reads (1, 2, 8, 7, 3, 6, 5, 4, 1). We also decide that in the next three iterations we are not permitted to swap the positions of these nodes. This means that in the next three iterations, swap (5, 6) is tabu. (The number of iterations during which the swap is in a tabu state has also been determined arbitrarily. Instead of value 3, we could have chosen value 2, or value 4, and so on.)

Table 8.14 Possible new routes in the first iteration

Swap					New route						Length	Gain
5	**6**	I	2	8	7	3	6	5	4	I	**34.38**	**3.46**
3	4	I	2	8	7	4	5	6	3	I	36.04	1.80
3	6	I	2	8	7	6	5	3	4	I	36.33	1.51
7	4	I	2	8	4	3	5	6	7	I	36.79	1.04
2	8	I	8	2	7	3	5	6	4	I	36.82	1.02
6	4	I	2	8	7	3	5	4	6	I	38.15	−0.31
8	7	I	2	7	8	3	5	6	4	I	38.32	−0.48
2	7	I	7	8	2	3	5	6	4	I	39.06	−1.23
5	4	I	2	8	7	3	4	6	5	I	39.93	−2.09
3	5	I	2	8	7	5	3	6	4	I	40.43	−2.60
2	4	I	4	8	7	3	5	6	2	I	41.37	−3.53
7	5	I	2	8	5	3	7	6	4	I	43.78	−5.94
2	3	I	3	8	7	2	5	6	4	I	43.97	−6.14
7	3	I	2	8	3	7	5	6	4	I	44.00	−6.17
8	3	I	2	3	7	8	5	6	4	I	44.50	−6.66
8	4	I	2	4	7	3	5	6	8	I	45.26	−7.42
7	6	I	2	8	6	3	5	7	4	I	46.44	−8.60
8	6	I	2	6	7	3	5	8	4	I	47.31	−9.47
2	6	I	6	8	7	3	5	2	4	I	51.63	−13.79
2	5	I	5	8	7	3	2	6	4	I	51.67	−13.83
8	5	I	2	5	7	3	8	6	4	I	53.32	−15.48

The tabu data structure, that represents the remaining tabu tenure XE "Tabu tenure" for node pairs, is given in Table 8.15.

The following are iterations 2, 3, 4, and 5:

Iteration 2:

From Table 8.16 we can see that the next swap is 2 and 8. This swap is not forbidden because, for this pair of nodes, the tabu value in Table 8.15 is 0. The new tabu list is given in Table 8.17.

Table 8.15 Tabu list after first iteration

	3	4	5	6	7	8
2	0	0	0	0	0	0
	3	0	0	0	0	0
		4	0	0	0	0
			5	3	0	0
				6	0	0
					7	0

Table 8.16 Possible new routes in the second iteration

Swap			New route								Length	Gain
2	**8**	**I**	**8**	**2**	**7**	**3**	**6**	**5**	**4**	**I**	**33.36**	**1.02**
3	6	I	2	8	7	6	3	5	4	I	34.59	−0.21
8	7	I	2	7	8	3	6	5	4	I	34.86	−0.48
2	7	I	7	8	2	3	6	5	4	I	35.61	−1.23
6	4	I	2	8	7	3	4	5	6	I	36.43	−2.05
6	5	I	2	8	7	3	5	6	4	I	37.84	−3.46
7	3	I	2	8	3	7	6	5	4	I	38.16	−3.78
5	4	I	2	8	7	3	6	4	5	I	38.19	−3.81
3	5	I	2	8	7	5	6	3	4	I	38.72	−4.34
7	4	I	2	8	4	3	6	5	7	I	39.18	−4.80
3	4	I	2	8	7	4	6	5	3	I	39.50	−5.12
2	3	I	3	8	7	2	6	5	4	I	39.71	−5.33
8	3	I	2	3	7	8	6	5	4	I	41.14	−6.76
2	4	I	4	8	7	3	6	5	2	I	42.18	−7.80
7	6	I	2	8	6	3	7	5	4	I	42.80	−8.42
7	5	I	2	8	5	3	6	7	4	I	43.95	−9.57
8	4	I	2	4	7	3	6	5	8	I	45.16	−10.78
8	6	I	2	6	7	3	8	5	4	I	46.57	−12.19
2	6	I	6	8	7	3	2	5	4	I	49.08	−14.70
8	5	I	2	5	7	3	6	8	4	I	50.60	−16.22
2	5	I	5	8	7	3	6	2	4	I	50.76	−16.38

Iteration 3:

The best swap is for the nodes 3 and 6 (see Table 8.18). This swap is not forbidden because, for this pair of nodes, the value in Table 8.17 is 0. The new tabu list is given in Table 8.19.

Iteration 4:

From Table 8.16, we can see that the best swap is for the nodes 6 and 3, but this swap is forbidden because, for this pair of nodes, the value in Table 8.19 is 3. The next best swap is 8 and 2. This swap is also forbidden

Table 8.17 Tabu list after second iteration

	3	4	5	6	7	8
2	0	0	0	0	0	3
	3	0	0	0	0	0
		4	0	0	0	0
			5	2	0	0
				6	0	0
					7	0

Table 8.18 Possible new routes in the third iteration

Swap			New route								Length	Gain
3	**6**	**I**	**8**	**2**	**7**	**6**	**3**	**5**	**4**	**I**	**33.57**	**−0.21**
2	7	I	8	7	2	3	6	5	4	I	34.30	−0.93
8	2	I	2	8	7	3	6	5	4	I	34.38	−1.02
8	7	I	7	2	8	3	6	5	4	I	35.15	−1.79
6	4	I	8	2	7	3	4	5	6	I	35.42	−2.05
6	5	I	8	2	7	3	5	6	4	I	36.82	−3.46
5	4	I	8	2	7	3	6	4	5	I	37.17	−3.81
7	3	I	8	2	3	7	6	5	4	I	37.60	−4.23
3	5	I	8	2	7	5	6	3	4	I	37.70	−4.34
3	4	I	8	2	7	4	6	5	3	I	38.48	−5.12
2	3	I	8	3	7	2	6	5	4	I	39.62	−6.25
8	3	I	3	2	7	8	6	5	4	I	40.22	−6.86
7	4	I	8	2	4	3	6	5	7	I	40.43	−7.06
7	6	I	8	2	6	3	7	5	4	I	41.73	−8.37
8	4	I	4	2	7	3	6	5	8	I	42.58	−9.21
2	4	I	8	4	7	3	6	5	2	I	43.74	−10.38
7	5	I	8	2	5	3	6	7	4	I	43.78	−10.42
2	6	I	8	6	7	3	2	5	4	I	46.91	−13.55
8	6	I	6	2	7	3	8	5	4	I	47.72	−14.36
8	5	I	5	2	7	3	6	8	4	I	49.41	−16.04
2	5	I	8	5	7	3	6	2	4	I	50.94	−17.58

(for this pair of nodes, the value in Table 8.19 is 2). The next best swap is for the nodes 3 and 4. This swap is not forbidden (for this pair of nodes, the tabu value is zero). The new route is (1, 8, 2, 7, 6, 4, 5, 3, 1) (Table 8.20). The new tabu list is given in Table 8.21.

Iteration 5:

In the fifth iteration, the best swap is for the nodes 4 and 5. This swap is allowed because, for the pair of nodes (4, 5), the tabu value in Table 8.21 is

Table 8.19 Tabu list after third iteration

	3	4	5	6	7	8
2	0	0	0	0	0	2
	3	0	0	3	0	0
		4	0	0	0	0
			5	I	0	0
				6	0	0
					7	0

Table 8.20 Possible new routes in the fourth iteration

Swap			New route									Length	Gain
6	3	I	8	2	7	3	6	5	4	I		33.36	0.21
8	2	I	2	8	7	6	3	5	4	I		34.59	−1.02
3	**4**	**I**	**8**	**2**	**7**	**6**	**4**	**5**	**3**	**I**		**35.06**	**−1.49**
3	5	I	8	2	7	6	5	3	4	I		35.32	−1.74
5	4	I	8	2	7	6	3	4	5	I		35.66	−2.09
2	7	I	8	7	2	6	3	5	4	I		36.04	−2.47
8	7	I	7	2	8	6	3	5	4	I		37.41	−3.84
6	4	I	8	2	7	4	3	5	6	I		39.05	−5.48
7	6	I	8	2	6	7	3	5	4	I		39.34	−5.77
6	5	I	8	2	7	5	3	6	4	I		39.42	−5.85
7	3	I	8	2	3	6	7	5	4	I		40.19	−6.62
2	6	I	8	6	7	2	3	5	4	I		40.33	−6.76
8	6	I	6	2	7	8	3	5	4	I		41.99	−8.42
7	4	I	8	2	4	6	3	5	7	I		42.14	−8.57
8	4	I	4	2	7	6	3	5	8	I		42.78	−9.21
7	5	I	8	2	5	6	3	7	4	I		43.58	−10.00
2	4	I	8	4	7	6	3	5	2	I		43.95	−10.38
8	3	I	3	2	7	6	8	5	4	I		46.16	−12.59
2	3	I	8	3	7	6	2	5	4	I		46.40	−12.83
8	5	I	5	2	7	6	3	8	4	I		47.36	−13.79
2	5	I	8	5	7	6	3	2	4	I		49.40	−15.83

Table 8.21 Tabu list after fourth iteration

	3	4	5	6	7	8
2	0	0	0	0	0	I
	3	3	0	2	0	0
		4	0	0	0	0
			5	0	0	0
				6	0	0
					7	0

zero. The new route is (1, 8, 2, 7, 6, 5, 4, 3, 1) (Table 8.22). The length of this route is 33.34 distance units. The new tabu list is given in Table 8.23.

Figure 8.9 shows the route length changes throughout the iterations.

Figure 8.10 shows the traveling salesman route after five tabu search iterations.

8.6 GENETIC ALGORITHMS

Genetic algorithms XE "Genetic algorithms" (GA) represent search techniques, based on the mechanics of natural selection, used to find optimal or near-optimal solutions of optimization and search problems.

Table 8.22 Possible new routes in the fifth iteration

Swap			New route								Length	Gain
4	**5**	**I**	**8**	**2**	**7**	**6**	**5**	**4**	**3**	**I**	**33.34**	**1.71**
4	3	I	8	2	7	6	3	5	4	I	33.57	1.49
6	3	I	8	2	7	3	4	5	6	I	35.42	−0.36
8	2	I	2	8	7	6	4	5	3	I	36.07	−1.02
6	5	I	8	2	7	5	4	6	3	I	37.45	−2.39
2	7	I	8	7	2	6	4	5	3	I	37.53	−2.47
6	4	I	8	2	7	4	6	5	3	I	38.48	−3.43
8	7	I	7	2	8	6	4	5	3	I	38.90	−3.84
5	3	I	8	2	7	6	4	3	5	I	39.12	−4.06
7	3	I	8	2	3	6	4	5	7	I	39.59	−4.54
8	6	I	6	2	7	8	4	5	3	I	40.76	−5.70
2	6	I	8	6	7	2	4	5	3	I	40.91	−5.85
7	6	I	8	2	6	7	4	5	3	I	41.01	−5.95
8	3	I	3	2	7	6	4	5	8	I	41.72	−6.66
2	3	I	8	3	7	6	4	5	2	I	43.04	−7.98
7	4	I	8	2	4	6	7	5	3	I	44.23	−9.17
7	5	I	8	2	5	6	4	7	3	I	45.06	−10.00
8	4	I	4	2	7	6	8	5	3	I	48.71	−13.65
2	4	I	8	4	7	6	2	5	3	I	48.80	−13.75
8	5	I	5	2	7	6	4	8	3	I	48.84	−13.79
2	5	I	8	5	7	6	4	2	3	I	50.89	−15.83

Table 8.23 Tabu list after fifth iteration

	3	4	5	6	7	8
2	0	0	0	0	0	0
3		2	0	1	0	0
4			3	0	0	0
5				0	0	0
6					0	0
7						0

These algorithms were developed by analogy with Darwin's theory of evolution and the basic principle of the "survival of the fittest." Charles Darwin indicated that "survival of the fittest" was the basis for the change of living things with time (evolution). As suggested by Darwin, the survival of the fittest should be understood as "preservation, during the battle for life, of varieties which possess any advantage in structure, constitution, or instinct."

The populations of lions, yaks, zebras, antelopes, giraffes, or hippos create the next generations of individuals through the process of mating,

Length of the route

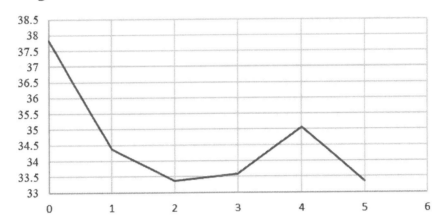

Figure 8.9 Route length changes through the iterations

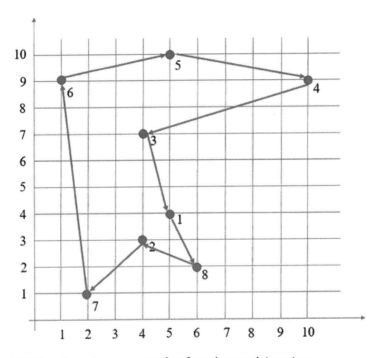

Figure 8.10 Traveling salesman route after five tabu search iterations

followed by selection. This process includes recombination of the genomes of the two parent individuals, by which offspring individuals inherit novel unique combinations of genes from the two parents. Every individual in nature has a specific degree of adaptation to environment. The fittest (best-adapted) individuals have a greater chance of surviving to participate in mating. The parents with better fitness create better offspring that have a greater chance of surviving in the environment. Offspring inherit the main characteristics of their parents. Additionally, some of the offspring's genes can be subjected to a novel mutation, albeit at a low probability, possibly resulting in a novel trait, advantageous or disadvantageous. Evolution of populations continuously happens in nature. Similarly, a genetic algorithm preserves a population of feasible solutions of the problem under investigation. By iteratively applying a set of stochastic operators, the analyst forces the population of feasible solutions to evolve through time. Table 8.24 indicates an analogy between the theory of evolution, the "survival of the fittest" principle, and the GA.

In the case of genetic algorithms, contradictory to conventional search techniques, the search for the optimal solution is run in parallel from a population of solutions. We describe GA stage by stage of their execution.

Initial population

In the first step, a group of feasible solutions of the problem in question are generated. These solutions form the initial population of the solutions. Every generated solution represents an individual that is characterized by a set of parameters (variables). We call these parameters "genes". Genes are grouped into a string. We call this string a chromosome. Every chromosome represents a feasible solution to the problem (Figure 8.11).

Let us assume that we want to find the maximum value of function $f(x) = x^3$ in the domain interval of x ranging from 0 to 15. By means of binary coding, the observed values of variable x can be presented in strings

Table 8.24 An analogy between the theory of evolutionary principle, the "survival of the fittest", and the GA

Nature	Genetic Algorithm
Environment	Optimization problem
Population of individuals that live in environment	A set of feasible solutions
Individual's degree of adaptation to environment	Solution quality (fitness function value)
Selection, recombination, and mutation of individuals in nature	Stochastic operators
Evolution of populations to match their environment	Applying stochastic operators on feasible solutions through iterations

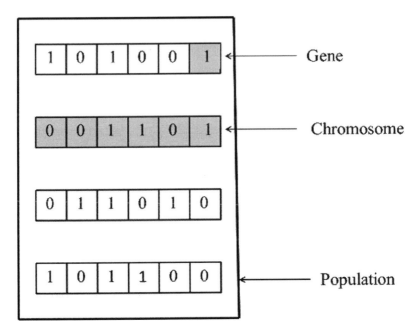

Figure 8.11 Gene, chromosome and population

of length 4 (since $2^4 = 16$). Table 8.25 shows sixteen strings with the corresponding decoded values.

Let us assume that, in the first step, the following four strings were randomly generated: 0011, 0110, 1010, and 1100 (Figure 8.12). These four strings form the initial population P(0).

Table 8.25 Sixteen strings with corresponding decoded values

String	Encoded value of variable x	String	Encoded value of variable x
0000	$0 = 0 \cdot 2^3 + 0 \cdot 2^2 + 0 \cdot 2^1 + 0 \cdot 2^0$	1000	$8 = 1 \cdot 2^3 + 0 \cdot 2^2 + 0 \cdot 2^1 + 0 \cdot 2^0$
0001	$1 = 0 \cdot 2^3 + 0 \cdot 2^2 + 0 \cdot 2^1 + 1 \cdot 2^0$	1001	$9 = 1 \cdot 2^3 + 0 \cdot 2^2 + 0 \cdot 2^1 + 1 \cdot 2^0$
0010	$2 = 0 \cdot 2^3 + 0 \cdot 2^2 + 1 \cdot 2^1 + 0 \cdot 2^0$	1010	$10 = 1 \cdot 2^3 + 0 \cdot 2^2 + 1 \cdot 2^1 + 0 \cdot 2^0$
0011	$3 = 0 \cdot 2^3 + 0 \cdot 2^2 + 1 \cdot 2^1 + 1 \cdot 2^0$	1011	$11 = 1 \cdot 2^3 + 0 \cdot 2^2 + 1 \cdot 2^1 + 1 \cdot 2^0$
0100	$4 = 0 \cdot 2^3 + 1 \cdot 2^2 + 0 \cdot 2^1 + 0 \cdot 2^0$	1100	$12 = 1 \cdot 2^3 + 1 \cdot 2^2 + 0 \cdot 2^1 + 0 \cdot 2^0$
0101	$5 = 0 \cdot 2^3 + 1 \cdot 2^2 + 0 \cdot 2^1 + 1 \cdot 2^0$	1101	$13 = 1 \cdot 2^3 + 1 \cdot 2^2 + 0 \cdot 2^1 + 1 \cdot 2^0$
0110	$6 = 0 \cdot 2^3 + 1 \cdot 2^2 + 1 \cdot 2^1 + 0 \cdot 2^0$	1110	$14 = 1 \cdot 2^3 + 1 \cdot 2^2 + 1 \cdot 2^1 + 0 \cdot 2^0$
0111	$7 = 0 \cdot 2^3 + 1 \cdot 2^2 + 1 \cdot 2^1 + 1 \cdot 2^0$	1111	$15 = 1 \cdot 2^3 + 1 \cdot 2^2 + 1 \cdot 2^1 + 1 \cdot 2^0$

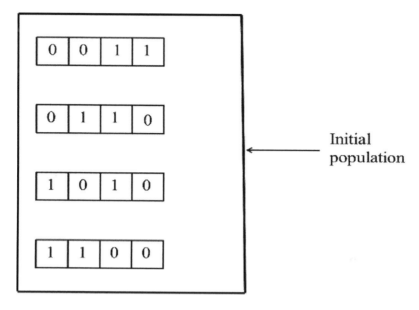

Figure 8.12 The initial population

Fitness function

In the next step, evaluation of the generated solutions is carried out. This includes the estimation of the individuals' fitness function values (fitness scores). The fitness function shows us how fit each individual is. In other words, the fitness score indicates the capability of an individual to compete with other individuals.

To make an estimation of the four generated strings, it is necessary to decode them. After decoding, we actually obtain the following four values of variable x: 3, 6, 10, and 12. The corresponding values of the function $f(x) = x^3$ are, respectively, equal to $f(3) = 27$, $f(6) = 216$, $f(10) = 1,000$, and $f(12) = 1,728$. As can be seen, string 1100 (decoded value 12) has the highest fitness value.

Selection

In nature, individuals that possess some advantage (constitution, instinct, speed, etc.) have higher chances of survival and reproduction. Similarly, in GA, individuals with high fitness scores have a higher chance of being selected in the mating pool. The future parents (some of the good solutions) are selected based on their fitness score. These chosen solutions undergo the phases of reproduction, crossover, and mutation. The remaining solutions are eliminated from consideration.

Let us denote by f_i the fitness score of string i. The probability $p(i)$ for string i to be selected for reproduction is equal to the ratio of f_i to the sum of all strings' fitness scores in the population:

$$p(i) = \frac{f_i}{\sum_{j=1}^{n} f_j}$$

where:

n – the number of individuals in the population

The chosen strings, 0011, 0110, 1010, and 1100, and the corresponding probabilities for each string to be selected for reproduction are given in Table 8.26.

The calculated probabilities are used to configure the roulette wheel. The probabilities correspond to the circle sector areas of the roulette wheel (Figure 8.13).

We use the roulette wheel to select the parents. The number of times the roulette wheel is spun is equal to the population size. In our case, the roulette wheel is spun four times. As we can see from Figure 8.13, the individuals with the higher fitness scores have the greatest chance of being selected for reproduction.

In our example, we use the following approach to simulate the roulette wheel selection. We denote by r the random number generated from the interval [0, 1]. Given that the [0, 1] random numbers are uniformly distributed in the interval [0, 1], we formulate the following rules for roulette wheel selection:

If	$0 \leq r \leq 0 + 0.01 = 0.01$	3 is chosen
If	$0.01 < r \leq 0.01 + 0.07 = 0.08$	6 is chosen
If	$0.08 < r \leq 0.08 + 0.34 = 0.42$	10 is chosen
If	$0.42 < r \leq 0.42 + 0.58 = 1$	12 is chosen

Table 8.26 Chosen strings and probabilities for strings to be selected for reproduction

Ordinary number i	String	Encoded value of the variable x	$f(x) = x^3$	The probability p(i) for string i to be selected for reproduction
1	0011	3	27	$\dfrac{27}{27 + 216 + 1,000 + 1,728} = 0.01$
2	0110	6	216	$\dfrac{216}{27 + 216 + 1,000 + 1,728} = 0.07$
3	1010	10	1,000	$\dfrac{1,000}{27 + 216 + 1,000 + 1,728} = 0.34$
4	1100	12	1,728	$\dfrac{1,728}{27 + 216 + 1,000 + 1,728} = 0.58$

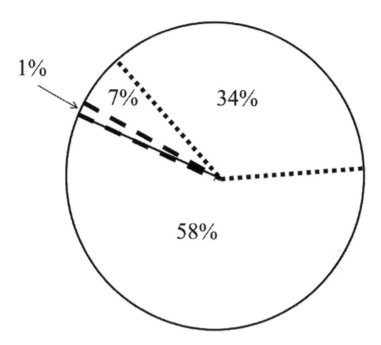

Figure 8.13 Roulette wheel and the sector areas

Let us assume that we generated the following four random numbers: 0.53, 0.23, 0.30, and 0.34. Consequently, strings 1100, 1010, 1010, and 1010 are chosen to be included in the mating pool (these strings correspond to the decoded values 12, 10, 10, and 10, respectively).

This type of selection for mating represents a proportional selection known as the "roulette wheel selection." In addition to "roulette wheel selection," several other ways of selection for mating have been suggested in the literature (rank selection, tournament selection, Boltzmann selection, etc.).

Parents are selected for mating among individuals in the mating pool. Parents could be selected sequentially (1–2, 3–4, 5–6, 7–8, etc.), or the analyst could perform random selection of the parents. There are various ways to produce offspring. All offspring could be the exact copy of the parents, all offspring could be made from parts of parents' chromosomes, specific numbers of the offspring could be exact copies of the parents, and the remaining number could be made from parts of parents' chromosomes, etc.

The probability of crossover (which achieves recombination of genes from the two parents in the creation of offspring) prescribes how frequently the analyst will perform the crossover. When crossover probability is equal to 1, then all offspring are made by the crossover operator. In the case when crossover probability is equal to 0, there is no crossover or recombination and all members of the new generation have exact copies of chromosomes from the previous population (parents), etc.

Crossover

Crossover XE "Crossover" operator is used to combine the genetic material of two parents to create new offspring with new genetic combinations. At the beginning, pairs of strings (parents) are chosen (sequentially, or by random selection) from a set of previously selected strings. In the next step, for each selected parent pair, the location for crossover is randomly chosen. Each pair of parents creates two offspring (Figure 8.14).

Combining the genetic material of the two parents (two solutions) corresponds to sexual reproduction. On the other hand, in some genetic algorithms, new solutions are created by cloning. This corresponds to asexual reproduction.

There are various crossover operators. Figure 8.15 shows the case of a single-point crossover operator. In this case, genes to the right of the crossover point are swapped between the two parents. In the case of a two-point crossover, two crossover points are generated. These points are randomly located. The genes between the two crossovers are swapped between the parents (Figure 8.15).

In the case of uniform crossover, the chromosomes are not divided into segments. Each gene is considered individually. In other words, for each gene, the decision is made whether or not the gene will be included in the offspring. This decision is made in a random manner, by flipping a coin. We

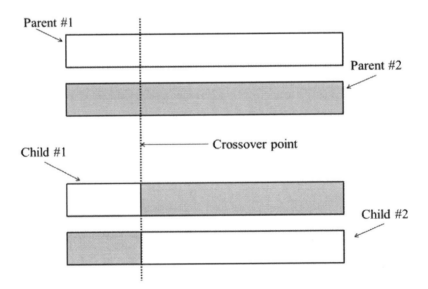

Figure 8.14 A single-point crossover operator: two parents and two created offspring

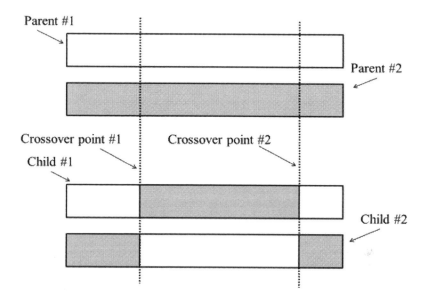

Figure 8.15 A two-point crossover operator: two parents and two created offspring

denote the generated random number from the interval [0, 1] by *r*. Given that the [0, 1] random numbers are uniformly distributed in the interval [0, 1], we formulate the following rules for the decision as to whether or not the specific gene will be included in the offspring:

If	$0 \leq r \leq 0.5$	the gene will not be included
If	$0.5 < r \leq 1$	the gene will be included

Figure 8.16 shows a uniform crossover operator.

The analyst, who is solving the problem, has complete freedom to decide to have, for example, more genetic material in the offspring from one parent. In such cases, the rules for the decision as to whether or not the specific gene will be included in the offspring could be, for example:

If	$0 \leq r \leq 0.2$	the gene will not be included
If	$0.2 < r \leq 1$	the gene will be included

Mutation

After completing crossover, the genetic operator mutation is used. In the case of binary coding, mutation XE "Mutation" of a certain number of genes refers to the change in value from 1 to 0 or *vice versa* (Figure 8.17). It should be noted that the probability of mutation is very low (of the order of magnitude 1/1000).

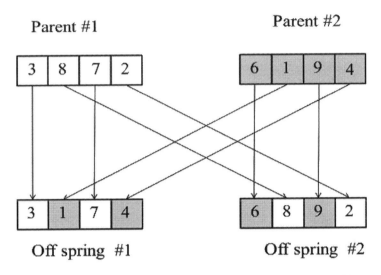

Figure 8.16 A uniform crossover operator: two parents and two created offspring

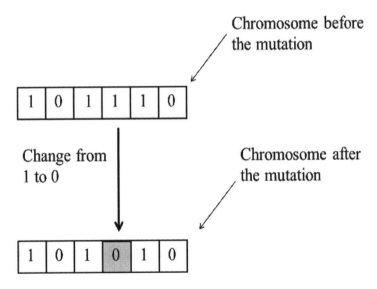

Figure 8.17 Mutation: Change from 1 to 0

The idea of the mutation operator is to prevent GA from being trapped into a local extreme. Mutation probability is an indicator that shows how frequently chromosome parts mutate. When the mutation probability equals 1, the complete chromosome is modified. When the mutation probability equals 0, nothing is changed. The purpose of mutation is to prevent an irretrievable loss of the genetic material at some point along the string. For example, in the overall population, a particularly significant gene might be missing (for example, none of the strings have 0 at the fourth location), which can considerably influence the determination of the optimal or near-optimal solution. Without mutation, none of the strings in all future populations could have 0 at the fourth location. This could cause, for example, premature convergence.

Termination of the algorithm

It has been shown that, at the start, the GA makes progress very quickly. Usually, in later stages, the improvements in fitness scores are very small. The following are typical GA termination conditions:

- When there has been no improvement in the fitness scores for the last few iterations
- When the prescribed number of generations is reached
- When the fitness scores have reached a specific prescribed value

Flow chart of the genetic algorithm

Having generated population P(1) (which has the same number of members as population P(0)), we proceed to use the reproduction, crossover, and mutation operators to generate a sequence of populations P(2), P(3), and so on. Modifications may occur in various applications of genetic algorithms (regarding the manner in which the strings for reproduction are selected, the manner of doing crossovers, the size of population that depend on the problem being optimized, and so on). The steps that characterize any genetic algorithm are shown in Figure 8.18.

8.7 ANT COLONY OPTIMIZATION

Ants which are successful at discovering food leave after them a pheromone trail that other ants can follow in order to reach the food. The arrival of other ants at the pheromone trail reinforces the pheromone signal. At the same time, the pheromone trail evaporates with time. The decision of one ant to follow a specific path to the food depends on the behavior of previous ants. At the same time, every ant increases the probability that the nestmates leaving the nest after him follow the same path. Through the pheromone trail, ants communicate among themselves. Ants are capable of solving complex problems. It has been shown by experiment that ants are able to identify the shortest path between two points in space.

Experiments with an obstacle (Figure 8.19) showed the ants' ability to detect the shorter path very quickly. In this experiment, the nest was connected with the food source. At the very beginning, all ants were

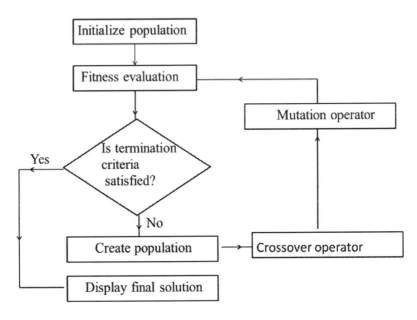

Figure 8.18 Flow chart of the genetic algorithm

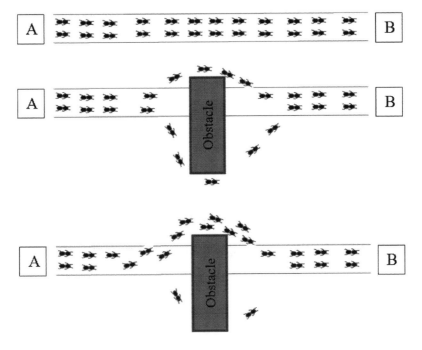

Figure 8.19 Experiment with the obstacle: (a) ants walk along the shortest path between the nest and the food source, (b) the obstacle forces ants to choose one of the two available paths, (c) most of the ants find out the new shortest path characterized by the very strong pheromone

located in the nest. No pheromone was deposited anywhere. Then, an obstacle was put in the path of ants. Every ant, in a probabilistic manner, chooses the path to reach the food source. Ants that randomly chose the shorter path arrive first at the food source. These ants go back over the path that they took to reach the food source. In this way, they reinforce the pheromones they deposited when traveling to the food source. They are, of course, also the first ants who return to the nest. The higher frequency of crossing the shorter path causes a higher pheromone concentration on the shorter path. In other words, the shorter path accumulates the pheromone quicker. Thus, the probability of choosing the shorter path constantly grows, and very quickly almost all the ants use the shorter path.

The ant colony optimization (ACO) represents a metaheuristic strategy capable of solving complex combinatorial optimization problems. The ACO is best illustrated when considering the traveling salesman problem (TSP). We describe the basic version of the ACO algorithm. When solving the TSP, artificial ants investigate the solution space, simulating real ants searching for food in the environment. The objective function values (the length of the TSP tour) correspond to the quality of food sources. The time is discrete in the artificial ants' environment.

At the beginning of the search process (time $t=0$), the artificial ants are located in different towns. Let us denote by $\tau_{ij}(t)$ the intensity of the pheromone trail on edge (i,j) at time t. At time $t=0$, the value of $\tau_{ij}(0)$ is equal to a small positive constant c. At time point t, each ant goes from the present town to the new town. After reaching the next town at time point $(t+1)$, each ant makes the next move towards the next (previously unvisited) town.

Being located in town i, ant k chooses the next town j to be visited, at time t, with the transition probability $p_{ij}^k(t)$ that is equal to:

$$p_{ij}^k(t) = \begin{cases} \dfrac{\left[\tau_{ij}(t)\right]^{\alpha} \cdot \left[\eta_{ij}(t)\right]^{\beta}}{\displaystyle\sum_{h \in \Omega_i^k(t)} \left[\tau_{ih}(t)\right]^{\alpha} \cdot \left[\eta_{ih}(t)\right]^{\beta}} & j \in \Omega_i^k(t) \\ 0 & j \notin \Omega_i^k(t) \end{cases}$$

where $\Omega_i^k(t)$ is the set of feasible nodes to be visited by ant k (the set of feasible nodes is updated for each ant after every move), d_{ij} is the Euclidean distance between node i and node j, $\eta_{ij} = \dfrac{1}{d_{ij}}$ is the "visibility", and α and β are parameters representing the relative importance of the trail intensity and visibility. The visibility is based on local information. The greater the importance the analyst assigns to visibility, the greater the probability that the neighboring towns will be selected. The greater the importance given to the link's trail intensity, the more desirable the link is, since many ants have already passed that way. The total of n moves is performed by n ants in the time interval $(t, t+1)$. Every ant finishes a traveling salesman tour after n iterations.

The m iterations of the algorithm are called a "cycle". The trail intensity $\tau_{ij}(t)$ is usually updated after each cycle in the following way:

$$\tau_{ij}(t) \leftarrow \rho \cdot \tau_{ij}(t) + \Delta\tau_{ij}(t)$$

where ρ is the coefficient $(0 < \rho < 1)$, such that $(1-\rho)$ represents evaporation of the trail in every cycle.

The total increase in trail intensity along the link (i,j) after one finished cycle is equal to:

$$\Delta\tau_{ij}(t) = \sum_{k=1}^{n} \Delta\tau_{ij}^{k}(t)$$

where $\Delta\tau_{ij}^{k}(t)$ is the amount of pheromone laid on the link (i,j) by the k-th ant during the cycle.

The pheromone quantity $\Delta\tau_{ij}^{k}(t)$ is calculated as $\Delta\tau_{ij}^{k}(t) = \dfrac{Q}{L_k(t)}$, if the k-th ant walks along the link (i,j) in its tour during the cycle. If not, the pheromone quantity equals: $\Delta\tau_{ij}^{k}(t) = 0$, where Q is a constant, and $L_k(t)$ is the tour length developed by the k-th ant within the cycle. As we can see, artificial ants collaborate among themselves in order to discover high-quality solutions. This collaboration is expressed through pheromone deposition.

8.8 BEE COLONY OPTIMIZATION

Bee colony optimization (BCO) is a biologically inspired method that uses principles of collective intelligence applied by the honey bees through the nectar-collecting process. The BCO uses an analogy between the way in which bees in nature search for food, and the way in which the optimization algorithm searches for an optimum solution to a particular combinatorial optimization problem. The BCO belongs to the class of population-based algorithms.

The BCO represents an artificial system made up of a number of agents (artificial bees). Artificial bees explore through the solution space, searching for feasible solutions. With the intention of finding better and better solutions, artificial bees collaborate and exchange information. By using collective knowledge and sharing information among themselves, artificial bees focus on more promising areas, and gradually discard less promising solutions in favor of the more promising ones. Slowly, artificial bees as a group generate and/or improve their solutions. The BCO search is run in iterations until some predefined stopping criterion is satisfied.

Two variations of the BCO algorithm can be differentiated: the *constructive* and *improvement* concepts. In the constructive version of the BCO algorithm, the BCO begins from scratch and, for each bee, builds a solution, step by step, by applying some stochastic problem-specific heuristics. Within each iteration, B solutions are generated, and the best of them is used for updating the current global best solution. The next iteration then results in B new solutions, among which we search for the new global best one.

There is also an improvement version of the BCO algorithm. In this case, the analyst starts from a complete solution. The complete solution could be generated randomly or by some heuristics. By perturbing that solution, artificial bees try to improve it. The scheme of improving solutions could be utilized in many different ways, and this approach may be very practical for solving difficult combinatorial optimization problems.

Artificial bees live in an environment characterized by discrete time. The population consists of B artificial bees that collaboratively search for the optimal solution.

Every artificial bee creates one solution to the problem. There are two alternating phases (the *forward pass* and the *backward pass*) that constitute a single step in the BCO algorithm. In each forward pass, every artificial bee explores the solution space. It applies a predefined number of moves, which construct and/or improve the solution, yielding a new solution. For example, let bees Bee 1, Bee 2, ..., Bee 5 participate in the decision-making process on *n* entities. At each forward pass, bees are supposed to select one entity. One of the possible situations after the third forward pass is illustrated in Figure 8.20.

After generating new partial solutions, the artificial bees fly back to the hive and begin the second phase, the so-called backward pass. In the backward pass, all artificial bees share information about the quality of their solutions. In nature, bees would return to the hive, perform a dancing ritual ("waggle dance"), which informs other bees about the amount of food they have discovered, and the proximity and direction of the patch to the hive. In the BCO algorithm, the bees make known the quality of the solution, i.e. the objective function value of the solution. After evaluation of all solutions, every bee decides with a certain probability whether it will stay loyal to its solution or not. The bees with better solutions have a greater probability to stay loyal and advertise their solutions. Artificial bees that are *loyal* to their discovered solutions are at the same time *recruiters*, i.e. their solutions would be considered by other bees. When the solution is discarded by a bee, the bee becomes *uncommitted* and has to choose one of the advertised solutions. This decision is taken with a certain probability

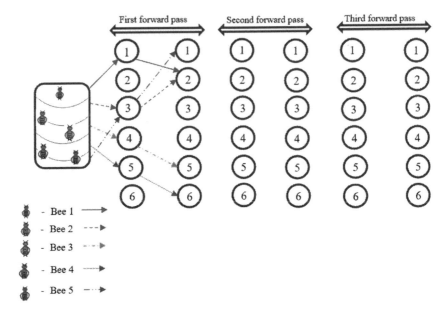

Figure 8.20 An illustration of the first forward pass

too, so that better solutions have a greater chance of being chosen for additional exploration. In such a way, within each backward pass, all bees are divided into two groups (R recruiters, and remaining $B-R$ uncommitted bees), as is shown in Figure 8.21. Values for R and $B-R$ change from one backward pass to another.

When evaluating all the partial solutions generated, Bee 3, from the prior example, makes a decision to leave behind its previously created solution and to join Bee 4. Also, Bee 5 makes a decision to join Bee 2 (see Figure 8.22).

Bee 2 and Bee 5 "fly together" along the path previously discovered by Bee 2. The partial solution generated by Bee 2 is associated (copied) to Bee 5, as well. When they arrive at the end of that common path, they are free to make an individual decision regarding the next constructive step to be made. Similarly, Bee 3 will "fly together" with Bee 4. Bee 1 will keep to its previously generated partial solution. Bee 1 was not chosen by any of the uncommitted hive-mates. In the second forward pass, they all fly independently. The described situation is illustrated in Figure 8.23.

After the second forward pass, a backward pass starts. Let us suppose that Bees 1, 2, and 4 stay loyal to their solutions, and Bees 3 and 5 do not. Now, Bee 1, Bee 2, and Bee 4 are recruiters and Bee 3 and Bee 5 are uncommitted followers (Figure 8.24). Let us suppose that Bee 3 will follow Bee 4 (Bee 3 will fly with Bee 4) and Bee 5 will follow Bee 1 (Bee 5 will fly with Bee 1). This is given in Figure 8.25.

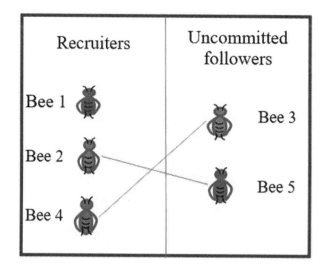

Figure 8.21 Recruitment of uncommitted followers

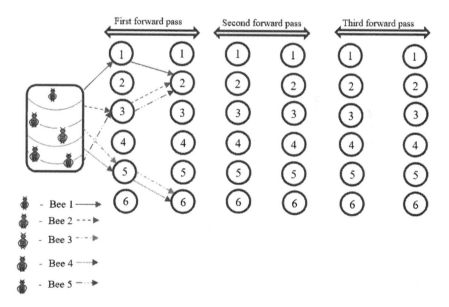

Figure 8.22 The possible result of a recruiting process within the first backward pass

When bees reach the last node from the second pass, they continue independently. In the third forward pass, they visit the last two nodes. In our example, this is illustrated in Figure 8.26.

The two phases of the BCO algorithm, the forward and backward passes, alternate in order to create all wanted feasible solutions (one for each bee).

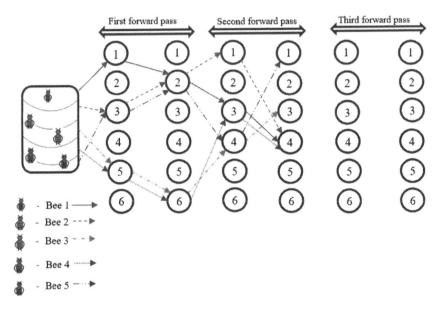

Figure 8.23 An illustration of the second forward pass

When all solutions are completed, the best one is recognized. It is used to update the global best solution. An iteration of the BCO is thus completed. At that moment, all B solutions are deleted, and the new iteration can start. The BCO runs iteration by iteration until a stopping condition is met. The stopping conditions could be, for example, the maximum total number of

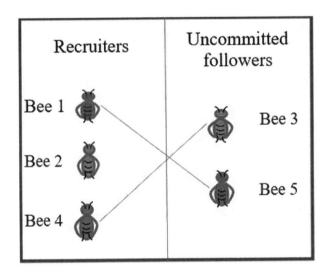

Figure 8.24 Recruitment of uncommitted followers in the second backward pass

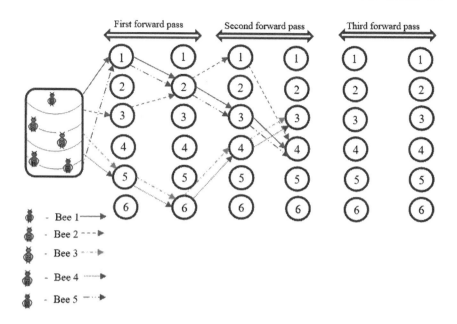

Figure 8.25 The result of a recruiting process within second backward pass

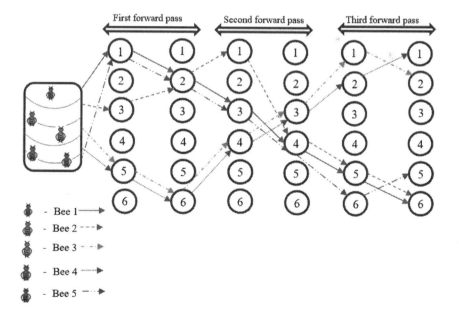

Figure 8.26 Bees' solutions after third forward pass

forward/backward passes, the maximum total number of forward/backward passes without the improvement of the objective function, the maximum allowed CPU time, etc. Finally, the best solution found (the so-called global best) is reported as the final one.

All the bees are in the hive at the beginning of the search. In line with the major idea of the BCO algorithm, the hive is an artificial object, with no precise location, and does not affect the algorithm execution. It is used merely to denote the synchronization points at which bees exchange information about the present state of the search.

The BCO algorithm parameters, the values of which must be set prior to the algorithm execution, are as follows:

#B	number of bees involved in the search,
#P	number of forward and backward passes in a single iteration,
#C	number of changes in one forward pass.

The analyst must also define the stopping criteria.

The pseudo code of the improvement version of the BCO algorithm can be given in the following way:

1. Generate initial solution
2. **do**
3.   Set the solution to all bees
4.   **for** $i = 1$ **to** #P
5.   Forward pass
6.   Backward pass
7.   **next**
8. **while** (stopping criteria are not satisfied)

During the forward pass, all bees modify their solutions. The pseudo code of the forward pass is the following:

1. **for** $i = 1$ **to** #C
2.   **for** $b = 1$ **to** #B
3.   Make one modification of the bee's b solution
4.   **next**
5.   Check if the new best solution has been discovered. If the new best solution has been discovered, save it.
6. **next**

In the BCO search algorithm, the bees announce the quality of the generated solutions. When all solutions have been evaluated, every bee decides with a certain probability whether it will stay loyal to its solution or not. The bees with better solutions have more chances to retain and advertise their solutions. Unlike bees in nature, the artificial bees that are loyal to their solutions are at the same time recruiters (their solutions would be considered by other bees). Once its solution is abandoned, a bee becomes uncommitted and must select one of the advertised solutions. This decision is taken with a probability that those selecting the better among the advertised solutions have a greater opportunity to be chosen for further exploration. The main steps of the backward pass are given with the following steps:

1. Normalization of the objective function values of the generated solutions
2. Loyalty decision for each bee
3. Assigning every uncommitted bee to one of the loyal bees

We normalize the objective function values of the solutions generated by artificial bees by using the following expressions:

$$O_b = \frac{T_b - T_{min}}{T_{max} - T_{min}} \qquad \forall b = 1, \cdots, B$$

if the goal is to maximize the objective function;

$$O_b = \frac{T_{max} - T_b}{T_{max} - T_{min}} \qquad \forall b = 1, \cdots, B$$

if the goal is to minimize the objective function,
where:
O_i – normalized bee's i objective function value
T_b – bee's b objective function value,
T_{min} – minimal objective function value considering all the bees
$$\left(T_{min} = \min_{b=1,\ldots,B}\{T_b\}\right),$$

T_{max}–maximal objective function value considering all the bees
$$\left(T_{max} = \max_{b=1,\ldots,B}\{T_b\}\right).$$

The probability that the b-th bee (at the beginning of the new forward pass) is loyal to its previously generated partial solution is expressed as follows:

$$p_b^{u+1} = e^{-\frac{O_{max} - O_b}{u}} \qquad b = 1,2,\ldots,B$$

where:
O_b – the normalized value for the objective function of partial solution created by the b-th bee;
O_{max} – maximum overall normalized values of partial solution to be compared;

u – the ordinary number of the forward pass (e.g. $u = 1$ for first forward pass, $u = 2$ for second forward pass, etc.).

Sometimes researchers do not use parameter u; in other words, they calculate the probability that the b-th bee stays loyal in the following way:
$p_b = e^{-(O_{max} - O_b)}$.

We conclude from the last relation that the better the partial solution generated (higher O_b value), the higher the probability that the bee b will be loyal to it. The greater the ordinary number of the forward pass, the greater the influence of the already-discovered partial solution. This is expressed by the term u in the denominator of the exponent. In other words, at the beginning of the search process, bees are "more courageous" to search the solution space. The more forward passes they make, the less courage the bees have to explore the solution space. The closer we are to approaching the end of the search process, the more focused the bees are on the already-known partial solution. By using the probability obtained for each artificial bee and a random number generator, it is decided whether to turn out to be an uncommitted follower, or to continue exploring a previously identified path.

Each uncommitted bee chooses, with a certain probability, a recruiter bee that it will follow. The probability that b's partial solution would be chosen by any uncommitted bee is equal to:

$$p_b = \frac{O_b}{\sum_{k=1}^{R} O_k} \qquad b = 1, 2, \ldots, R$$

where O_k represents the normalized value for the objective function of the k-th advertised partial solution, and R denotes the number of recruiters. By using the last relation and a random number generator, each uncommitted follower joins one recruiter.

8.9 EXAMPLE

Demand in transportation networks is described by the origin-destination matrix. A traffic assignment problem could be described in the following way: for an analyzed transportation network described by its set of nodes, a set of links, link orientation, node connections, and link performance functions, and, for a known origin-destination matrix, calculate the link flows and link travel times.

Via the traffic assignment process, we try to answer the following question: how are network users distributed through the traffic network? Clearly, network users could be distributed in numerous ways through the network paths and links.

System optimal route choice results in the minimum total travel costs of all users. The system optimal route choice problem is described by the following mathematical program:

Minimize

$$F = \sum_{a \in A} x_a t_a$$

subject to:

$$\sum_{p \in P_{rs}} f_p^{rs} = q_{rs} \qquad \forall r \in R, s \in S$$

$$f_p^{rs} \geq 0 \qquad \forall p \in P_{rs}, r \in R, s \in S$$

where:

R – set of origins,

S – set of destinations,

A – set of network links,

$r \in R$ – origin,

$s \in S$ – destination,

q_{rs} – total number of travelers from node r to node s,

P_{rs} – set of all paths between node r and node s,

x_a – flow along link a,

t_a – travel time along link a,

f_p^{rs} – flow along path p that leads from node r to node s.

Traffic assignment, which is the result of this program, usually cannot be achieved in practice. The objective function represents the total travel time (total travel costs) of all users spent in the transportation network.

Let us demonstrate how we could use the genetic algorithm for solving the system optimal route choice problem. We analyze a simple network shown in Figure 8.27.

There are two freeways between nodes A and B. In total, there are 10,000 trips that should be distributed on these two freeways. Capacity of the first freeway is $c_1 = 4,000$ vehicle/hour, and capacity of the second link is $c_2 = 5,700$ vehicle/hour. Free flow time on the first link is $t_1^0 = 15$ minutes and on the second link is $t_2^0 = 25$ minutes. Travel time on the first link is calculated as:

$$t_1 = t_1^0 \cdot \left[1 + 0.15 \cdot \left(\frac{x_1}{c_1} \right)^4 \right]$$

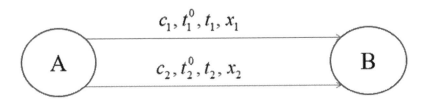

Figure 8.27 System optimal route choice in the case of two alternative routes

Travel time on the second link is:

$$t_2 = t_2^0 \cdot \left[1 + 0.15 \cdot \left(\frac{x_2}{c_2}\right)^4\right]$$

Distribute the trips to the link, i.e. determine the values of x_1 and x_2 ($x_1 + x_2 = 10,000$), in such a way as to minimize the total travel time of all users T_t. The total travel time T_t equals:

$$T_t = T_1 + T_2 = x_1 \cdot t_1 + x_2 \cdot t_2$$

Solve the system optimal route choice problem by using the following metaheuristics:

a) Genetic algorithm,
b) Simulated annealing,
c) Tabu search,
d) Bee colony optimization and
e) Ant colony optimization.

Solution:

We have to determine the values of x_1 and x_2. Since $x_1 + x_2 = 10,000$ it is enough to determine, for example, x_1. Once the x_1 is known, the x_2 is easily calculated x_2 as $x_2 = 10,000 - x_1$. Genetic algorithms usually work with binary strings. We represent x_1 as a string of 10 binary (0, 1) digits. We use the following expression to map the binary string into a decimal number:

$$d = \sum_{i=1}^{n} k_i \cdot 2^{i-1}$$

In the case when all digits in the binary string are equal to 1, then $d = 16,383$. We see that, in this case, the d value is bigger than the total number of trips (10,000). In order to have all flow values in the interval [0,10,000] we use the following expression:

$$y = \frac{10,000}{16,383} \cdot d = 0.610389 \cdot d$$

After rounding, the total number of vehicles x_1 on the first freeway equals:

$$x_1 = \lfloor y \rfloor$$

a) Genetic Algorithm

We use the following parameter values in the genetic algorithm:

- Number of chromosomes in population: 10
- We use a single-point crossover operator. The probability of crossover equals 0.25.
- The probability of mutation equals 0.01.

Generation I

We generate the initial set of chromosomes in a random manner. These chromosomes are shown in Table 8.27. Each chromosome represents one solution. Table 8.28 shows the characteristics of these solutions. All travel times are expressed in hours.

Table 8.27 GA: The initial set of chromosomes

Chr 1	0	1	0	0	1	1	1	0	0	1	0	1	1	0
Chr 2	1	1	0	1	1	1	1	0	0	1	0	0	1	1
Chr 3	1	0	0	0	0	1	0	0	0	1	1	0	1	1
Chr 4	0	1	0	0	0	0	1	0	1	1	0	0	0	1
Chr 5	0	1	0	0	0	1	1	1	0	1	1	0	1	1
Chr 6	0	0	0	1	1	0	1	1	1	1	0	0	1	1
Chr 7	0	1	0	0	0	0	0	0	1	0	0	0	1	1
Chr 8	0	1	0	1	1	0	1	0	1	0	1	1	0	1
Chr 9	0	0	0	0	0	0	1	1	0	1	0	0	1	1
Chr 10	0	1	0	0	1	1	1	0	0	0	1	0	1	0

Table 8.28 GA: Characteristics of the chromosomes

	d	y	x_1	t_1	x_2	t_2	T_t
Chr 1	5014	3,060.49	3,060	0.263	6,940	0.5540	4,649.15
Chr 2	14,227	8,684.00	8,684	1.083	1,316	0.4168	9,953.76
Chr 3	8,475	5,173.05	5,173	0.355	4,827	0.4488	4,002.29
Chr 4	4,273	2,608.19	2,608	0.257	7,392	0.5934	5,056.42
Chr 5	4,571	2,790.09	2,790	0.259	7,210	0.5767	4,880.03
Chr 6	1,779	1,085.88	1,085	0.250	8,915	0.7907	7,320.22
Chr 7	4,131	2,521.52	2,521	0.256	7,479	0.6019	5,146.89
Chr 8	5,805	3,543.31	3,543	0.273	6,457	0.5196	4,322.51
Chr 9	211	128.79	128	0.250	9,872	0.9790	9,696.78
Chr 10	5,002	3,053.16	3,053	0.263	6,947	0.5546	4,654.69

The smaller the total travel time of all users, the better the chromosome.

Selection

We use the roulette wheel to select the parents (Figure 8.28). Let us calculate the probability p_1 for the first string to be selected for reproduction. This probability is equal to the ratio of $1/T_{t,1_i}$ to the sum of all strings' fitness scores in the population, i.e.:

$$p_1 = \frac{1/T_{t,1}}{\sum_{i=1}^{14} 1/T_{t,i}} = \frac{0.00022}{0.01398} = 0.1164$$

In the same way, the probability p_2 for the second string to be selected for reproduction equals:

$$p_2 = \frac{1/T_{t,2}}{\sum_{i=1}^{14} 1/T_{t,i}} = \frac{0.0001}{0.01398} = 0.0544$$

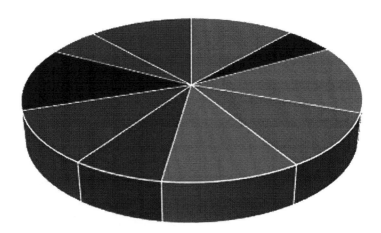

■ Ch. 1   ■ Ch. 2   ■ Ch. 3   ■ Ch. 4   ■ Ch. 5

■ Ch. 6   ■ Ch. 7   ■ Ch. 8   ■ Ch. 9   ■ Ch. 10

Figure 8.28 GA: Roulette wheel

Table 8.29 GA: Probabilities for selection process

	T_t	$1/T_t$	p	Cumulative probabilities
Chr 1	4,649.154	0.00022	0.1164	0.1164
Chr 2	9,953.765	0.0001	0.0544	0.1707
Chr 3	4,002.285	0.00025	0.1352	0.3059
Chr 4	5,056.417	0.0002	0.1070	0.4129
Chr 5	4,880.035	0.0002	0.1109	0.5238
Chr 6	7,320.220	0.00014	0.0739	0.5977
Chr 7	5,146.890	0.00019	0.1051	0.7028
Chr 8	4,322.507	0.00023	0.1252	0.8280
Chr 9	9,696.782	0.0001	0.0558	0.8838
Chr 10	4,654.690	0.00021	0.1162	1.0000
	Sum	0.01398		

The calculated probabilities for all 10 chromosomes are given in the fourth column of Table 8.29. The cumulative probability values are shown in the fifth column of Table 8.29.

We generate 10 random numbers between 0 and 1, and determine which chromosome is selected for reproduction. This procedure is shown in Table 8.30. The selected chromosomes are shown in Table 8.31.

Crossover

The chromosome participates in a crossover with the probability that is equal to 0.25. For each chromosome, we make one simulation test to make

Table 8.30 GA: Simulation of the selection process

i	Generated random number (r_i)	Simulation	Selected chromosome
1	0.948	$0.8838 < r_1 < 1$	Chr 10
2	0.390	$0.3059 < r_2 < 0.4129$	Chr 4
3	0.037	$0 < r_3 < 0.1164$	Chr 1
4	0.950	$0.8838 < r_4 < 1$	Chr 10
5	0.832	$0.8280 < r_5 < 0.8838$	Chr 9
6	0.642	$0.5977 < r_6 < 0.7028$	Chr 7
7	0.895	$0.8838 < r_7 < 1$	Chr 10
8	0.628	$0.5977 < r_8 < 0.7028$	Chr 7
9	0.287	$0.1707 < r_9 < 0.3059$	Chr 3
10	0.119	$0.1164 < r_{10} < 0.1707$	Chr 2

Table 8.31 GA: The chromosomes selected for reproduction

Chr 10	0	I	0	0	I	I	I	0	0	0	I	0	I	0
Chr 4	0	I	0	0	0	0	I	0	I	I	0	0	0	I
Chr 1	0	I	0	0	I	I	I	0	0	I	0	I	I	0
Chr 10	0	I	0	0	I	I	I	0	0	0	I	0	I	0
Chr 9	0	0	0	0	0	0	I	I	0	I	0	0	I	I
Chr 7	0	I	0	0	0	0	0	0	I	0	0	0	I	I
Chr 10	0	I	0	0	I	I	I	0	0	0	I	0	I	0
Chr 7	0	I	0	0	0	0	0	0	I	0	0	0	I	I
Chr 3	I	0	0	0	0	I	0	0	0	I	I	0	I	I
Chr 2	I	I	0	I	I	I	I	0	0	I	0	0	I	I

Table 8.32 GA: Chromosome selection for crossover

	Generated random number r_i	$r_i < 0.25$
Chr 10	0.4950	FALSE
Chr 4	0.6525	FALSE
Chr 1	0.8964	FALSE
Chr 10	0.4579	FALSE
Chr 9	0.7217	FALSE
Chr 7	0.6698	FALSE
Chr 10	0.4633	FALSE
Chr 7	0.2273	TRUE
Chr 3	0.1402	TRUE
Chr 2	0.5555	FALSE

Table 8.33 GA: Chromosome 7

0	I	0	0	0	0	0	0	I	0	0	0	I	I

Table 8.34 GA: Chromosome 3

I	0	0	0	0	I	0	0	0	I	I	0	I	I

a decision about participation in the crossover. The obtained results are shown in Table 8.32. The random numbers generated are given in the second column. The values TRUE in the third column are given to the chromosomes that participate in the crossover. The values FALSE are assigned to the chromosomes that do not participate in the crossover.

We see from Table 8.32 that chromosome 7 (Table 8.33) and chromosome 3 (Table 8.34) participate in the crossover:

Since we use a single-point crossover operator, we generate one random number between 1 and 14. We generated number 4. We created two new chromosomes. The first chromosome, called chromosome 731 (Table 8.35), consists of the first 4 genes from chromosome 7 and the last 10 genes from

Table 8.35 GA: Chromosome 731

0	1	0	0	0	1	0	0	0	1	1	0	1	1

Table 8.36 GA: Chromosome 732

1	0	0	0	0	0	0	0	1	0	0	0	1	1

Table 8.37 GA: The set of chromosomes generated after the crossover operator

Chr 10	0	1	0	0	1	1	1	0	0	0	1	0	1	0
Chr 4	0	1	0	0	0	0	1	0	1	1	0	0	0	1
Chr 1	0	1	0	0	1	1	1	0	0	1	0	1	1	0
Chr 10	0	1	0	0	1	1	1	0	0	0	1	0	1	0
Chr 9	0	0	0	0	0	0	1	1	0	1	0	0	1	1
Chr 7	0	1	0	0	0	0	0	0	1	0	0	0	1	1
Chr 10	0	1	0	0	1	1	1	0	0	0	1	0	1	0
Chr 731	**0**	**1**	**0**	**0**	**0**	**1**	**0**	**0**	**0**	**1**	**1**	**0**	**1**	**1**
Chr 732	**1**	**0**	**0**	**0**	**0**	**0**	**0**	**0**	**1**	**0**	**0**	**0**	**1**	**1**
Chr 2	1	1	0	1	1	1	1	0	0	1	0	0	1	1

chromosome 3. The second new chromosome, called chromosome 732 (Table 8.36), consist of the first 4 genes of chromosome 3, and the last 10 genes from chromosome 7.

The set of chromosomes generated after the crossover operator is given in Table 8.37.

Chromosome 7 and chromosome 3 are removed from the set. Their offspring, chromosome 731, and chromosome 732 (denoted in bold letters) are the new members of the sets of chromosomes.

Mutation

We generate one random number for each gene in every chromosome. The generated random numbers are shown in Table 8.38. We note that only the random number for gene 3 on chromosome 9 is smaller than 0.01 (denoted in bold). This means that the value of gene 3 in chromosome 9 is changed from 0 to 1.

Generation 2

We start creation of the second generation from the solution given in Table 8.39. The characteristics of these solutions are given in the Table 8.40.

Selection

Table 8.41 shows total travel times, values of fitness functions, probabilities, and cumulative probabilities for each chromosome. According to cumulative probabilities in Table 8.41 and random numbers given in the second column of Table 8.42, we make selections of chromosomes. The

Table 8.38 GA: Generated random numbers for mutation

	Gene 1	Gene 2	Gene 3	Gene 4	Gene 5	Gene 6	Gene 7	Gene 8	Gene 9	Gene 10	Gene 11	Gene 12	Gene 13	Gene 14
Chr 10	0.4126	0.9528	0.2878	0.6170	0.7987	0.6795	0.7163	0.5683	0.4124	0.3823	0.0933	0.0678	0.8750	0.0274
Chr 4	0.7725	0.3474	0.4353	0.9302	0.3465	0.3784	0.7640	0.4015	0.3984	0.6757	0.0262	0.7507	0.4437	0.2182
Chr 1	0.6800	0.8980	0.2066	0.3850	0.9357	0.8758	0.5283	0.9872	0.9208	0.2671	0.5480	0.8943	0.6812	0.2429
Chr 10	0.5594	0.6855	0.2918	0.4963	0.0927	0.8676	0.1407	0.3460	0.6203	0.8904	0.7339	0.4256	0.2165	0.2300
Chr 9	0.5048	0.4459	0.0013	0.9351	0.0985	0.6067	0.4034	0.3323	0.4401	0.2755	0.1522	0.6021	0.6694	0.5140
Chr 7	0.8436	0.0647	0.8591	0.0993	0.7921	0.4094	0.7646	0.4226	0.7470	0.7147	0.5333	0.5317	0.2637	0.1142
Chr 10	0.3182	0.5549	0.9239	0.8668	0.5789	0.4623	0.3936	0.7343	0.6176	0.3861	0.1918	0.5800	0.1757	0.3115
Chr 731	0.4669	0.2807	0.1621	0.7178	0.9941	0.7646	0.9655	0.4264	0.5227	0.5308	0.8905	0.1567	0.0723	0.2640
Chr 732	0.7967	0.9134	0.9239	0.4758	0.7052	0.2398	0.3460	0.4515	0.3701	0.0531	0.5003	0.9052	0.3034	0.2940
Chr 2	0.4905	0.0578	0.5314	0.0934	0.9142	0.3689	0.2661	0.0861	0.4093	0.1704	0.1497	0.0519	0.3189	0.1147

Table 8.39 GA: Second generation: The initial set of chromosomes

Chr 10	0	1	0	0	1	1	1	0	0	0	1	0	1	0
Chr 4	0	1	0	0	0	0	1	0	1	1	0	0	0	1
Chr 1	0	1	0	0	1	1	1	0	0	1	0	1	1	0
Chr 10	0	1	0	0	1	1	1	0	0	0	1	0	1	0
Chr 9	0	0	1	0	0	0	1	1	0	1	0	0	1	1
Chr 7	0	1	0	0	0	0	0	0	1	0	0	0	1	1
Chr 10	0	1	0	0	1	1	1	0	0	0	1	0	1	0
Chr 731	0	1	0	0	0	1	0	0	0	1	1	0	1	1
Chr 732	1	0	0	0	0	0	0	0	1	0	0	0	1	1
Chr 2	1	1	0	1	1	1	1	0	0	1	0	0	1	1

Table 8.40 GA: Second generation: The characteristics of the solutions in the initial set

	d	y	x_1	t_1	x_2	t_2	T_t
Chr 10	5,002	3,053.16	3,053	0.263	6,947	0.5546	4,654.69
Chr 4	4,273	2,608.19	2,608	0.257	7,392	0.5934	5,056.42
Chr 1	5,014	3,060.49	3,060	0.263	6,940	0.5540	4,649.15
Chr 10	5,002	3,053.16	3,053	0.263	6,947	0.5546	4,654.69
Chr 9	2,259	1,378.87	1,378	0.251	8,622	0.7439	6,758.84
Chr 7	4,131	2,521.52	2,521	0.256	7,479	0.6019	5,146.89
Chr 10	5,002	3,053.16	3,053	0.263	6,947	0.5546	4,654.69
Chr 731	4,379	2,672.89	2,672	0.257	7,328	0.5874	4,992.43
Chr 732	8,227	5,021.67	5,021	0.343	4,979	0.4531	3,978.46
Chr 2	14,227	8684	8,684	1.083	1,316	0.4168	9,953.76

Table 8.41 GA: Second generation: Probabilities for selection process

	T_t	$1/T_t$	p	Cumulative probabilities
Chr 10	4,654.690	0.00021	0.1101	0.1101
Chr 4	5,056.417	0.0002	0.1013	0.2114
Chr 1	4,649.154	0.00022	0.1102	0.3216
Chr 10	4,654.690	0.00021	0.1101	0.4317
Chr 9	6,758.842	0.00015	0.0758	0.5075
Chr 7	5,146.890	0.00019	0.0995	0.6070
Chr 10	4,654.690	0.00021	0.1101	0.7171
Chr731	4,992.430	0.0002	0.1026	0.8197
Chr 732	3,978.463	0.00025	0.1288	0.9485
Chr 2	9,953.765	0.0001	0.0515	1.0000
	Sum	0.00195		

Table 8.42 GA: Simulation of the selection process

i	Random number (r_i)	Simulation	Chromosome
1	0.903	$0.8197 < r_1 < 1$	Chr 732
2	0.513	$0.5075 < r_2 < 0.6070$	Chr 7
3	0.717	$0.6070 \leq r_3 < 0.7171$	Chr 10
4	0.136	$0.1101 \leq r_4 < 0.2114$	Chr 4
5	0.006	$0 \leq r_5 < 0.1101$	Chr 10
6	0.627	$0.6070 \leq r_6 < 0.7171$	Chr 10
7	0.140	$0.1101 \leq r_7 < 0.2114$	Chr 4
8	0.074	$0 \leq r_8 < 0.1101$	Chr 10
9	0.254	$0.2114 < r_9 < 3216$	Chr 1
10	0.154	$0.1101 \leq r_{10} < 0.2114$	Chr 4

Table 8.43 GA: Second generation: The chromosomes selected for reproduction

Chr 732	1	0	0	0	0	0	0	0	1	0	0	0	1	1
Chr 7	0	1	0	0	0	0	0	0	1	0	0	0	1	1
Chr 10	0	1	0	0	1	1	1	0	0	0	1	0	1	0
Chr 4	0	1	0	0	0	0	1	0	1	1	0	0	0	1
Chr 10	0	1	0	0	1	1	1	0	0	0	1	0	1	0
Chr 10	0	1	0	0	1	1	1	0	0	0	1	0	1	0
Chr 4	0	1	0	0	0	0	1	0	1	1	0	0	0	1
Chr 10	0	1	0	0	1	1	1	0	0	0	1	0	1	0
Chr 1	0	1	0	0	1	1	1	0	0	1	0	1	1	0
Chr 4	0	1	0	0	0	0	1	0	1	1	0	0	0	1

third column of Table 8.42 shows the selected chromosomes. These chromosomes are given in Table 8.43.

Crossover

We see from Table 8.44 that chromosomes 7, 10, 4, and 1 participate in the crossover. In other words, parents are selected for pairing among chromosomes 7, 10, 4, and 1. We decide to select parents sequentially, i.e. 7–10, and 4–1.

Tables 8.45 and 8.46 show chromosomes 7 and 10.

Since we use a single-point crossover operator, we generate one random number between 1 and 14. We generated number 11. We create two new chromosomes. The first chromosome, called chromosome 7101, consists of the first 11 genes from chromosome 7 and the last 3 genes from chromosome 10 (Table 8.47). The second new chromosome, called chromosome 7102, consists of the first 11 genes of chromosome 10, and the last 3 genes from chromosome 7 (Table 8.48). The new chromosomes, created by crossover, are:

Chromosomes 4 and 1 are given in Tables 8.49 and 8.50.

Table 8.44 GA: Second generation: Chromosome selection for crossover

	r_i	$r_i < 0.25$
Chr 732	0.5893	FALSE
Chr 7	0.1549	TRUE
Chr 10	0.1948	TRUE
Chr 4	0.2497	TRUE
Chr 10	0.5375	FALSE
Chr 10	0.6056	FALSE
Chr 4	0.4112	FALSE
Chr 10	0.6582	FALSE
Chr 1	0.1859	TRUE
Chr 4	0.3871	FALSE

Table 8.45 GA: Chromosome 7

0	1	0	0	0	0	0	0	1	0	0	0	1	1

Table 8.46 GA: Chromosome 10

0	1	0	0	1	1	1	0	0	0	1	0	1	0

Table 8.47 GA: Chromosome 7101

0	1	0	0	0	0	0	0	1	0	0	0	1	0

Table 8.48 Chromosome 7102

0	1	0	0	1	1	1	0	0	0	1	0	1	1

Table 8.49 GA: Chromosome 4

0	1	0	0	0	0	1	0	1	1	0	0	0	1

Table 8.50 GA: Chromosome I

0	I	0	0	I	I	I	0	0	I	0	I	I	0

Table 8.51 GA: Chromosome 411

0	I	0	0	I	I	I	0	0	I	0	I	I	0

Table 8.52 GA: Chromosome 412

0	I	0	0	0	0	I	0	I	I	0	0	0	I

Table 8.53 GA: The set of chromosomes of the second generation created after the crossover operator

Chr 732	I	0	0	0	0	0	0	0	I	0	0	0	I	I
Chr 7101	0	I	0	0	0	0	0	0	I	0	0	0	I	0
Chr 7102	0	I	0	0	I	I	I	0	0	0	I	0	I	I
Chr 411	0	I	0	0	I	I	I	0	0	I	0	I	I	0
Chr 10	0	I	0	0	I	I	I	0	0	0	I	0	I	0
Chr 10	0	I	0	0	I	I	I	0	0	0	I	0	I	0
Chr 4	0	I	0	0	0	0	I	0	I	I	0	0	0	I
Chr 10	0	I	0	0	I	I	I	0	0	0	I	0	I	0
Chr 412	0	I	0	0	0	0	I	0	I	I	0	0	0	I
Chr 4	0	I	0	0	0	0	I	0	I	I	0	0	0	I

We again generate one random number between 1 and 14. We generated number 4. We create two new chromosomes. The first chromosome, called chromosome 411, consists of the first 4 genes from chromosome 4 and the last 10 genes from chromosome 1 (Table 8.51). The second new chromosome, called chromosome 412, consist of the first 4 genes of chromosome 1, and the last 10 genes from chromosome 4 (Table 8.52). The new chromosomes, created by crossover, are:

The set of chromosomes of the second generation, created after the crossover operator, is given in Table 8.53.

To perform mutation in the second generation, we also generated one random number for each gene in every chromosome. The generated random numbers are shown in Table 8.54. We note that only the random number for gene 2 of chromosome 7101 is smaller than 0.01 (denoted in bold). This means that the value of gene 2 in chromosome 7101 is changed from 1 to 0.

Table 8.54 GA: Second generation: Generated random numbers for mutation

	Gene 1	Gene 2	Gene 3	Gene 4	Gene 5	Gene 6	Gene 7	Gene 8	Gene 9	Gene 10	Gene 11	Gene 12	Gene 13	Gene 14
Chr 732	0.8919	0.9930	0.4534	0.5792	0.8444	0.6095	0.2208	0.7360	0.6379	0.7573	0.2903	0.6300	0.4407	0.5093
Chr 7101	0.1642	0.0053	0.6957	0.0741	0.5366	0.3653	0.3286	0.3129	0.2196	0.4949	0.2019	0.6478	0.7536	0.6544
Chr 7102	0.7585	0.4528	0.0856	0.3067	0.7215	0.5707	0.0728	0.2640	0.0838	0.1368	0.8381	0.8474	0.1939	0.8567
Chr 411	0.3015	0.1614	0.1692	0.1451	0.5209	0.8257	0.1953	0.8630	0.7556	0.9148	0.6130	0.7983	0.1471	0.5806
Chr 10	0.6727	0.3546	0.8134	0.3553	0.6873	0.1187	0.6543	0.2993	0.1759	0.6105	0.2931	0.0029	0.1392	0.1997
Chr 10	0.6668	0.7829	0.6913	0.7851	0.1709	0.8450	0.0600	0.5373	0.3855	0.8189	0.8260	0.9802	0.0417	0.4088
Chr 4	0.0480	0.4332	0.6509	0.3009	0.3019	0.6392	0.8470	0.2594	0.7604	0.2943	0.8247	0.0949	0.1421	0.7149
Chr 10	0.2212	0.5511	0.4680	0.3229	0.6172	0.3990	0.1359	0.9502	0.7565	0.6343	0.3845	0.7986	0.5559	0.4627
Chr 412	0.7914	0.5213	0.0499	0.0043	0.3349	0.3731	0.5968	0.8243	0.3929	0.7588	0.5281	0.5686	0.3646	0.3967
Chr 4	0.2298	0.3652	0.2308	0.2474	0.2878	0.8903	0.8419	0.0807	0.9528	0.9804	0.0570	0.5317	0.5584	0.4785

Mutation

Table 8.54 shows random numbers generated for the mutation process. We can note from the table that the second gene of chromosome 7101 should be changed. That gene was 1, and now it should be zero.

The solutions in the second generation are shown in Table 8.55.

The characteristics of the solutions in the second generation are given in Table 8.56.

Imagine that we want to finish our search for the best solution. We should take the solution with the smallest value of total travel time, obtained in any generation. In this example, the solution from chromosome 732, with the total travel time $T_t = 3,978.46$ [hours], should be taken for the final one. This means that 5,021 vehicles should be assigned to the first freeway and the remaining 4,979 vehicles to the second freeway.

Table 8.55 GA: The solutions in the second generation

Chr 732	1	0	0	0	0	0	0	0	1	0	0	0	1	1
Chr 7101	0	0	0	0	0	0	0	0	1	0	0	0	1	0
Chr 7102	0	1	0	0	1	1	1	0	0	0	1	0	1	1
Chr 411	0	1	0	0	1	1	1	0	0	1	0	1	1	0
Chr 10	0	1	0	0	1	1	1	0	0	0	1	0	1	0
Chr 10	0	1	0	0	1	1	1	0	0	0	1	0	1	0
Chr 4	0	1	0	0	0	0	1	0	1	1	0	0	0	1
Chr 10	0	1	0	0	1	1	1	0	0	0	1	0	1	0
Chr 412	0	1	0	1	0	0	1	0	1	1	0	0	0	1
Chr 4	0	1	0	0	0	0	1	0	1	1	0	0	0	1

Table 8.56 GA: The characteristics of the solutions in the second generation

	d	y	x_1	t_1	x_2	t_2	T_t
Chr 732	8,227	5,021.67	5,021	0.343	4,979	0.45305	3,978.46
Chr 7101	34	20.7532	20	0.25	9,980	1.00402	10,025.2
Chr 7102	5,003	3,053.78	3,053	0.263	6,947	0.55457	4,654.69
Chr 411	5,014	3,060.49	3,060	0.263	6,940	0.55401	4,649.15
Chr 10	5,002	3,053.16	3,053	0.263	6,947	0.55457	4,654.69
Chr 10	5,002	3,053.16	3,053	0.263	6,947	0.55457	4,654.69
Chr 4	4,273	2,608.19	2,608	0.257	7,392	0.59344	5,056.42
Chr 10	5,002	3,053.16	3,053	0.263	6,947	0.55457	4,654.69
Chr 412	5,297	3,233.23	3,233	0.266	6,767	0.54082	4,519.73
Chr 4	4,273	2,608.19	2,608	0.257	7,392	0.59344	5,056.42

In this example, due to limited space, we created only two generations of solutions. When solving real-life examples, analysts could use 100, 1,000, or 10,000 generations. Increasing the number of generations generally improves the final result, while simultaneously increasing computer time. It is also worthwhile noting that a larger population requires a larger number of generations.

b) Simulated annealing

We choose the temperature $T = 100$ for the initial temperature. We also decide to perform three iterations at the same temperature. The new temperature is calculated in the following way: $T_{new} = 0.9 \cdot T_{old}$. We make one modification to the solution in each iteration. In each iteration, we change the values of two genes. Genes, whose values will be changed, are chosen in a random manner. We generate two random numbers between 1 and 14 and choose these two genes. We also generate the initial solution in a random manner. Table 8.57 shows the initial solution that was randomly generated. The initial solution generated has the characteristics given in Table 8.58.

Iteration 1:

We generate two random numbers, namely $r_1 = 6$ and $r_2 = 13$. The new solution is obtained by changing values for genes 6 and 13 in the initial solution. The new solution is described in Table 8.59.

The characteristics of the new solution are described in Table 8.60.

Table 8.57 SA: Initial solution

0	0	0	1	1	1	1	0	1	1	1	0	1	0

Table 8.58 SA: Characteristics of the initial solution

d	y	x_1	t_1	T_1	x_2	t_2	T_2	T_t
1,978	1,207.35	1207	0.2503	302.125	8793	0.7706	6,775.9	7,078.06

Table 8.59 SA: The new solution obtained by changing values for genes 6 and 13 in the initial solution

0	0	0	1	1	0	1	0	1	1	1	0	0	0

Table 8.60 SA: Characteristics of the new solution obtained by changing values for genes 6 and 13 in the initial solution

d	y	x_1	t_1	T_1	x_2	t_2	T_2	T_t
1720	1049.87	1049	0.2502	262.436	8951	0.7967	7131.6	7394.05

In the next step we calculate the value of ΔE. This value equals:

$$\Delta E = 7,394.05 - 7,078.06 = 315.99$$

Since $\Delta E > 0$, we should calculate the probability of accepting the new solution. This probability equals:

$$p = e^{-(\Delta E/T)} = e^{-(315.99/100)} = 0.0424$$

Now, we generate the random number between 0 and 1:

$$r = 0.2219$$

Since $r > p$, we reject the new solution.

Iteration 2:

The generated random numbers are: $r_1 = 4$ and $r_2 = 12$. Table 8.61 shows the new solution.

The characteristics of the new solution obtained by changing values for genes 4 and 12 in the initial solution are presented in Table 8.62.

We calculate the value of ΔE. This value equals:

$$\Delta E = 8,305.7 - 7,078.06 = 1,373.24$$

Since $\Delta E > 0$, we should calculate the probability of accepting the new solution. This probability equals:

$$p = e^{-(\Delta E/T)} = e^{-(1,373.24/100)} = 1.08 \cdot 10^{-6}$$

Table 8.61 SA: The new solution obtained by changing values for genes 4 and 12 in the initial solution

0	0	0	0	1	1	1	0	1	1	1	1	1	0

Table 8.62 SA: Characteristics of the new solution

d	y	x_1	t_1	x_2	t_2	T_t
958	584.752	584	0.25	9,416	0.8821	8,451.75

We generate a random number between 0 and 1:

$r = 0.2219$

Since $r > p$, the new solution is rejected.

Iteration 3:

The new random numbers are: $r_1 = 14$ and $r_2 = 8$. New solution is described in Table 8.63, and the characteristics of that solution are given in Table 8.64.

$$\Delta E = 7{,}001.32 - 7{,}078.06 = -76.74$$

Since $\Delta E < 0$ we accept the new solution.

Iteration 4:

We change the temperature in this iteration. The new value for the temperature is:

$$T_{new} = 0.9 \cdot T_{old} = 0.9 \cdot 100 = 90$$

Random numbers are respectively: $r_1 = 9$ and $r_2 = 10$. Table 8.65 shows the new solution and Table 8.66 shows characteristics of that solution.

$$\Delta E = 7{,}058.76 - 7{,}001.32 = 57.44$$

$$p = e^{-(\Delta E/T)} = e^{-(57.44/90)} = 0.528$$

Table 8.63 SA: The new solution obtained by changing values for genes 14 and 8 in the initial solution

0	0	0	I	I	I	I	I	I	I	I	0	I	I

Table 8.64 SA: Characteristics of the new solution

d	y	x_1	t_1	x_2	t_2	T_t
2,043	1,247.02	1,247	0.2504	8,753	0.7642	7,001.32

Table 8.65 SA: The new solution obtained by changing values for genes 9 and 10 in the initial solution

0	0	0	I	I	I	I	I	0	0	I	0	I	I

$$r = 0.941$$

Since $r > p$, the new solution is not accepted.

Iteration 5:

The new random numbers are $r_1 = 6$ and $r_2 = 11$. This solution is described in Table 8.67, and the characteristics of that solution are given in Table 8.68.

$$\Delta E = 7,320.22 - 7,001.32 = 318.9$$

Since $\Delta E > 0$, we calculate the probability of accepting the new solution:

$$p = e^{-(\Delta E/T)} = e^{-(318.9/90)} = 0.0289$$

$$r = 0.623$$

Since $r > p$, the new solution is not accepted.

Iteration 6:

Random numbers are: $r_1 = 10$ and $r_2 = 3$. Table 8.69 describes the new solution and Table 8.70 shows characteristics of this solution.

$$\Delta E = 5,183.36 - 7,001.32 = -1,817.96$$

We accept the new solution.

Table 8.66 SA: Characteristics of the new solution

d	y	x_1	t_1	T_1	x_2	t_2	T_2	T_t
1,995	1,217.73	1,217	0.2503	304.641	8,783	0.769	6,754.1	7,058.76

Table 8.67 SA: The new solution obtained by changing values for genes 6 and 11 in the initial solution

0	0	0	I	I	0	I	I	I	I	0	0	I	I

Table 8.68 SA: Characteristics of the new solution

d	y	x_1	t_1	T_1	x_2	t_2	T_2	T_t
1,779	1,085.88	1,085	0.2502	271.47	8,915	0.7907	7,048.8	7,320.22

Table 8.69 SA: The new solution obtained by changing values for genes 10 and 3 in the initial solution

0	0	I	I	I	I	I	I	I	0	I	0	I	I

Table 8.70 SA: Characteristics of the new solution

d	y	x_1	t_1	x_2	t_2	T_t
4,075	2,487.33	2,487	0.2556	7,513	0.6053	5,183.36

Iteration 7:

The new temperature is: $T_{new} = 0.9 \cdot T_{old} = 0.9 \cdot 90 = 81$.

The new numbers are: $r_1 = 2$ and $r_2 = 5$. Now, the description of the solution is given in Table 8.71. Table 8.72 shows characteristics of that solution.

$$\Delta E = 3,968.17 - 5,183.36 = -1,215.19$$

Since $\Delta E < 0$, we accept the new solution.

Iteration 8:

The new random numbers are: $r_1 = 10$ and $r_2 = 5$. The new solution is described in Table 8.73, and the characteristics of that solution are given in Table 8.74.

$$\Delta E = 3,975.81 - 3,968.17 = 7.64$$

Table 8.71 SA: The new solution obtained by changing values for genes 2 and 5 in the initial solution

0	1	1	1	0	1	1	1	1	0	1	0	1	1

Table 8.72 SA: Characteristics of the new solution

d	y	x_1	t_1	x_2	t_2	T_t
7,659	4,674.97	4,674	0.3199	5,326	0.4643	3,968.17

Table 8.73 SA: The new solution obtained by changing values for genes 10 and 5 in the initial solution

0	1	1	1	1	1	1	1	1	1	1	0	1	1

Table 8.74 SA: Characteristics of the new solution

d	y	x_1	t_1	x_2	t_2	T_t
8,187	4,997.25	4,997	0.3413	5,003	0.4538	3,975.81

Because $\delta > 0$, we calculate the probability:

$$p = e^{-(\Delta E/T)} = e^{-(7.64/81)} = 0.91$$

$$r = 0.584$$

Since $r < p$, the new solution is accepted.

Iteration 9:

The random numbers in this iteration are $r_1 = 7$ and $r_2 = 8$. The solution obtained is shown in Table 8.75, and the solution's characteristics are presented in Table 8.76.

$$\Delta E = 3{,}967.06 - 3{,}975.81 = -8.75$$

Since $\Delta E < 0$, we accept the new solution.

Iteration 10:

The new temperature is: $T_{new} = 0.9 \cdot T_{old} = 0.9 \cdot 81 = 72.9$.

The new random numbers are $r_1 = 4$ and $r_2 = 1$. Table 8.77 shows this solution, the characteristics of which are given in Table 8.78.

Table 8.75 SA: The new solution obtained by changing values for genes 7 and 8 in the initial solution

0	1	1	1	1	1	0	0	1	1	1	0	1	1

Table 8.76 SA: Characteristics of the new solution

d	y	x_1	t_1	x_2	t_2	T_t
7,995	4,880.06	4,880	0.3331	5,120	0.4574	3,967.06

Table 8.77 SA: The new solution obtained by changing values for genes 7 and 8 in the initial solution

1	1	1	0	1	1	0	0	1	1	1	0	1	1

Table 8.78 SA: Characteristics of the new solution

d	y	x_1	t_1	x_2	t_2	T_t
15,163	9,255.33	9,255	1.3247	745	0.4167	12,570.8

$$\Delta E = 12,570.8 - 3,967.06 = 8,603.69$$

Since $\Delta E > 0$, we calculate the probability:

$$p = e^{-(\Delta E/T)} = e^{-(8,603.69/72.5)} = 5.55 \cdot 10^{-52}$$

$$r = 0.088$$

Since $r > p$, the new solution is not accepted.

Figure 8.29 shows the changes in the objective function values. The best solution obtained by simulated annealing, after ten iterations, has the total travel time equal to 3,967.06 hours.

c) Tabu search

We generated the initial solution in a random manner. That initial solution is described in Table 8.79 and the characteristics of that solution are given in Table 8.80. We decided that, when one gene is changed, it is forbidden to change that gene again during the subsequent four iterations.

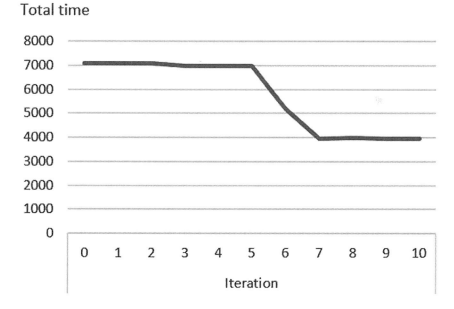

Total time

Figure 8.29 SA: Changes in the objective function values

Table 8.79 TS: The initial solution

0	0	I	0	I	0	0	I	I	0	I	I	I	I

Table 8.80 TS: Characteristics of the initial solution

d	y	x_1	t_1	x_2	t_2	T_t
2,671	1,630.35	1,630	0.251	8,370	0.707	6,328.93

Table 8.81 TS: 14 solutions in the initial solution's neighborhood

Sol. 1	1	0	1	0	1	0	0	1	1	0	1	1	1	1
Sol. 2	0	1	1	0	1	0	0	1	1	0	1	1	1	1
Sol. 3	0	0	0	0	1	0	0	1	1	0	1	1	1	1
Sol. 4	0	0	1	1	1	0	0	1	1	0	1	1	1	1
Sol. 5	0	0	1	0	0	0	0	1	1	0	1	1	1	1
Sol. 6	0	0	1	0	1	1	0	1	1	0	1	1	1	1
Sol. 7	0	0	1	0	1	0	1	1	1	0	1	1	1	1
Sol. 8	0	0	1	0	1	0	0	0	1	0	1	1	1	1
Sol. 9	0	0	1	0	1	0	0	1	0	0	1	1	1	1
Sol. 10	0	0	1	0	1	0	0	1	1	1	1	1	1	1
Sol. 11	0	0	1	0	1	0	0	1	1	0	0	1	1	1
Sol. 12	0	0	1	0	1	0	0	1	1	0	1	0	1	1
Sol. 13	0	0	1	0	1	0	0	1	1	0	1	1	0	1
Sol. 14	0	0	1	0	1	0	0	1	1	0	1	1	1	0

Iteration 1

The tabu search is characterized by an exploration of the solution space in the neighborhood of the current solution. Every step in the tabu search represents a move from the current solution to a new, potentially better solution in the neighborhood of the current solution.

Table 8.81 contains 14 possible modifications of the initial solution. Modification consists of a change of one gene. In the first solution, we modify the first digit. In the second solution, we change the value for the second digit, etc. The total number of possible changes is equal to 14 (Table 8.81). The characteristics of the 14 solutions are given in Table 8.82.

We sort the solutions in ascending order of total travel time values (Table 8.83). Solution 2 is the first solution in this order. Considering that tabu for this solution (see the third column in Table 8.83) equals zero, we can accept this solution.

Solution 2 is described in Table 8.84.

Iteration 2:

Now we consider the solution at the end of the first iteration (Table 8.85).

We consider possible changes to this solution in the same way as we did in iteration 1. It is possible to make changes in 14 genes. Table 8.86 shows

Table 8.82 TS: The characteristics of the 14 solutions in the initial solution's neighborhood

	d	y	x_1	t_1	x_2	t_2	T_t
Sol. 1	10,863	6,630.65	6,630	0.533	3,370	0.424	4,963.95
Sol. 2	6,767	4,130.50	4,130	0.293	5,870	0.487	4,066.98
Sol. 3	623	380.27	380	0.250	9,620	0.924	8,981.49
Sol. 4	3,695	2,255.39	2,255	0.254	7,745	0.630	5,449.39
Sol. 5	2,159	1,317.83	1,317	0.250	8,683	0.753	6,870.08
Sol. 6	2,927	1,786.61	1,786	0.251	8,214	0.686	6,085.54
Sol. 7	2,799	1,708.48	1,708	0.251	8,292	0.697	6,205.14
Sol. 8	2,607	1,591.28	1,591	0.251	8,409	0.713	6,392.44
Sol. 9	2,639	1,610.82	1,610	0.251	8,390	0.710	6,361.36
Sol. 10	2,687	1,640.11	1,640	0.251	8,360	0.706	6,312.82
Sol. 11	2,663	1,625.47	1,625	0.251	8,375	0.708	6,337.01
Sol. 12	2,667	1,627.91	1,627	0.251	8,373	0.708	6,333.78
Sol. 13	2,669	1,629.13	1,629	0.251	8,371	0.707	6,330.55
Sol. 14	2,670	1,629.74	1,629	0.251	8,371	0.707	6,330.55

Table 8.83 TS: 14 solutions in the initial solution's neighborhood sorted in ascending order of total travel time values

Solution	T_t	Tabu
Solution 2	**4,066.98**	**0**
Solution 1	4,963.95	0
Solution 4	5,449.39	0
Solution 6	6,085.54	0
Solution 7	6,205.14	0
Solution 10	6,312.82	0
Solution 13	6,330.55	0
Solution 14	6,330.55	0
Solution 12	6,333.78	0
Solution 11	6,337.01	0
Solution 9	6,361.36	0
Solution 8	6,392.44	0
Solution 5	6,870.08	0
Solution 3	8,981.49	0

Table 8.84 TS: Solution 2

Solution 2	0	1	1	0	1	0	0	1	1	0	1	1	1	1

Table 8.85 TS: Solution 2 (solution at the end of the first iteration)

0	1	1	0	1	0	0	1	1	0	1	1	1	1

Table 8.86 TS: 14 solutions in the solution's 2neighborhood sorted in ascending order of total travel time values

Solution	T_t	Tabu
Solution 4	**3,965.27**	**0**
Solution 6	4,024.84	0
Solution 7	4,044.55	0
Solution 10	4,063.96	0
Solution 13	4,067.29	0
Solution 14	4,067.29	0
Solution 12	4,067.60	0
Solution 11	4,068.52	0
Solution 9	4,073.18	0
Solution 8	4,079.22	0
Solution 5	4,184.22	0
Solution 3	4,799.07	0
Solution 2	6,328.93	4
Solution 1	11,937.80	0

Table 8.87 TS: Solution 4

Solution 4	0	1	1	1	1	0	0	1	1	0	1	1	1	1

Table 8.88 TS: The solution at the beginning of the third iteration (solution 4)

0	1	1	1	1	0	0	1	1	0	1	1	1	1

the new solutions arranged in ascending order of the objective function values (total travel time).

Solution 4 has the smallest value of the total travel time. Since this solution also has a tabu value equal to 0, this solution is accepted. This solution is described in Table 8.87.

Iteration 3:

We begin our search from solution 4 (Table 8.88).

The 14 solutions in the neighborhood of solution 4 were sorted in ascending order of total travel time values and are shown in Table 8.89.

Solution 10 has the smallest value of the total travel time. This solution also has a tabu value equal to zero. We accept solution 10, as described in Table 8.90.

Iteration 4:

We begin our search from Solution 10 (Table 8.91):

Table 8.89 TS: 14 solutions in the solution's 4neighborhood sorted in ascending order of total travel time values

Solution	T_t	Tabu
Solution 10	**3,965.13**	**0**
Solution 13	3,965.28	0
Solution 14	3,965.28	0
Solution 12	3,965.30	0
Solution 11	3,965.35	0
Solution 7	3,965.48	0
Solution 9	3,965.66	0
Solution 8	3,966.27	0
Solution 6	3,968.70	0
Solution 5	3,993.53	0
Solution 4	4,066.98	4
Solution 3	4,344.33	0
Solution 2	5,449.39	3
Solution 1	15,480.64	0

Table 8.90 TS: Solution 10

Solution 10	0	1	1	1	1	0	0	1	1	1	1	1	1	1

Table 8.91 TS: The solution at the beginning of the fourth iteration (solution 10)

0	1	1	1	1	0	0	1	1	1	1	1	1	1

The solutions in the neighborhood of Solution 10 sorted in ascending order of total travel time values are shown in Table 8.92.

Solution 13 has the smallest value of the total travel time. This solution also has a tabu value equal to zero. We accept Solution 13 (Table 8.93).

Iteration 5:

We begin our search from Solution 13 (Table 8.94):

The 14 solutions in the neighborhood of Solution 13 neighborhood are listed in ascending order of total travel time values, and are shown in Table 8.95.

It can be noted that Solution 13 has the best value for the total travel time. However, this solution has a tabu value equal to 4, which means that this solution cannot be changed in the next 4 iterations. Because of that, we consider the next solution, Solution 14. This solution has a tabu value of 0, which means that it can be accepted. We accept this solution, which is described in (Table 8.96).

Table 8.92 TS: 14 solutions in the solution's 10 neighborhood sorted in ascending order of total travel time values

Solution	T_t	Tabu
Solution 13	**3,965.14**	**0**
Solution 14	3,965.14	0
Solution 12	3,965.16	0
Solution 11	3,965.19	0
Solution 10	3,965.27	4
Solution 9	3,965.45	0
Solution 7	3,965.73	0
Solution 8	3,965.94	0
Solution 6	3,969.34	0
Solution 5	3,992.08	0
Solution 4	4,063.96	3
Solution 3	4,338.53	0
Solution 2	5,437.29	2
Solution 1	1,5545.4	0

Table 8.93 TS: Solution 13

Solution 13	0	1	1	1	1	0	0	1	1	1	1	1	0	1

Table 8.94 TS: The solution at the beginning of the fifth iteration (solution 13)

0	1	1	1	1	0	0	1	1	1	1	1	0	1

Table 8.95 TS: 14 solutions in the solution's 13 neighborhood sorted in ascending order of total travel time values

Solution	T_t	Tabu
Solution 13	3,965.13	4
Solution 14	**3,965.15**	**0**
Solution 12	3,965.18	0
Solution 11	3,965.20	0
Solution 10	3,965.28	3
Solution 9	3,965.47	0
Solution 7	3,965.70	0
Solution 8	3,965.97	0
Solution 6	3,969.27	0
Solution 5	3,992.24	0
Solution 4	4,064.26	2
Solution 3	4,339.11	0
Solution 2	5,439.70	1
Solution 1	15,538.94	0

Table 8.96 TS: Solution 14

Solution 14	0	1	1	1	1	0	0	1	1	1	1	1	0	0

Iteration 6:

We begin our search from Solution 14 (Table 8.97):

New solutions in the neighborhood of Solution 14 were sorted in ascending order of total travel time values, and are shown in Table 8.98.

We can note that Solutions 13 and 14 cannot be accepted since tabu values for these solutions are greater than 0. Because of that, we accept Solution 12 (Table 8.99).

Table 8.97 TS: The solution at the beginning of the sixth iteration (solution 14)

0	1	1	1	1	0	0	1	1	1	1	1	0	0

Table 8.98 TS: 14 solutions in the solution's 13 neighborhood sorted in ascending order of total travel time values

Solution	T_t	Tabu
Solution 13	3,965.14	3
Solution 14	3,965.14	4
Solution 12	**3,965.18**	**0**
Solution 11	3,965.22	0
Solution 10	3,965.30	2
Solution 9	3,965.50	0
Solution 7	3,965.67	0
Solution 8	3,966.00	0
Solution 6	3,969.21	0
Solution 5	3,992.40	0
Solution 4	4,064.56	1
Solution 3	4,339.69	0
Solution 2	5,439.70	0
Solution 1	15,532.45	0

Table 8.99 TS: Solution 12

Solution 12	0	1	1	1	1	0	0	1	1	1	1	0	0	0

Iteration 7:

We begin our search from Solution 12 (Table 8.100).

Table 8.101 shows the new solutions.

From Table 8.102 we see that the best solution with the zero tabu value is Solution 11 (Table 8.102).

Figure 8.30 shows how the objective function value changes, through 7 iterations, generated by the tabu search technique.

d) Bee colony optimization (BCO)

Let us suppose that we have the following BCO parameters:

- number of bees: $\#B = 8$
- number of forward/backward passes: $\#P = 3$
- number of changes in each forward pass $\#C = 1$

In this example we will make one complete iteration.

Table 8.100 TS: The solution at the beginning of the seventh iteration (solution 12)

0	1	1	1	1	0	0	1	1	1	1	0	0	0

Table 8.101 TS: 14 solutions in the solution's 14 neighborhood sorted in ascending order of total travel time values

Solution	T_t	Tabu
Solution 12	3,965.15	4
Solution 13	3,965.16	2
Solution 14	3,965.18	3
Solution 11	**3,965.25**	**0**
Solution 10	3,965.33	1
Solution 9	3,965.54	0
Solution 7	3,965.62	0
Solution 8	3,966.10	0
Solution 6	3,969.08	0
Solution 5	3,992.72	0
Solution 4	4,065.47	0
Solution 3	4,341.43	0
Solution 2	5,443.33	0
Solution 1	15,519.48	0

Table 8.102 TS: Solution 11

Solution 11	0	1	1	1	1	0	0	1	1	1	0	0	0	0

At the beginning, we generate the initial solution. The initial solution is described in Table 8.103. Characteristics of this solution are given in Table 8.104. This solution should be saved temporarily as the best-known solution.

First iteration

At the beginning of the iteration, we assign the initial solution to all bees (Table 8.105).

First forward pass

We generated the following eight random numbers (one random number for each bee): $r_1 = 12, r_2 = 11, r_3 = 9, r_4 = 3, r_5 = 13, r_6 = 3, r_7 = 14$ and $r_8 = 4$.

The random numbers generated indicate positions of the genes that should be changed. For instance, the random number $r_1 = 12$ indicates that

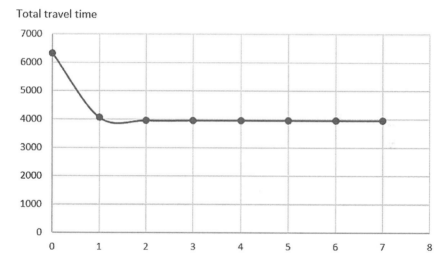

Total travel time

Figure 8.30 TS: The objective function value changes, through 7 iterations, generated by the tabu search technique

Table 8.103 BCO: The initial solution

0	0	0	1	1	0	0	0	0	0	0	1	0	1

Table 8.104 BCO: Characteristics of the initial solution

d	y	x_1	t_1	x_2	t_2	T_t
1,541	940.609	940	0.25011	9,060	0.81559	7,624.39

Table 8.105 BCO: All bees have the same initial solution

Bee 1	0	0	0	1	1	0	0	0	0	0	0	1	0	1
Bee 2	0	0	0	1	1	0	0	0	0	0	0	1	0	1
Bee 3	0	0	0	1	1	0	0	0	0	0	0	1	0	1
Bee 4	0	0	0	1	1	0	0	0	0	0	0	1	0	1
Bee 5	0	0	0	1	1	0	0	0	0	0	0	1	0	1
Bee 6	0	0	0	1	1	0	0	0	0	0	0	1	0	1
Bee 7	0	0	0	1	1	0	0	0	0	0	0	1	0	1
Bee 8	0	0	0	1	1	0	0	0	0	0	0	1	0	1

Table 8.106 BCO: Bees' solutions after the first forward pass

Bee 1	0	0	0	1	1	0	0	0	0	0	0	**0**	0	1
Bee 2	0	0	0	1	1	0	0	0	0	0	**1**	1	0	1
Bee 3	0	0	0	1	1	0	0	0	**1**	0	0	1	0	1
Bee 4	0	0	**1**	1	1	0	0	0	0	0	0	1	0	1
Bee 5	0	0	0	1	1	0	0	0	0	0	0	1	**1**	1
Bee 6	0	0	**1**	1	1	0	0	0	0	0	0	1	0	1
Bee 7	0	0	0	1	1	0	0	0	0	0	0	1	0	**0**
Bee 8	0	0	0	**0**	1	0	0	0	0	0	0	1	0	1

Table 8.107 BCO: Characteristics of bees' solutions after first forward pass

	d	y	x_1	t_1	x_2	t_2	T_t
Bee 1	1,537	938.168	938	0.25011	9,062	0.81595	7,628.71
Bee 2	1,549	945.492	945	0.25012	9,055	0.81471	7,613.59
Bee 3	1,573	960.142	960	0.25012	9,040	0.81208	7,581.35
Bee 4	3,589	2,190.69	2,190	0.25337	7,810	0.63695	5,529.47
Bee 5	1,543	941.83	941	0.25011	9,059	0.81542	7,622.23
Bee 6	3,589	2,190.69	2,190	0.25337	7,810	0.63695	5,529.47
Bee 7	1,540	939.999	939	0.25011	9,061	0.81577	7,626.55
Bee 8	517	315.571	315	0.25	9,685	0.9376	9,159.37

the twelfth gene in the Bee 1 solution should be changed. We replaced 1 in this position by 0. The bees' solutions after the first forward pass are shown in Table 8.106. The new gene values are indicated in bold letters.

The characteristics of these solutions are given in Table 8.107.

Now, we check if any of the generated solutions are better than the best-known solution. We calculate:

$$\min\{7,628.71; 7,613.59; 7,581.35; 5,529.47;$$

$$7,622.23; 5,529.47; 7,626.55; 9,159.37\} = 5,529.47$$

Table 8.108 BCO: The new best-known solution

0	0	1	1	1	0	0	0	0	0	0	1	0	1

Since:

$$5,529.47 < 7,624.39$$

we save the solution of Bee 4 (or Bee 6), as the best-known solution, and the best solution found is described in Table 8.108.

First backward pass

We normalize objective function values of the solutions generated by artificial bees by using the following expressions:

$$O_b = \frac{T_{max} - T_b}{T_{max} - T_{min}}$$

where:

O_b – normalized bee's b objective function value

T_b – bee's b objective function value,

T_{min} – minimal objective function value, considering all the bees,

T_{max} – maximal objective function value, considering all the bees.

We calculate the minimal and maximal objective function values:

$$\min\{7,628.71; 7,613.59; 7,581.35; 5,529.47;$$

$$7,622.23; 5,529.47; 7,626.55; 9,159.37\} = 5529.47$$

$$\max\{7,628.71; 7,613.59; 7,581.35; 5,529.47;$$

$$7,622.23; 5,529.47; 7,626.55; 9,159.37\} = 9159.37$$

The normalized values of the objective function are given in the second column of Table 8.109.

Table 8.109 BCO: The loyalty decision in first backward pass

	Normalized value	p	r	Bee is loyal
Bee 1	0.422	0.561	0.310	TRUE
Bee 2	0.426	0.563	0.788	FALSE
Bee 3	0.435	0.568	0.600	FALSE
Bee 4	1.000	1.000	0.842	TRUE
Bee 5	0.423	0.562	0.919	FALSE
Bee 6	1.000	1.000	0.391	TRUE
Bee 7	0.422	0.561	0.917	FALSE
Bee 8	0.000	0.368	0.554	FALSE

Each bee has to make a decision as to whether it will stay loyal to its previous solution or not. We will use the following expression to calculate this probability: $p = e^{-(O_{max} - O_{min})}$. The probability that Bee 1 will stay loyal to its solution equals:

$$p = e^{-(O_{max} - O_1)} = e^{-(1-0.422)} = e^{-0.578} = 0.561$$

Taking a random number between 0 and 1, we make a decision. Let us suppose that the random number equals $r = 0.310$. We can notice that:

$$p = 0.561 > 0.310 = r$$

As a consequence, Bee 1 will be loyal to its previously generated solution. The following is satisfied in the case of Bee 2:

$$p = e^{-(O_{max} - O_2)} = e^{-(1-0.426)} = e^{-0.574} = 0.563$$

$$r = 0.788$$

$$p = 0.563 < 0.788 = r$$

As a consequence, Bee 2 will become an uncommitted follower.

The bees' loyalty decisions in the first backward passes are shown in the last column of Table 8.109.

The sum of the normalized objective function values of the solutions discovered by all loyal bees (Bee 1, Bee 4, and Bee 6) equals:

$$\sum_{k=1}^{3} O_k = 0.422 + 1 + 1 = 2.422$$

Each uncommitted bee chooses, with a certain probability, the recruiter that it will follow. The probability that b's partial solution would be chosen by any uncommitted bee is equal to:

$$p_b = \frac{O_b}{\sum_{k=1}^{R} O_k} \qquad b = 1, 2, \cdots, R$$

where O_k represents the normalized value for the objective function of the k-th advertised partial solution, and R denotes the number of recruiters. In our case, $R = 3$. The calculated probabilities are shown in Table 8.110.

By using probabilities p_b and a random number generator, each uncommitted follower bee joins one recruiter.

Table 8.110 BCO: Probabilities that specific recruiters will be followed

	Normalized value	Probability that the bee will be followed	Cumulative probability
Bee 1	0.422	0.174	0.174
Bee 4	1	0.413	0.587
Bee 6	1	0.413	1

Table 8.111 BCO: Joining uncommitted followers to recruiters in the first backward pass

Bee	Generated random number		Uncommitted followers join recruiters
Bee 2	0.319	$0.174 < 0.319 < 0.587$	Bee 2 will follow Bee 4
Bee 3	0.330	$0.174 < 0.330 < 0.587$	Bee 3 will follow Bee 4
Bee 5	0.792	$0.587 < 0.792 < 1$	Bee 5 will follow Bee 6
Bee 7	0.657	$0.587 < 0.657 < 1$	Bee 7 will follow Bee 6
Bee 8	0.202	$0.174 < 0.202 < 0.587$	Bee 8 will follow Bee 4

Table 8.112 BCO: Bees' solution at the end of the first backward pass

Bee 1	0	0	0	1	1	0	0	0	0	0	0	0	0	1
Bee 2	0	0	1	1	1	0	0	0	0	0	0	1	0	1
Bee 3	0	0	1	1	1	0	0	0	0	0	0	1	0	1
Bee 4	0	0	1	1	1	0	0	0	0	0	0	1	0	1
Bee 5	0	0	1	1	1	0	0	0	0	0	0	1	0	1
Bee 6	0	0	1	1	1	0	0	0	0	0	0	1	0	1
Bee 7	0	0	1	1	1	0	0	0	0	0	0	1	0	1
Bee 8	0	0	1	1	1	0	0	0	0	0	0	1	0	1

In order to simulate the decision of Bee 2, we generate a random number $r = 0.319$. Since $0.174 < r < 0.587$ we conclude that Bee 2 will follow Bee 4. The joining of uncommitted followers to specific recruiters is listed in Table 8.111.

Bees' solutions at the end of the first backward pass are given in Table 8.112.

We begin the second forward pass, and we generate new random numbers: $r_1 = 14, r_2 = 6, r_3 = 3, r_4 = 2, r_5 = 10, r_6 = 12, r_7 = 7$ and $r_8 = 9$. The random numbers generated indicate positions of the genes that should be changed. The modified bees' solutions are given in Table 8.113. Characteristics of these solutions are given in Table 8.114.

We see that:

$$\min_{i=1,\dots,8} \{T_t(\text{bee } i)\} = 3,967.35$$

Table 8.113 BCO: Bees' solutions after the second forward pass

Bee 1	0	0	0	1	1	0	0	0	0	0	0	0	0	0
Bee 2	0	0	1	1	1	1	0	0	0	0	0	1	0	1
Bee 3	0	0	0	1	1	0	0	0	0	0	0	1	0	1
Bee 4	0	1	1	1	1	0	0	0	0	0	0	1	0	1
Bee 5	0	0	1	1	1	0	0	0	0	1	0	1	0	1
Bee 6	0	0	1	1	1	0	0	0	0	0	0	0	0	1
Bee 7	0	0	1	1	1	0	1	0	0	0	0	1	0	1
Bee 8	0	0	1	1	1	0	0	0	1	0	0	1	0	1

Table 8.114 BCO: Characteristics of bees' solutions after second forward pass

	d	y	x_1	t_1	x_2	t_2	T_t
Bee 1	1,536	937.557	937	0.25011	9,063	0.81612	7,630.87
Bee 2	3,845	2,346.95	2,346	0.25444	7,654	0.61987	5,341.41
Bee 3	1,541	940.609	940	0.25011	9,060	0.81559	7,624.39
Bee 4	7,685	4,690.84	4,690	0.32087	5,310	0.46374	3,967.35
Bee 5	3,605	2,200.45	2,200	0.25343	7,800	0.63583	5,516.99
Bee 6	3,585	2,188.24	2,188	0.25336	7,812	0.63718	5,531.97
Bee 7	3,717	2,268.82	2,268	0.25388	7,732	0.62828	5,433.67
Bee 8	3,621	2,210.22	2,210	0.25349	7,790	0.6347	5,504.56

Table 8.115 BCO: The best-known solution after two forward passes

0	1	1	1	1	0	0	0	0	0	0	1	0	1

Since $3,967.35 < 5,529.47$, we save the solution of Bee 4 as the best-known solution, in Table 8.115.

Second backward pass

We found:

$$\min_{i=1,\dots,8}\{T_t(\text{bee}\,i)\} = 3,967.35$$

$$\max_{i=1,\dots,8}\{T_t(\text{bee}\,i)\} = 7,630.87$$

The bee's loyalty decision in the second backward pass is shown in the last column of Table 8.116.

We calculate:

$$\sum_{l\in L} O_l = 0.368 + 0.687 + 0.369 + 1 + 0.655 + 0.652 + 0.670 + 0.657 = 2.777$$

Table 8.116 BCO: The loyalty decision in second backward pass

	Normalized value	p	r	Bee is loyal
Bee 1	0	0.368	0.072	TRUE
Bee 2	0.625	0.687	0.443	TRUE
Bee 3	0.002	0.369	0.036	TRUE
Bee 4	1	1	0.382	TRUE
Bee 5	0.577	0.655	0.118	TRUE
Bee 6	0.573	0.652	0.135	TRUE
Bee 7	0.600	0.670	0.730	FALSE
Bee 8	0.580	0.657	0.893	FALSE

Table 8.117 BCO: The probabilities that specific recruiters will be followed in the second backward pass

	Normalized value	Probability to follow	Cumulative probability
Bee 1	0	0	0
Bee 2	0.625	0.225	0.225
Bee 3	0.002	0.001	0.226
Bee 4	1	0.360	0.586
Bee 5	0.577	0.208	0.794
Bee 6	0.573	0.206	1.000

Table 8.118 BCO: Joining uncommitted followers to recruiters in the second backward pass

	r	Simulation	Follows
Bee 7	0.620	$0.586 < r < 0.794$	Bee 5
Bee 8	0.412	$0.226 < r < 0.586$	Bee 4

The probabilities that specific recruiters will be followed in the second backward pass are shown in Table 8.117.

Joining uncommitted followers to recruiters is shown in Table 8.118.

Bees' solution at the end of the second backward pass is given in Table 8.119.

Third forward pass

The generated random numbers are: $r_1 = 12, r_2 = 10, r_3 = 5, r_4 = 7, r_5 = 9,$ $r_6 = 4, r_7 = 1$ and $r_8 = 12$. Modified bees' solutions are given in Table 8.120. Characteristics of these solutions are given in Table 8.121.

Table 8.119 BCO: Bees' solution at the end of the second backward pass

Bee 1	0	0	0	1	1	0	0	0	0	0	0	0	0	0
Bee 2	0	0	1	1	1	1	0	0	0	0	0	1	0	1
Bee 3	0	0	0	1	1	0	0	0	0	0	0	1	0	1
Bee 4	0	1	1	1	1	0	0	0	0	0	0	1	0	1
Bee 5	0	0	1	1	1	0	0	0	0	1	0	1	0	1
Bee 6	0	0	1	1	1	0	0	0	0	0	0	0	0	1
Bee 7	0	0	1	1	1	0	0	0	0	1	0	1	0	1
Bee 8	0	1	1	1	1	0	0	0	0	0	0	1	0	1

Table 8.120 BCO: Bees' solutions after the third forward pass

Bee 1	0	0	0	1	1	0	0	0	0	0	0	1	0	0
Bee 2	0	0	1	1	1	1	0	0	0	1	0	1	0	1
Bee 3	0	0	0	1	0	0	0	0	0	0	0	1	0	1
Bee 4	0	1	1	1	1	0	1	0	0	0	0	1	0	1
Bee 5	0	0	1	1	1	0	0	0	1	1	0	1	0	1
Bee 6	0	0	1	0	1	0	0	0	0	0	0	0	0	1
Bee 7	1	0	1	1	1	0	0	0	0	1	0	1	0	1
Bee 8	0	1	1	1	1	0	0	0	0	0	0	0	0	1

Table 8.121 BCO: Characteristics of bees' solutions after third forward pass

	d	y	x_1	t_1	x_2	t_2	T_t
Bee 1	1,540	939.999	939	0.25011	9,061	0.81577	7,626.55
Bee 2	3,861	2,356.71	2,356	0.25451	7,644	0.61881	5,329.83
Bee 3	1,029	628.09	628	0.25002	9,372	0.87345	8,342.98
Bee 4	7,813	4,768.97	4,768	0.32571	5,232	0.46103	3,965.09
Bee 5	3,637	2,219.98	2,219	0.25355	7,781	0.6337	5,493.43
Bee 6	2,561	1,563.21	1,563	0.25087	8,437	0.71667	6438.7
Bee 7	11,797	7,200.76	7,200	0.64366	2,800	0.42031	5,811.21
Bee 8	7,681	4,688.4	4,688	0.32075	5,312	0.46381	3,967.44

Table 8.122 BCO: New best-known solution found in third iteration

0	1	1	1	1	0	1	0	0	0	0	1	0	1

We see that:

$$\min_{i=1,\ldots,8}\left\{T_t\left(\mathrm{bee}\,i\right)\right\} = 3,965.095$$

Since $3,965.095 < 3,967.35$, we save the solution of Bee 4 as the best-known solution, which is given in Table 8.122.

Table 8.123 BCO: The loyalty decision in third backward pass

	Normalized value	p	r	Bee is loyal
Bee 1	0.16365	0.433	0.629	FALSE
Bee 2	0.688	0.732	0.590	TRUE
Bee 3	0	0.368	0.873	FALSE
Bee 4	1	1	0.677	TRUE
Bee 5	0.651	0.705	0.502	TRUE
Bee 6	0.435	0.568	0.246	TRUE
Bee 7	0.578	0.656	0.502	TRUE
Bee 8	0.999	0.999	0.518	TRUE

Third backward pass

We found:

$$\min_{i=1,\ldots,8}\left\{T_t\left(\text{bee}\,i\right)\right\} = 3,965.095$$

$$\max_{i=1,\ldots,8}\left\{T_t\left(\text{bee}\,i\right)\right\} = 8,342.982$$

The bee's loyalty decision in the third backward pass is shown in the last column of Table 8.123.
We calculate:

$$\sum_{l \in L} O_l = 4.352$$

The probabilities that specific recruiters will be followed in the third backward pass are shown in Table 8.124.
Joining uncommitted followers to recruiters is shown in Table 8.125.
Bees' solutions at the end of the third backward pass are given in Table 8.126.

Table 8.124 BCO: The probabilities that specific recruiters will be followed in the third backward pass

	Normalized value	Probability to follow	Cumulative probability
Bee 2	0.688	0.158	0.158
Bee 4	1.000	0.230	0.388
Bee 5	0.651	0.150	0.538
Bee 6	0.435	0.100	0.637
Bee 7	0.578	0.133	0.770
Bee 8	0.999	0.230	1.000

Table 8.125 BCO: Joining uncommitted followers to recruiters in the third backward pass

	r	Simulation	Follows
Bee 1	0.238	$0.158 < r < 0.388$	Bee 4
Bee 3	0.999	$0.770 < r < 1$	Bee 8

Table 8.126 BCO: Bees' solution at the end of the third backward pass

Bee 1	0	1	1	1	1	0	1	0	0	0	0	1	0	1
Bee 2	0	0	1	1	1	1	0	0	0	1	0	1	0	1
Bee 3	0	1	1	1	1	0	0	0	0	0	0	0	0	1
Bee 4	0	1	1	1	1	0	1	0	0	0	0	1	0	1
Bee 5	0	0	1	1	1	0	0	0	1	1	0	1	0	1
Bee 6	0	0	1	0	1	0	0	0	0	0	0	0	0	1
Bee 7	1	0	1	1	1	0	0	0	0	1	0	1	0	1
Bee 8	0	1	1	1	1	0	0	0	0	0	0	0	0	1

Table 8.127 BCO: Bees' solutions at the beginning of second iteration

Bee 1	0	1	1	1	1	0	1	0	0	0	0	1	0	1
Bee 2	0	1	1	1	1	0	1	0	0	0	0	1	0	1
Bee 3	0	1	1	1	1	0	1	0	0	0	0	1	0	1
Bee 4	0	1	1	1	1	0	1	0	0	0	0	1	0	1
Bee 5	0	1	1	1	1	0	1	0	0	0	0	1	0	1
Bee 6	0	1	1	1	1	0	1	0	0	0	0	1	0	1
Bee 7	0	1	1	1	1	0	1	0	0	0	0	1	0	1
Bee 8	0	1	1	1	1	0	1	0	0	0	0	1	0	1

Iteration 2.

To each bee, we set up the best-found solution. Because of that, all bees have the same solution as described in Table 8.127.

Bee 1	0	1	1	1	1	0	1	0	0	0	0	1	0	1
Bee 2	0	1	1	1	1	0	1	0	0	0	0	1	0	1
Bee 3	0	1	1	1	1	0	1	0	0	0	0	1	0	1
Bee 4	0	1	1	1	1	0	1	0	0	0	0	1	0	1
Bee 5	0	1	1	1	1	0	1	0	0	0	0	1	0	1
Bee 6	0	1	1	1	1	0	1	0	0	0	0	1	0	1
Bee 7	0	1	1	1	1	0	1	0	0	0	0	1	0	1
Bee 8	0	1	1	1	1	0	1	0	0	0	0	1	0	1

e) Ant colony optimization (ACO)

In this example, we use the ant system (AS) algorithm. The AS is the first version of the ACO algorithms, with all other ACO algorithms being based on the AS algorithm. In this example, we use the following parameters:

- number of ants is 5,
- $\alpha = 1$,
- $\beta = 1$,
- $\rho = 0.5$,
- initial pheromone quantity = 10^{-4}.

In order to solve the problem in question, we created a graph (Figure 8.31). Each ant should make a trip from stage 0 to stage 14 (see Figure 8.31). Stages from 1 to 14 represent digits in a binary string. Each stage from 1 to 14 contains two nodes: 0 and 1. At the beginning of each iteration, all ants are located at an artificial node 0 in stage 0. An ant located in stage i goes with a certain probability to node 0 or to node 1 at step $i+1$. To calculate these probabilities, we have first to calculate T_t for both decisions. For that purpose, we have to determine the binary string completely, and then to calculate the number of vehicles. Let us use $d(j)$ to denote the node that the ant took in step j. If the ant is in stage i and should then go to stage $i+1$, then the string for the decision to go to node 0 in step $i+1$ is:

$$d(0) = 0, d(1), d(2), \ldots, d(i), d(i+1) = 0, d(i+2) = 0, \ldots, d(14) = 0,$$

and the string for the decision to go to node 1 in step $i+1$ is:

$$d(0) = 0, d(1), d(2), \ldots, d(i), d(i+1) = 1, d(i+2) = 0, \ldots, d(14) = 0.$$

In this example, we take into consideration our previous decisions (from stage 0 to stage i), and the decision in stage i, whereas, for all other stages, we suppose that the ant takes value 0.

After one ant finishes the trip from stage 0 to stage 14, the next ant starts. When all ants finish their trips, we calculate the characteristics of the ants' solutions and check whether a better solution than the previously best-known solution was found. If the answer is yes, we save that solution. After that, we update the pheromone quantity and we start the next iteration. At the beginning of each iteration, ants are based at node 0 in stage 0 and all of them have empty solutions.

In this example, we started our search with the initial pheromone values that are shown in Table 8.128.

Iteration I

At the beginning of iteration 1, all ants have empty solutions and they are located at node 0 in stage 0. Let us explain how the first ant makes a decision

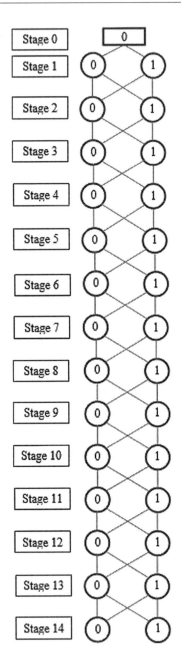

Figure 8.31 ACO: Each ant should make a trip from stage 0 to stage 14

Table 8.128 ACO: The initial pheromone values on the links

Stage i		Next stage (Stage i+1)	
		0	1
Stage 0	0	10^{-4}	10^{-4}
Stage 1	0	10^{-4}	10^{-4}
	1	10^{-4}	10^{-4}
Stage 2	0	10^{-4}	10^{-4}
	1	10^{-4}	10^{-4}
Stage 3	0	10^{-4}	10^{-4}
	1	10^{-4}	10^{-4}
Stage 4	0	10^{-4}	10^{-4}
	1	10^{-4}	10^{-4}
Stage 5	0	10^{-4}	10^{-4}
	1	10^{-4}	10^{-4}
Stage 6	0	10^{-4}	10^{-4}
	1	10^{-4}	10^{-4}
Stage 7	0	10^{-4}	10^{-4}
	1	10^{-4}	10^{-4}
Stage 8	0	10^{-4}	10^{-4}
	1	10^{-4}	10^{-4}
Stage 9	0	10^{-4}	10^{-4}
	1	10^{-4}	10^{-4}
Stage 10	0	10^{-4}	10^{-4}
	1	10^{-4}	10^{-4}
Stage 11	0	10^{-4}	10^{-4}
	1	10^{-4}	10^{-4}
Stage 12	0	10^{-4}	10^{-4}
	1	10^{-4}	10^{-4}
Stage 13	0	10^{-4}	10^{-4}
	1	10^{-4}	10^{-4}

as to where it will go in the next step. Ant 1 can go to node 0 or to node 1 in stage 1. If it goes to 0, the binary string is: 0,0,0,0,0,0,0,0,0,0,0,0,0,0. For this string, $T_t = 10,087.47$. If the ant goes to 1, the binary string is: 1,0,0,0,0,0,0,0,0,0,0,0,0,0. For this string, $T_t = 3,976.12$. The probability that the first ant will choose 0 is:

$$p_0 = \frac{\left(\frac{1}{10,087.47}\right)^{\alpha} \cdot \left(10^{-4}\right)^{\beta}}{\left(\frac{1}{10,087.47}\right)^{\alpha} \cdot \left(10^{-4}\right)^{\beta} + \left(\frac{1}{3,976.12}\right)^{\alpha} \cdot \left(10^{-4}\right)^{\beta}} = 0.283$$

The probability that the first ant will choose 1 is:

$$p_1 = \frac{\left(\dfrac{1}{3976.12}\right)^{\alpha} \cdot \left(10^{-4}\right)^{\beta}}{\left(\dfrac{1}{10,087.47}\right)^{\alpha} \cdot \left(10^{-4}\right)^{\beta} + \left(\dfrac{1}{3,976.12}\right)^{\alpha} \cdot \left(10^{-4}\right)^{\beta}} = 1 - 0.283 = 0.717$$

We generate random number r. In the case when $r < p_0$, the ant goes to node 0, otherwise it will go to node 1. In our case, we generate random number $r = 0.331$. Since $r > p_0$, the ant will go to node 1.

Table 8.129 shows the movements of the first three ants. The column $d(i)$ shows the node (decision) where the ant goes in that step. The quality of the solutions generated is shown in Table 8.130. Ants' solutions generated after the first iteration are graphically illustrated in Figure 8.32.

When all the ants have created their solutions, we have to calculate the new values of the pheromone. Let us consider the link between nodes 0 in stage 0 and node 0 in stage 1. Ants 2 and 3 passed through this link. Their values of total travel time are respectively $T_{t,2} = 4,148.90$ distance units and $T_{t,3} = 4,327.05$ distance units. The pheromone value equals:

$$\tau_{0(0),0(1),new} = \rho \cdot \tau_{0(0),0(1),old} + \Delta\tau_{0(0),0(1)}$$

$$\Delta\tau_{0(0),0(1)} = \frac{1}{T_{t,2}} + \frac{1}{T_{t,3}} = \frac{1}{4148.9} + \frac{1}{4327.05}$$

$$= 2.41 \cdot 10^{-4} + 2.31 \cdot 10^{-4} = 4.72 \cdot 10^{-4}$$

$$\tau_{0(0),0(1),new} = 0.5 \cdot 10^{-4} + 4.72 \cdot 10^{-4} = 5.22 \cdot 10^{-4}$$

In a similar way, we calculated the other pheromone values. These values are given in Table 8.131.

Tables 8.132–8.134 and Figure 8.33 show the results obtained within the second iteration.

PROBLEMS

8.1. A vehicle delivers goods to small shops. The total demand of all the shops is less than the vehicle capacity. The vehicle starts the trip from the warehouse. After the last delivery, the vehicle returns to the warehouse. The shortest distances between the nodes (node 0 is the warehouse, the shops are denoted as nodes 1, 2, ..., 5) are given in Table 8.135.

Table 8.129 ACO: Ants construct their solutions in the first iteration

	Ant 1			Ant 2			Ant 3			Ant 4			Ant 5		
	p_0	r	$d(i)$	p_0	r	$d(i)$	p_0	r	$d(i)$	p_0	r	$d(i)$	p_0	r	$d(i)$
Stage1	0.283	0.331	—	0.283	0.143	0	0.283	0.031	0	0.283	0.804	—	0.283	0.556	—
Stage2	0.617	0.102	0	0.339	0.371	—	0.339	0.940	—	0.617	0.024	0	0.617	0.573	0
Stage3	0.535	0.001	0	0.449	0.584	—	0.449	0.035	0	0.535	0.880	—	0.535	0.730	—
Stage4	0.511	0.426	0	0.487	0.467	0	0.471	0.680	—	0.537	0.542	—	0.537	0.279	0
Stage5	0.504	0.712	—	0.492	0.108	0	0.488	0.868	—	0.523	0.725	—	0.517	0.995	—
Stage6	0.503	0.463	0	0.496	0.203	0	0.495	0.199	0	0.512	0.808	—	0.510	0.547	—
Stage7	0.501	0.397	0	0.498	0.737	—	0.497	0.657	—	0.506	0.765	—	0.505	0.586	—
Stage8	0.501	0.165	0	0.499	0.792	—	0.499	0.305	0	0.503	0.593	—	0.503	0.754	—
Stage9	0.500	0.400	0	0.500	0.759	—	0.499	0.570	—	0.502	0.411	0	0.501	0.752	—
Stage10	0.500	0.656	—	0.500	0.844	—	0.500	0.342	0	0.501	0.008	0	0.501	0.372	0
Stage11	0.500	0.110	0	0.500	0.258	0	0.500	0.460	0	0.500	0.466	0	0.500	0.224	0
Stage12	0.500	0.719	—	0.500	0.906	—	0.500	0.100	0	0.500	0.797	—	0.500	0.464	0
Stage13	0.500	0.001	0	0.500	0.639	—	0.500	0.236	0	0.500	0.128	0	0.500	0.920	—
Stage14	0.500	0.664	—	0.500	0.011	0	0.500	0.327	0	0.500	0.837	—	0.500	0.006	0

Table 8.130 ACO: The quality of the generated solutions in the first iteration

	d	y	x_1	t_1	x_2	t_2	T_t
Ant 1	8,725	5,325.64	5,325	0.368	4,675	0.445	4,038.56
Ant 2	6,390	3,900.38	3,900	0.284	6,100	0.499	4,148.90
Ant 3	5,792	3,535.37	3,535	0.273	6,465	0.520	4,327.05
Ant 4	12,229	7,464.44	7,464	0.705	2,536	0.419	6,322.39
Ant 5	11,234	6,857.11	6,857	0.574	3,143	0.422	5,262.55

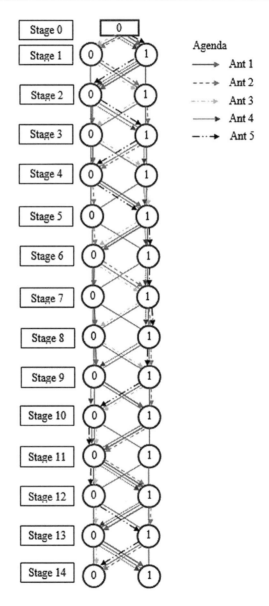

Figure 8.32 ACO: Paths of the ants generated at the end of the first iteration

Table 8.131 ACO: The new pheromone values after the first iteration

Stage i		Next stage (Stage i + 1)	
		0	1
Stage 0	0	$5.22 \cdot 10^{-4}$	$6.46 \cdot 10^{-4}$
Stage 1	0	$5.00 \cdot 10^{-5}$	$5.22 \cdot 10^{-4}$
	1	$6.46 \cdot 10^{-4}$	$5.00 \cdot 10^{-5}$
Stage 2	0	$2.98 \cdot 10^{-4}$	$3.98 \cdot 10^{-4}$
	1	$2.81 \cdot 10^{-4}$	$2.91 \cdot 10^{-4}$
Stage 3	0	$2.98 \cdot 10^{-4}$	$2.81 \cdot 10^{-4}$
	1	$4.81 \cdot 10^{-4}$	$2.08 \cdot 10^{-4}$
Stage 4	0	$2.91 \cdot 10^{-4}$	$4.88 \cdot 10^{-4}$
	1	$5.00 \cdot 10^{-5}$	$4.39 \cdot 10^{-4}$
Stage 5	0	$2.91 \cdot 10^{-4}$	$5.00 \cdot 10^{-5}$
	1	$5.29 \cdot 10^{-4}$	$3.98 \cdot 10^{-4}$
Stage 6	0	$2.98 \cdot 10^{-4}$	$5.22 \cdot 10^{-4}$
	1	$5.00 \cdot 10^{-5}$	$3.98 \cdot 10^{-4}$
Stage 7	0	$2.98 \cdot 10^{-4}$	$5.00 \cdot 10^{-5}$
	1	$2.81 \cdot 10^{-4}$	$6.39 \cdot 10^{-4}$
Stage 8	0	$2.98 \cdot 10^{-4}$	$2.81 \cdot 10^{-4}$
	1	$2.08 \cdot 10^{-4}$	$4.81 \cdot 10^{-4}$
Stage 9	0	$2.08 \cdot 10^{-4}$	$2.98 \cdot 10^{-4}$
	1	$4.71 \cdot 10^{-4}$	$2.91 \cdot 10^{-4}$
Stage 10	0	$6.29 \cdot 10^{-4}$	$5.00 \cdot 10^{-5}$
	1	$5.39 \cdot 10^{-4}$	$5.00 \cdot 10^{-5}$
Stage 11	0	$4.71 \cdot 10^{-4}$	$6.97 \cdot 10^{-4}$
	1	$5.00 \cdot 10^{-5}$	$5.00 \cdot 10^{-5}$
Stage 12	0	$2.81 \cdot 10^{-4}$	$2.40 \cdot 10^{-4}$
	1	$4.56 \cdot 10^{-4}$	$2.91 \cdot 10^{-4}$
Stage 13	0	$5.00 \cdot 10^{-5}$	$5.00 \cdot 10^{-5}$
	1	$5.00 \cdot 10^{-5}$	$5.00 \cdot 10^{-5}$

Determine the vehicle's route by the greedy nearest-neighbor heuristic algorithm for the traveling salesman problem.

8.2. If chromosome 1 has the following genes: 0, 0, 1, 1, 0, 1, 0, 1, 1, 0, and chromosome 2 has the following genes: 1, 0, 1, 0, 0, 1, 1, 0, 1, 1, then:

a) determine the new chromosomes obtained by a single-point cross-over operation, where the point is the fourth gene.

b) determine the new chromosomes obtained by a two-point cross-over operation, where the points are the third and the seventh gene.

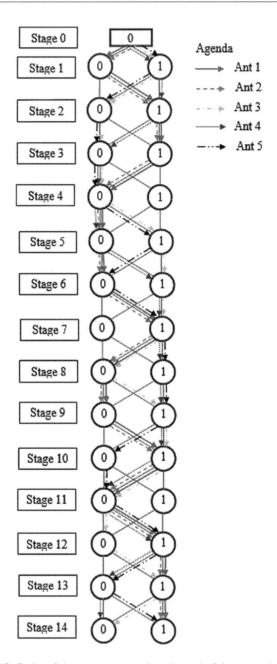

Figure 8.33 ACO: Paths of the ants generated at the end of the second iteration

Table 8.132 ACO: Ants construct their solutions in the second iteration

	Ant 1			Ant 2			Ant 3			Ant 4			Ant 5		
	p_o	r	$d(i)$	p_o	r	$d(i)$	p_o	r	$d(i)$	p_o	r	$d(i)$	p_o	r	$d(i)$
Stage1	0.242	0.240	0	0.242	0.124	0	0.242	0.545	1	0.242	0.888	1	0.242	0.255	1
Stage2	0.047	0.365	1	0.047	0.998	1	0.954	0.922	0	0.954	0.981	1	0.954	0.524	0
Stage3	0.441	0.614	1	0.441	0.622	1	0.462	0.295	0	0.607	0.547	0	0.462	0.006	0
Stage4	0.687	0.377	0	0.687	0.037	0	0.525	0.269	0	0.570	0.064	0	0.525	0.383	0
Stage5	0.367	0.046	0	0.367	0.232	0	0.377	0.900	1	0.399	0.387	0	0.377	0.620	1
Stage6	0.851	0.620	0	0.851	0.829	0	0.573	0.742	1	0.860	0.868	1	0.573	0.370	0
Stage7	0.361	0.215	0	0.361	0.709	1	0.112	0.913	1	0.114	0.383	1	0.364	0.683	1
Stage8	0.856	0.198	1	0.305	0.273	0	0.306	0.073	0	0.308	0.740	1	0.306	0.756	1
Stage9	0.514	0.876	0	0.514	0.095	0	0.515	0.881	1	0.304	0.598	1	0.302	0.893	1
Stage10	0.618	0.290	0	0.411	0.997	1	0.618	0.874	1	0.619	0.912	1	0.618	0.348	0
Stage11	0.926	0.138	1	0.915	0.658	0	0.915	0.372	0	0.915	0.383	0	0.926	0.432	0
Stage12	0.403	0.963	0	0.403	0.465	1	0.403	0.326	0	0.404	0.972	1	0.403	0.819	1
Stage13	0.610	0.013	1	0.610	0.751	1	0.539	0.896	1	0.610	0.047	0	0.610	0.214	0
Stage14	0.500	0.966	1	0.500	0.913	1	0.500	0.207	0	0.500	0.106	0	0.500	0.505	1

Table 8.133 ACO: The quality of the generated solutions in the second iteration

	d	y	x_1	t_1	x_2	t_2	T_t
Ant 1	6,181	3,772.81	3,772	0.280	6,228	0.506	4,204.64
Ant 2	6,295	3,842.40	3,842	0.282	6,158	0.502	4,173.26
Ant 3	9,138	5,577.73	5,577	0.392	4,423	0.439	4,127.69
Ant 4	12,788	7,805.65	7,805	0.794	2,195	0.418	7,111.68
Ant 5	8,933	5,452.60	5,452	0.379	4,548	0.442	4,078.83

Table 8.134 ACO: The new pheromone values after the second iteration

		Next stage (Stage $i+1$)	
Stage i		0	1
Stage 0	0	$7.39 \cdot 10^{-4}$	$9.51 \cdot 10^{-4}$
Stage 1	0	$2.50 \cdot 10^{-5}$	$7.39 \cdot 10^{-4}$
	1	$8.10 \cdot 10^{-4}$	$1.66 \cdot 10^{-4}$
Stage 2	0	$6.36 \cdot 10^{-4}$	$1.99 \cdot 10^{-4}$
	1	$2.81 \cdot 10^{-4}$	$6.23 \cdot 10^{-4}$
Stage 3	0	$7.77 \cdot 10^{-4}$	$1.41 \cdot 10^{-4}$
	1	$7.18 \cdot 10^{-4}$	$1.04 \cdot 10^{-4}$
Stage 4	0	$7.64 \cdot 10^{-4}$	$7.31 \cdot 10^{-4}$
	1	$2.50 \cdot 10^{-5}$	$2.20 \cdot 10^{-4}$
Stage 5	0	$6.23 \cdot 10^{-4}$	$1.66 \cdot 10^{-4}$
	1	$5.10 \cdot 10^{-4}$	$4.41 \cdot 10^{-4}$
Stage 6	0	$3.87 \cdot 10^{-4}$	$7.46 \cdot 10^{-4}$
	1	$2.50 \cdot 10^{-5}$	$5.82 \cdot 10^{-4}$
Stage 7	0	$3.87 \cdot 10^{-4}$	$2.50 \cdot 10^{-5}$
	1	$6.22 \cdot 10^{-4}$	$7.05 \cdot 10^{-4}$
Stage 8	0	$3.88 \cdot 10^{-4}$	$6.21 \cdot 10^{-4}$
	1	$1.04 \cdot 10^{-4}$	$6.26 \cdot 10^{-4}$
Stage 9	0	$1.04 \cdot 10^{-4}$	$3.88 \cdot 10^{-4}$
	1	$7.19 \cdot 10^{-4}$	$5.28 \cdot 10^{-4}$
Stage 10	0	$7.98 \cdot 10^{-4}$	$2.50 \cdot 10^{-5}$
	1	$8.92 \cdot 10^{-4}$	$2.50 \cdot 10^{-5}$
Stage 11	0	$4.78 \cdot 10^{-4}$	$1.21 \cdot 10^{-3}$
	1	$2.50 \cdot 10^{-5}$	$2.50 \cdot 10^{-5}$
Stage 12	0	$1.41 \cdot 10^{-4}$	$3.62 \cdot 10^{-4}$
	1	$8.52 \cdot 10^{-4}$	$3.85 \cdot 10^{-4}$
Stage 13	0	$2.50 \cdot 10^{-5}$	$2.50 \cdot 10^{-5}$
	1	$2.50 \cdot 10^{-5}$	$2.50 \cdot 10^{-5}$

Table 8.135 Distances between the nodes

	0	1	2	3	4	5
0	0	42.43	49.50	35.36	28.28	28.28
1	42.43	0	91.92	77.78	14.14	70.71
2	49.50	91.92	0	14.14	77.78	21.21
3	35.36	77.78	14.14	0	63.64	7.07
4	28.28	14.14	77.78	63.64	0	56.57
5	28.28	70.71	21.21	7.07	56.57	0

8.3. If the previous solution in the simulated annealing strategy has the value 950, the new solution has the value 980, the temperature is 80 and the random number is 0.42, do we accept the new solution or should it be rejected?

8.4. After the forward pass in the BCO algorithm, 5 bees have solutions with the following objective function values: $T_1 = 1,200$, $T_2 = 1,000$, $T_3 = 1,250$, $T_4 = 980$, and $T_5 = 1,100$. The aim in this problem is to find a solution with the minimum objective function. Determine the normalized values of the bees' objective functions and the probabilities that bees will stay loyal to their solutions.

8.5. An ant should go from node 1 to one of three nodes (2, 3, or 4). Calculate the probabilities that the ant will choose any of these nodes according to the following values:
   a) $\tau_{12} = 0.001$, $\tau_{13} = 0.015$, $\tau_{14} = 0.025$, $d_{12} = 50$, $d_{13} = 65$, $d_{14} = 55$, $\alpha = 1$, and $\beta = 1$.
   b) $\tau_{12} = 0.001$, $\tau_{13} = 0.015$, $\tau_{14} = 0.025$, $d_{12} = 50$, $d_{13} = 65$, $d_{14} = 55$, $\alpha = 1$, and $\beta = 5$.

8.6. Within the iteration, two ants went through the link (2,5). The objective function values of the ants' solutions were 130 and 180, respectively, the previous value of pheromone through the link (2,5) was 0.00125, the parameter ρ equals 0.8, and Q equals 1. Determine the new value of the pheromone through the observed link, if our goal in the problem in question is to find the solution that has the minimum objective function value.

BIBLIOGRAPHY

Beni, G., The concept of cellular robotic system. In *Proceedings of the 1988 IEEE International Symposium on Intelligent Control*, IEEE Computer Society Press, Los Alamitos, CA, 57–62, 1988.

Beni, G. and Hackwood, S., Stationary Waves in Cyclic Warms. In *Proceedings of the1992 International Symposium on Intelligent Control*, IEEE Computer Society Press, Los Alamitos, CA, 234–242, 1992.

Beni, G. and Wang, J., Swarm Intelligence. In *Proceedings of the Seventh Annual Meeting of the Robotics Society of Japan*, RSJ Press, Tokyo, 425–428, 1989.

Bonabeau, E., Dorigo, M., and Theraulaz, G., *Swarm Intelligence*, Oxford University Press, Oxford, 1999.

Černy, V., A Thermodynamical Approach to the Travelling Salesman Problem: An Efficient Simulation Algorithm, *Journal of Optimization Theory and Applications*, 45, 41–51, 1985.

Colorni, A., Dorigo, M., and Maniezzo, V., Distributed Optimization by Ant Colonies. In F. Varela and P. Bourgine (Eds.), *Proceedings of the First European Conference on Artificial Life*, Elsevier, Paris, France, 134–142, 1991.

Crainic, T. G., M. Toulouse and M. Gendreau, Toward a Taxonomy of Parallel Tabu Search Heuristics, *INFORMS Journal on Computing*, 9, 61–72, 1997.

Davidović, T., Ramljak, D., Šelmić, M., and Teodorović, D., Bee Colony Optimization for the *p*-center Problem, *Computers & Operations Research*, 38, 1367–1376, 2011.

Dorigo, M. and DiCaro, G., The Ant Colony Optimization Metaheuristic. In D. Corne, M. Dorigo, and F. Glover (Eds.), *New Ideas in Optimization*, McGraw-Hill, London, 1999.

Dorigo, M. and Gambardella, L. M., Ant Colonies for the Traveling Salesman Problem, *BioSystems*, 43, 73–81, 1997a.

Dorigo, M. and Gambardella, L. M., Ant Colony System: A Cooperative Learning Approach to the Travelling Salesman Problem, *IEEE Transactions on Evolutionary Computation* 1, 53–66, 1997b.

Dorigo, M., Maniezzo, V., and Colorni, A., The Ant System: An Autocatalytic Optimizing Process. Technical Report No. 91–016 Revised. Politecnico di Milano, Milano, Italy, 1991.

Dorigo, M., Maniezzo, V., and Colorni, A., Ant System: Optimization by a Colony of Cooperating Agents, *IEEE Transactions on Systems, Man and Cybernetics*, 26 (Part B), 29–41, 1996.

Eglese, R. W., Simulated Annealing: A Tool for Operational Research, *European Journal of Operational Research*, 46, 271–281, 1990.

Gendreau, M., Hertz, A., and Laporte, G., A Tabu Search Heuristic for the Vehicle Routing Problem, *Management Science*, 40, 1276–1290, 1994.

Glover, F., and Laguna, M., *Tabu Search*, Kluwer Academic Publishers, Norwell, MA, 1997.

Glover, F., Tabu Search – Part I, *ORSA Journal on Computing* 1, 190–206, 1989.

Glover, F., Tabu Search – Part II, *ORSA Journal on Computing* 2, 4–32, 1990.

Glover, F., Taillard, E., and de Werra, D., A User's Guide to Tabu Search, *Annals of Operations Research*, 41, 3–28, 1993.

Goldberg D., *Genetic Algorithms in Search, Optimization, and Machine Learning*, Addison Wesley Publishing Company Inc, New York, 1989.

Holland, J. H., *Adaptation in Natural and Artificial Systems*, University of Michigan Press, Ann Arbor, MI, 1975.

Kennedy, J. and Eberhart, R. C. 1995. Particle Swarm Optimization. In *Proceedings of the IEEE International Conference on Neural Networks*, Piscataway, NJ, 1942–1948.

Kennedy, J., Eberhart, R. C. and Shi, Y., *Swarm Intelligence*, Morgan Kaufmann Publishers, San Francisco, 2001.

Kirkpatrick, S., Gelatt Jr., C. D., and Vecchi, M. P., Optimization by Simulated Annealing, *Science*, 220, 671–680, 1983.

Lučić, P. and Teodorović, D., Bee System: Modeling Combinatorial Optimization Transportation Engineering Problems by Swarm Intelligence. In *Preprints of the TRISTAN IV Triennial Symposium on Transportation Analysis*, Sao Miguel, Azores Islands, Portugal, 441–445, 2001.

Lučić, P., Teodorović, Transportation Modeling: An Artificial Life Approach. In *Proceedings of the 14th IEEE International Conference on Tools with Artificial Intelligence*, Washington, DC, 216–223, 2002.

Lučić, P., Teodorović, Computing with Bees: Attacking Complex Transportation Engineering Problems, *International Journal on Artificial Intelligence Tools*, 12, 375–394, 2003a.

Lučić, P., Teodorović, Vehicle Routing Problem with Uncertain Demand at Nodes: the Bee System and Fuzzy Logic Approach. In J. L. Verdegay (Ed.), *Fuzzy Sets in Optimization*, Springer, Heidelberg, Berlin, 67–82, 2003b.

Metropolis, N., Rosenbluth, A. W., Rosenbluth, M. N., Teller, A. H., and Teller, E. Equations of State Calculations by Fast Computing Machines, *Journal of Chemical Physics*, 21, 1087–1092, 1953.

Mitchell, M., *An Introduction to Genetic Algorithms*, MIT Press, Cambridge, 1996.

Nikolić, M. and Teodorović, D., Transit Network Design by Bee Colony Optimization, *Expert Systems with Applications*, 40, 5945–5955, 2013a.

Nikolić, M. and Teodorović, D., Empirical Study of the Bee Colony Optimization (BCO) Algorithm, *Expert Systems with Applications*, 40, 4609–4620, 2013b.

Nikolić, M. and Teodorović, D., A Simultaneous Transit Network Design and Frequency Setting: Computing with Bees, *Expert Systems with Applications*, 41, 7200–7209, 2014.

Nikolić, M. and Teodorović, D., Mitigation of Disruptions in Public Transit by Bee Colony Optimization, *Transportation Planning and Technology*, 42, 573–586, 2019.

Nikolić, M., Teodorović, D., and Vukadinović, K., Disruption Management in Public Transit: The Bee Colony Optimization (BCO) Approach, *Transportation Planning and Technology*, 38, 162–180, 2015.

Ruthenbar, R., Simulated Annealing Algorithms: An Overview, *IEEE Circuits and Devices*, 5, 19–26, 1989.

Simon, D., *Evolutionary Optimization Algorithms: Biologically-Inspired and Population-Based Approaches to Computer Intelligence*, Wiley, Hoboken, NJ, 2013.

Teodorović, D., Swarm Intelligence Systems for Transportation Engineering: Principles and Applications, *Transportation Research C*, 16, 651–782, 2008.

Teodorović, D., Bee Colony Optimization (BCO). In C. P. Lim, L. C. Jain, and S. Dehuri (Eds.), *Innovations in Swarm Intelligence*, Springer, Berlin, Heidelberg, 39–60, 2009.

Teodorović, D. and Dell'Orco, M., Bee Colony Optimization – A Cooperative Learning Approach to Complex Transportation Problems. In: *Advanced OR and AI Methods in Transportation: Proceedings of the 10th Meeting of the EURO Working Group on Transportation*, Poznan, Poland, 51–60, 2005.

Teodorović, D., and Dell'Orco, M., Mitigating Traffic Congestion: Solving the Ride-Matching Problem by Bee Colony Optimization, *Transportation Planning and Technology* 31, 135–152, 2008.

Teodorović, D. and Lučić, P., Schedule Synchronization in Public Transit by Fuzzy Ant System, *Transportation Planning and Technology*, 28, 47–77, 2005.

Tovey, G., Simulated Annealing, *American Journal of Mathematical and Management Sciences*, 8, 389–407, 1988.

Appendix

Table A.1 Values of cumulative probabilities of standard normal distribution

z	0	0.01	0.02	0.03	0.04	0.05	0.06	0.07	0.08	0.09
−3.4	0.0003	0.0003	0.0003	0.0003	0.0003	0.0003	0.0003	0.0003	0.0003	0.0002
−3.3	0.0005	0.0005	0.0005	0.0004	0.0004	0.0004	0.0004	0.0004	0.0004	0.0003
−3.2	0.0007	0.0007	0.0006	0.0006	0.0006	0.0006	0.0006	0.0005	0.0005	0.0005
−3.1	0.001	0.0009	0.0009	0.0009	0.0008	0.0008	0.0008	0.0008	0.0007	0.0007
−3	0.0013	0.0013	0.0013	0.0012	0.0012	0.0011	0.0011	0.0011	0.001	0.001
−2.9	0.0019	0.0018	0.0018	0.0017	0.0016	0.0016	0.0015	0.0015	0.0014	0.0014
−2.8	0.0026	0.0025	0.0024	0.0023	0.0023	0.0022	0.0021	0.0021	0.002	0.0019
−2.7	0.0035	0.0034	0.0033	0.0032	0.0031	0.003	0.0029	0.0028	0.0027	0.0026
−2.6	0.0047	0.0045	0.0044	0.0043	0.0041	0.004	0.0039	0.0038	0.0037	0.0036
−2.5	0.0062	0.006	0.0059	0.0057	0.0055	0.0054	0.0052	0.0051	0.0049	0.0048
−2.4	0.0082	0.008	0.0078	0.0075	0.0073	0.0071	0.0069	0.0068	0.0066	0.0064
−2.3	0.0107	0.0104	0.0102	0.0099	0.0096	0.0094	0.0091	0.0089	0.0087	0.0084
−2.2	0.0139	0.0136	0.0132	0.0129	0.0125	0.0122	0.0119	0.0116	0.0113	0.011
−2.1	0.0179	0.0174	0.017	0.0166	0.0162	0.0158	0.0154	0.015	0.0146	0.0143
−2	0.0228	0.0222	0.0217	0.0212	0.0207	0.0202	0.0197	0.0192	0.0188	0.0183
−1.9	0.0287	0.0281	0.0274	0.0268	0.0262	0.0256	0.025	0.0244	0.0239	0.0233
−1.8	0.0359	0.0351	0.0344	0.0336	0.0329	0.0322	0.0314	0.0307	0.0301	0.0294
−1.7	0.0446	0.0436	0.0427	0.0418	0.0409	0.0401	0.0392	0.0384	0.0375	0.0367
−1.6	0.0548	0.0537	0.0526	0.0516	0.0505	0.0495	0.0485	0.0475	0.0465	0.0455
−1.5	0.0668	0.0655	0.0643	0.063	0.0618	0.0606	0.0594	0.0582	0.0571	0.0559
−1.4	0.0808	0.0793	0.0778	0.0764	0.0749	0.0735	0.0721	0.0708	0.0694	0.0681
−1.3	0.0968	0.0951	0.0934	0.0918	0.0901	0.0885	0.0869	0.0853	0.0838	0.0823
−1.2	0.1151	0.1131	0.1112	0.1093	0.1075	0.1056	0.1038	0.102	0.1003	0.0985
−1.1	0.1357	0.1335	0.1314	0.1292	0.1271	0.1251	0.123	0.121	0.119	0.117
−1	0.1587	0.1562	0.1539	0.1515	0.1492	0.1469	0.1446	0.1423	0.1401	0.1379
−0.9	0.1841	0.1814	0.1788	0.1762	0.1736	0.1711	0.1685	0.166	0.1635	0.1611
−0.8	0.2119	0.209	0.2061	0.2033	0.2005	0.1977	0.1949	0.1922	0.1894	0.1867
−0.7	0.242	0.2389	0.2358	0.2327	0.2296	0.2266	0.2236	0.2206	0.2177	0.2148
−0.6	0.2743	0.2709	0.2676	0.2643	0.2611	0.2578	0.2546	0.2514	0.2483	0.2451

(Continued)

Table A.1 (Continued) Values of cumulative probabilities of standard normal distribution

z	0	0.01	0.02	0.03	0.04	0.05	0.06	0.07	0.08	0.09
−0.5	0.3085	0.305	0.3015	0.2981	0.2946	0.2912	0.2877	0.2843	0.281	0.2776
−0.4	0.3446	0.3409	0.3372	0.3336	0.33	0.3264	0.3228	0.3192	0.3156	0.3121
−0.3	0.3821	0.3783	0.3745	0.3707	0.3669	0.3632	0.3594	0.3557	0.352	0.3483
−0.2	0.4207	0.4168	0.4129	0.409	0.4052	0.4013	0.3974	0.3936	0.3897	0.3859
−0.1	0.4602	0.4562	0.4522	0.4483	0.4443	0.4404	0.4364	0.4325	0.4286	0.4247
−0	0.5	0.496	0.492	0.488	0.484	0.4801	0.4761	0.4721	0.4681	0.4641
0	0.5	0.504	0.508	0.512	0.516	0.5199	0.5239	0.5279	0.5319	0.5359
0.1	0.5398	0.5438	0.5478	0.5517	0.5557	0.5596	0.5636	0.5675	0.5714	0.5753
0.2	0.5793	0.5832	0.5871	0.591	0.5948	0.5987	0.6026	0.6064	0.6103	0.6141
0.3	0.6179	0.6217	0.6255	0.6293	0.6331	0.6368	0.6406	0.6443	0.648	0.6517
0.4	0.6554	0.6591	0.6628	0.6664	0.67	0.6736	0.6772	0.6808	0.6844	0.6879
0.5	0.6915	0.695	0.6985	0.7019	0.7054	0.7088	0.7123	0.7157	0.719	0.7224
0.6	0.7257	0.7291	0.7324	0.7357	0.7389	0.7422	0.7454	0.7486	0.7517	0.7549
0.7	0.758	0.7611	0.7642	0.7673	0.7704	0.7734	0.7764	0.7794	0.7823	0.7852
0.8	0.7881	0.791	0.7939	0.7967	0.7995	0.8023	0.8051	0.8078	0.8106	0.8133
0.9	0.8159	0.8186	0.8212	0.8238	0.8264	0.8289	0.8315	0.834	0.8365	0.8389
1	0.8413	0.8438	0.8461	0.8485	0.8508	0.8531	0.8554	0.8577	0.8599	0.8621
1.1	0.8643	0.8665	0.8686	0.8708	0.8729	0.8749	0.877	0.879	0.881	0.883
1.2	0.8849	0.8869	0.8888	0.8907	0.8925	0.8944	0.8962	0.898	0.8997	0.9015
1.3	0.9032	0.9049	0.9066	0.9082	0.9099	0.9115	0.9131	0.9147	0.9162	0.9177
1.4	0.9192	0.9207	0.9222	0.9236	0.9251	0.9265	0.9279	0.9292	0.9306	0.9319
1.5	0.9332	0.9345	0.9357	0.937	0.9382	0.9394	0.9406	0.9418	0.9429	0.9441
1.6	0.9452	0.9463	0.9474	0.9484	0.9495	0.9505	0.9515	0.9525	0.9535	0.9545
1.7	0.9554	0.9564	0.9573	0.9582	0.9591	0.9599	0.9608	0.9616	0.9625	0.9633
1.8	0.9641	0.9649	0.9656	0.9664	0.9671	0.9678	0.9686	0.9693	0.9699	0.9706
1.9	0.9713	0.9719	0.9726	0.9732	0.9738	0.9744	0.975	0.9756	0.9761	0.9767
2	0.9772	0.9778	0.9783	0.9788	0.9793	0.9798	0.9803	0.9808	0.9812	0.9817
2.1	0.9821	0.9826	0.983	0.9834	0.9838	0.9842	0.9846	0.985	0.9854	0.9857
2.2	0.9861	0.9864	0.9868	0.9871	0.9875	0.9878	0.9881	0.9884	0.9887	0.989
2.3	0.9893	0.9896	0.9898	0.9901	0.9904	0.9906	0.9909	0.9911	0.9913	0.9916
2.4	0.9918	0.992	0.9922	0.9925	0.9927	0.9929	0.9931	0.9932	0.9934	0.9936
2.5	0.9938	0.994	0.9941	0.9943	0.9945	0.9946	0.9948	0.9949	0.9951	0.9952
2.6	0.9953	0.9955	0.9956	0.9957	0.9959	0.996	0.9961	0.9962	0.9963	0.9964
2.7	0.9965	0.9966	0.9967	0.9968	0.9969	0.997	0.9971	0.9972	0.9973	0.9974
2.8	0.9974	0.9975	0.9976	0.9977	0.9977	0.9978	0.9979	0.9979	0.998	0.9981
2.9	0.9981	0.9982	0.9982	0.9983	0.9984	0.9984	0.9985	0.9985	0.9986	0.9986
3	0.9987	0.9987	0.9987	0.9988	0.9988	0.9989	0.9989	0.9989	0.999	0.999
3.1	0.999	0.9991	0.9991	0.9991	0.9992	0.9992	0.9992	0.9992	0.9993	0.9993
3.2	0.9993	0.9993	0.9994	0.9994	0.9994	0.9994	0.9994	0.9995	0.9995	0.9995
3.3	0.9995	0.9995	0.9995	0.9996	0.9996	0.9996	0.9996	0.9996	0.9996	0.9997
3.4	0.9997	0.9997	0.9997	0.9997	0.9997	0.9997	0.9997	0.9997	0.9997	0.9998

(Continued)

Table A.1 (Continued) Values of cumulative probabilities of standard normal distribution

z	0	0.01	0.02	0.03	0.04	0.05	0.06	0.07	0.08	0.09
3.5	0.9998	0.9998	0.9998	0.9998	0.9998	0.9998	0.9998	0.9998	0.9998	0.9998
3.6	0.9998	0.9998	0.9999	0.9999	0.9999	0.9999	0.9999	0.9999	0.9999	0.9999

Table A.2 Values for the chi-squared distribution

n/α	0.01	0.025	0.05	0.1	0.9	0.95	0.975	0.99
1	6.6349	5.0239	3.8415	2.7055	0.0158	0.0039	0.001	0.0002
2	9.2103	7.3778	5.9915	4.6052	0.2107	0.1026	0.0506	0.0201
3	11.3449	9.3484	7.8147	6.2514	0.5844	0.3518	0.2158	0.1148
4	13.2767	11.1433	9.4877	7.7794	1.0636	0.7107	0.4844	0.2971
5	15.0863	12.8325	11.0705	9.2364	1.6103	1.1455	0.8312	0.5543
6	16.8119	14.4494	12.5916	10.6446	2.2041	1.6354	1.2373	0.8721
7	18.4753	16.0128	14.0671	12.017	2.8331	2.1673	1.6899	1.239
8	20.0902	17.5345	15.5073	13.3616	3.4895	2.7326	2.1797	1.6465
9	21.666	19.0228	16.919	14.6837	4.1682	3.3251	2.7004	2.0879
10	23.2093	20.4832	18.307	15.9872	4.8652	3.9403	3.247	2.5582
11	24.725	21.92	19.6751	17.275	5.5778	4.5748	3.8157	3.0535
12	26.217	23.3367	21.0261	18.5493	6.3038	5.226	4.4038	3.5706
13	27.6882	24.7356	22.362	19.8119	7.0415	5.8919	5.0088	4.1069
14	29.1412	26.1189	23.6848	21.0641	7.7895	6.5706	5.6287	4.6604
15	30.5779	27.4884	24.9958	22.3071	8.5468	7.2609	6.2621	5.2293
16	31.9999	28.8454	26.2962	23.5418	9.3122	7.9616	6.9077	5.8122
17	33.4087	30.191	27.5871	24.769	10.0852	8.6718	7.5642	6.4078
18	34.8053	31.5264	28.8693	25.9894	10.8649	9.3905	8.2307	7.0149
19	36.1909	32.8523	30.1435	27.2036	11.6509	10.117	8.9065	7.6327
20	37.5662	34.1696	31.4104	28.412	12.4426	10.851	9.5908	8.2604
21	38.9322	35.4789	32.6706	29.6151	13.2396	11.591	10.283	8.8972
22	40.2894	36.7807	33.9244	30.8133	14.0415	12.338	10.982	9.5425
23	41.6384	38.0756	35.1725	32.0069	14.848	13.091	11.689	10.196
24	42.9798	39.3641	36.415	33.1962	15.6587	13.848	12.401	10.856
25	44.3141	40.6465	37.6525	34.3816	16.4734	14.611	13.12	11.524
26	45.6417	41.9232	38.8851	35.5632	17.2919	15.379	13.844	12.198
27	46.9629	43.1945	40.1133	36.7412	18.1139	16.151	14.573	12.879
28	48.2782	44.4608	41.3371	37.9159	18.9392	16.928	15.308	13.565
29	49.5879	45.7223	42.557	39.0875	19.7677	17.708	16.047	14.257
30	50.8922	46.9792	43.773	40.256	20.5992	18.493	16.791	14.954
31	52.1914	48.2319	44.9853	41.4217	21.4336	19.281	17.539	15.656

(Continued)

Table A.2 (Continued) Values for the chi-squared distribution

n/α	0.01	0.025	0.05	0.1	0.9	0.95	0.975	0.99
32	53.4858	49.4804	46.1943	42.5847	22.2706	20.072	18.291	16.362
33	54.7755	50.7251	47.3999	43.7452	23.1102	20.867	19.047	17.074
34	56.0609	51.966	48.6024	44.9032	23.9523	21.664	19.806	17.789
35	57.3421	53.2033	49.8018	46.0588	24.7967	22.465	20.569	18.509
36	58.6192	54.4373	50.9985	47.2122	25.6433	23.269	21.336	19.233
37	59.8925	55.668	52.1923	48.3634	26.4921	24.075	22.106	19.96
38	61.1621	56.8955	53.3835	49.5126	27.343	24.884	22.879	20.691
39	62.4281	58.1201	54.5722	50.6598	28.1958	25.695	23.654	21.426
40	63.6907	59.3417	55.7585	51.8051	29.0505	26.509	24.433	22.164
50	76.1539	71.4202	67.5048	63.1671	37.6886	34.764	32.357	29.707
60	88.3794	83.2977	79.0819	74.397	46.4589	43.188	40.482	37.485
70	100.425	95.0232	90.5312	85.527	55.3289	51.739	48.758	45.442
80	112.329	106.629	101.88	96.5782	64.2778	60.392	57.153	53.54
90	124.116	118.136	113.145	107.565	73.2911	69.126	65.647	61.754
100	135.807	129.561	124.342	118.498	82.3581	77.93	74.222	70.065

Index